李 杰/编 著

程序思维之前端框架实战系列

uni-app多端跨平台开发

从入门到企业级实战 ProMAX版

中国水利水电出版社
www.waterpub.com.cn

· 北京 ·

内 容 提 要

《uni-app多端跨平台开发从入门到企业级实战》是一本系统介绍热门前端多端框架uni-app的实战教程。全书由基础知识和实战两大部分组成，包括初识uni-app，uni-app环境搭建，生命周期，尺寸单位、Flex布局与背景图片，pages.json配置，manifest.json配置，路由，判断运行环境和平台，常用组件，常用API以及仿美团点餐小程序客户端开发。

本书以面试和实战为基础，通过阅读本书，读者可快速学会uni-app的相关知识并将其应用到实战项目中。例如，优化性能、解决iPhone X 的“刘海”兼容性问题、实现微信支付与微信授权登录、获取位置并在地图上显示、获取用户所在的城市和街道信息、发布与审核微信小程序等。对于想要学习更多框架知识，并且想要快速将这些知识应用到实战项目中的你来说，这本书可谓是不二之选。

图书在版编目（CIP）数据

uni-app 多端跨平台开发从入门到企业级实战 / 李杰编著 . —北京 : 中国水利水电出版社 , 2022.6(2024.6重印)

ISBN 978-7-5226-0348-3

Ⅰ. ① u… Ⅱ. ①李… Ⅲ. ①网页制作工具—程序设计 Ⅳ. ① TP392.092.2

中国版本图书馆 CIP 数据核字 (2021) 第 280450 号

书　　名	uni-app 多端跨平台开发从入门到企业级实战 uni-app DUODUAN KUA PINGTAI KAIFA CONG RUMEN DAO QIYE JI SHIZHAN
作　　者	李杰　编著
出版发行	中国水利水电出版社 （北京市海淀区玉渊潭南路 1 号 D 座 100038） 网址：www.waterpub.com.cn E-mail：zhiboshangshu@163.com 电话：（010）62572966-2205/2266/2201（营销中心）
经　　售	北京科水图书销售有限公司 电话：（010）68545874、63202643 全国各地新华书店和相关出版物销售网点
排　　版	北京智博尚书文化传媒有限公司
印　　刷	北京富博印刷有限公司
规　　格	190mm×235mm　16 开本　28 印张　716 千字
版　　次	2022 年 6 月第 1 版　2024 年 6 月第 4 次印刷
印　　数	8001—10000册
定　　价	108.00 元

前　言

前端工程师是目前非常热门和受欢迎的职位，可与大数据、人工智能、云计算等火爆职位相媲美。其入门简单、学习人群广泛、工资高、就业好。

前端技能包括HTML5、CSS3、JavaScript、Node，目前较火的前端框架包括Vue、React，流行的多端开发框架有uni-app、Taro等。

月活8.4亿的多端开发框架uni-app，在不到两年的时间内，迅速受到广大开发者的青睐和推崇。uni-app之所以如此火爆，与其特性密切相关：第一，月活8.4亿，热度高；第二，具有颠覆性的优势——快，快到可以节省7套代码，解决方案可跨多端，可发布到iOS、Android、H5及各种小程序等多个平台；第三，学习门槛低，基于通用的前端技术，采用Vue语法+微信小程序API，无额外学习成本；第四，通过条件编译+平台特有API调用，可以为某平台编写个性代码，支持原生代码混写和原生SDK集成；第五，周边生态丰富，插件市场中拥有数千款插件，支持NPM，支持小程序组件和SDK，兼容mpvue组件和项目，兼容Weex组件。

学习技巧

如果你是职场精英，当你所在的公司需要使用uni-app开发项目时，那么你可以从本书中快速学习开发项目的技巧并应用到企业实战项目中。

如果你是职场小白，那么你可以从基础开始学习直到开发出企业级实战项目，在学习过程中，本书可以帮助你提升自学的能力、解决问题的能力、二次开发的能力，以及举一反三的能力，这些能力有助于你找到更理想的工作。

在学习过程中，如果遇到不理解的地方，可以扫描二维码查看视频详解。

本书内容组织

uni-app是基于Vue语法开发出来的前端框架，一个人一套代码可以开发出多端应用，包括Android版App，iOS版App，H5网站，微信小程序、支付宝小程序、百度小程序、头条小程序、QQ小程序、快应用、360小程序。通过学习本书，读者最终可以开发出一套类似美团的企业级点餐小程序。

本书由基础知识和实战两大部分组成，如图0-1所示。

基础知识部分是1~10章，包括初识uni-app，uni-app环境搭建，生命周期，尺寸单位、Flex布局与背景图片，pages.json配置，manifest.json配置，路由，判断运行环境和平台，常用组件，常用API。其中，读者可以重点关注Hbuilder X的安装与使用方法，APP、小程序和H5网站的调试与运行方法，多平台开发的注意事项，常用组件和常用API的使用方法，获取手机号码一站式登录的方法，以及微信支付等高级功能的开发方法。此外，可以跟着本书学习快速读懂uni-app官方文档的方法，掌握快速学习的技巧，为日后学习其他技能打下坚实的基础。

实战项目是第11章，包括仿美团点餐小程序客户端开发。该实战项目是本书的重点内容，也是本书的制胜法宝。

本书附赠资源中有用于开发实战项目的服务端接口文档，不仅能让巩固读者所学知识点，而且能让读者的前端技术有一个质的提高。

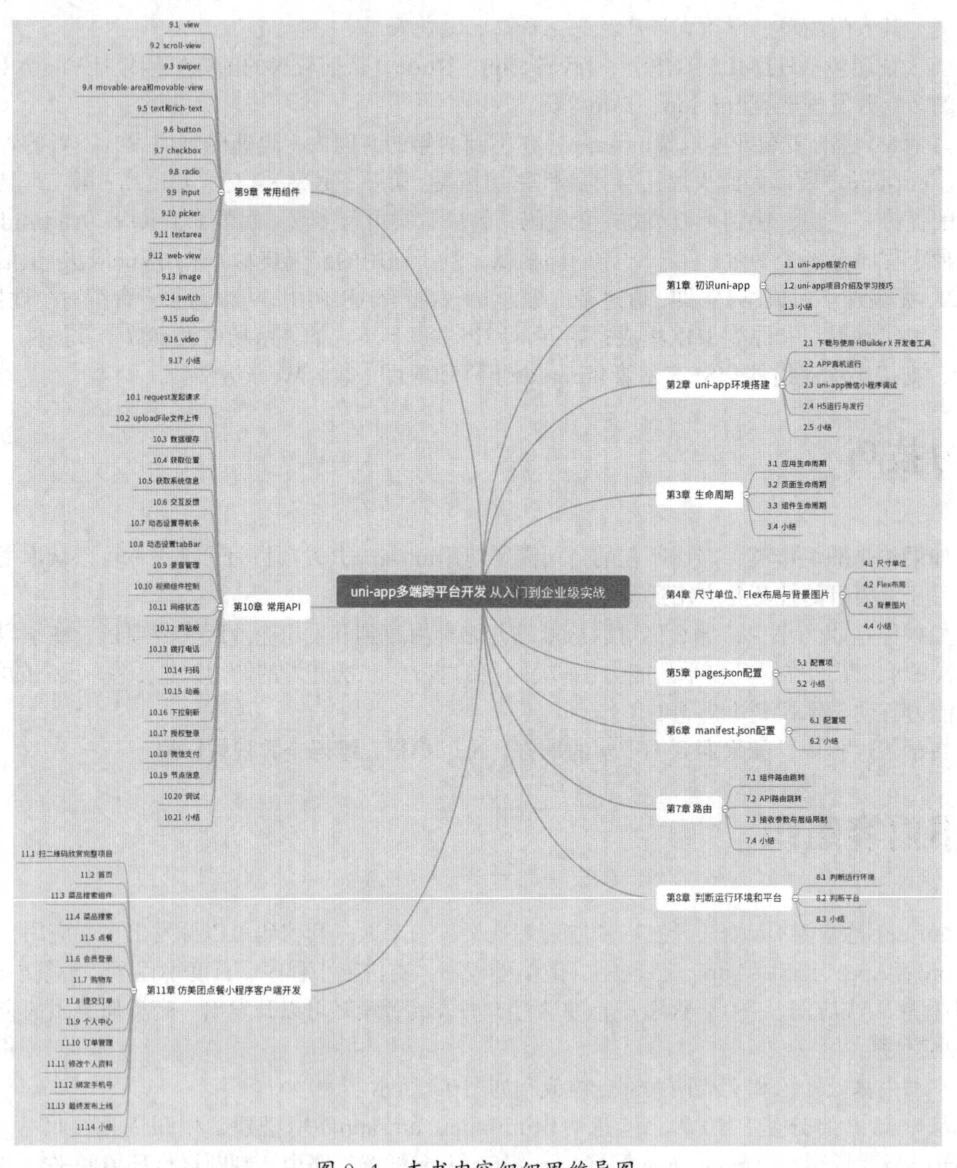

图 0-1 本书内容组织思维导图

本书显著特色

1. 视频讲解

本书关键章节配有二维码，微信扫一扫，可以随时随地观看讲解视频。一方面，视频讲解可以帮助读者增强逻辑思维，提高举一反三、自我学习、解决问题的能力；另一方面，可以增加读者在工作中的实战技巧，提升修补各种漏洞的能力，开发出高性能、高兼容、高维护的项目。此外，笔者用幽默风趣的语言将抽象的问题具体化，使每一个知识点都对应实战开发项目。

2. 实战项目

真实的企业级仿美团点餐小程序客户端开发实战，完整的接口文档，实用的购物车、在线支付开发等开发人员必会功能。

3. 资源放送

本书提供的配套学习资源如下：

- 98集视频讲解；
- 96个源代码包；
- 精品PPT课件；
- 企业级完整接口文档。

另外，本书附赠以下学习资源，供读者在工作与学习中进一步掌握相关知识。

- 赠送价值5000元的仿美团微信点餐小程序；
- 赠送前台和后台架构模板；
- 特别奉送上百道前端面试题，助读者面试一臂之力；
- 可以向笔者咨询工作中的问题，未必局限于本书。

4. 在线服务

本书提供售后疑难解答服务，有以下3种方式：

（1）扫描下方二维码，关注微信公众号“html5程序思维”，在后台发送“uniapp”，添加笔者QQ或微信进行远程答疑；

（2）加入QQ群：211851273，或关注抖音号（html5程序思维）、快手号（html5程序思维）与笔者或同行在线交流；

（3）扫描下方二维码，加入“本书专属读者在线服务交流圈”，与本书其他读者一起，分享读书心得、提出对本书的建议，以及咨询笔者问题等。

html5 程序思维 公众号

本书专属读者在线服务交流圈

本书资源下载

本书提供全部配套视频和源代码包，有以下2种下载方式：

（1）扫描下方二维码，关注微信公众号“html5程序思维”，在后台发送“uniapp”，获得本书资源下载链接，然后将此链接复制到计算机浏览器的地址栏中，根据提示下载即可；

（2）扫描下方二维码，加入“本书专属读者在线服务交流圈”，在置顶的动态中获得本书资源下载链接，然后将此链接复制到计算机浏览器的地址栏中，根据提示下载即可。

html5 程序思维 公众号

本书专属读者在线服务交流圈

关于作者

李杰，资深前端工程师、技术总监，拥有10年以上一线开发经验，以及丰富的Web前端和移动端开发经验，擅长HTML5、JavaScript、Vue、uni-app、React、PHP、MySQL、混合式App开发。

程序思维创始人，曾经是八维教育实训主任、千锋教育高级HTML5前端讲师、尚品中国联合创始人、学橙教育联合创始人，曾与许多大型企业合作开发项目，拥有百余客户、千余案例。

网易签约讲师，学生千余名；CSDN签约讲师，学生万余名；51CTO签约讲师，学生千余名。各大平台视频课程销量前茅，学生高薪就业，工作稳定，好评过万。

2018年创业开发的学橙教育App入围APICloud教育行业优秀案例。

2020年畅销React视频课程受CSDN邀请入选平台“前端入门到全栈一卡通”套餐课，课程大卖，好评不断。

作　者

2022年5月

目　录

第1章

初识 uni-app

uni-app是一个使用 Vue.js 开发所有前端应用的框架，开发者编写一套代码，即可发布到iOS、Android、H5及各种小程序（如微信、支付宝、百度、头条、QQ、钉钉、淘宝等）、快应用等多个平台。

扫一扫，看视频

1.1 uni-app 框架介绍

本节介绍uni-app的概念，以及选择uni-app的原因。

1.1.1 uni-app 概述

uni，译音为“优你”，含义是统一。uni-app是DCloud公司开发的一个基于Vue.js 开发所有前端应用的框架。

uni-app官网:https://uniapp.dcloud.io。

将一套代码发布到10个平台上并不是梦想。眼见为实，各位读者可扫描以下10个二维码，亲自体验最全面的跨平台效果，如图1-1所示。

图 1-1　10 个平台案例展示

1.1.2 选择 uni-app 的原因

uni-app在开发者数量、案例数量、跨平台能力、扩展灵活性、性能体验、周边生态、学习成本、开发成本八大关键指标上拥有巨大的优势。

1. 开发者 / 案例数量更多

uni-app涉及几十万应用、开发者数是上亿、有多个社群在讨论uni-app。

2. 跨平台能力及扩展灵活性更强

(1)开发者编写一套代码，可发布到iOS、Android及各种小程序、快应用等多个平台。

(2)在跨平台的同时，通过条件编译和平台特有API调用，可以“优雅”地为某平台编写个性化代码，调用专有能力而不影响其他平台。

(3)支持原生代码混写和原生SDK集成。

3. 性能体验优秀

(1)体验更好的Hybrid框架，加载新页面时速度更快。

(2) App端支持Weex原生渲染，可支撑更流畅的用户体验。

(3)小程序端的性能优于市场其他框架。

4. 周边生态丰富

（1）插件市场拥有数千款插件。插件市场地址为https://ext.dcloud.net.cn。

（2）支持NPM（Node Package Manager，Node.js包管理和分发工具），支持小程序组件和SDK，兼容mpvue组件和项目，兼容Weex组件。

（3）微信生态的各种SDK可直接用于跨平台App。

5. 学习成本低

基于通用的前端技术栈，采用Vue语法和微信小程序API，无额外学习成本。

6. 开发成本低

（1）招聘、管理、测试各方面的成本都大幅下降。

（2）HBuilderX是高效开发神器，熟练掌握后研发效率至少翻倍。

uni-app的功能框架如图1-2所示。

图 1-2　uni-app 的功能框架

从图1-2可以看出，uni-app在跨平台过程中不牺牲平台特色，可“优雅”地调用平台专有能力，真正做到海纳百川、各取所长。

1.2　uni-app 项目介绍及学习技巧

1. 项目展示及功能介绍

实战项目是一个仿美团点餐小程序客户端开发，扫描图1-3所示的二维码可以欣赏案例。

图 1-3 好运买点餐小程序二维码

该项目的核心功能如下：附近商家展示、搜索菜品、菜品展示、套餐、购物车、堂内点餐、自提、外卖配送、积分商城、地图定位显示商家位置、会员登录自动获取手机号、充值、优惠券、微信支付、退款等。

2. 学习流程和学习技巧

学习uni-app必须有Vue基础，最好有Vue项目开发经验，建议读者在学习uni-app之前先学习一下Vue。在学习uni-app时一定要会查看官方的技术文档，在开发中遇到问题时要会使用官方的技术文档解决问题。

1.3 小结

uni-app是一个使用Vue.js开发前端应用的框架，开发一套代码可以发布到10个平台，如iOS、Android、Web（响应式）以及各种小程序（微信、支付宝、百度、头条、QQ等）、快应用等多个平台。uni-app在跨平台的过程中，不牺牲平台特色，可简单地调用平台专有技术，真正做到海纳百川、各取所长。本章主要介绍了uni-app的特性及优势，可以使读者更深入地了解uni-app，以及更熟练地掌握项目演示和开发技巧，在学习uni-app之前一定要先学习Vue，更要学会查看uni-app的官方文档。

第2章

uni-app 环境搭建

扫一扫，看视频

要使用uni-app，需要安装HBuilder X，在HBuilder X软件中运行项目，编写代码。当然，也可以使用Vscode、WebStorm等软件编写代码，编写完成后用HBuilder X软件运行项目。

本章推荐配合视频教程学习。

2.1 下载与使用 HBuilder X 开发者工具

在使用uni-app项目之前，必须使用HBuilder X开发者工具运行项目和生成生产环境的代码。

2.1.1 下载 HBuilder X

HBuilder X下载地址：https://www.dcloud.io/hbuilderx.html

光标移动到more下三角按钮，会出现一个下拉菜单，如图2-1 所示。

图 2-1 HBuilder X 下载窗口

此处有两个版本：正式版和Alpha版（内部测试版）。选择正式版，根据自己的操作系统选择是下载Windows版还是MacOS版。其中，Windows版是免安装绿色版，下载后需要使用解压缩软件解压，打开解压后的目录，运行HBuilderX.exe文件即可运行软件；MacOS版直接双击HBuilderX.3.6.4.20220922.dmg进行安装。

2.1.2 创建项目

使用HBuilder X创建项目有两种方式。

1. 通过 HBuilder X 可视化界面创建项目

这里以MacOS版为例演示，Windows版和MacOS版的操作方式是一样的。打开HBuilder X，选择“文件”→“新建”→“1.项目”命令，如图2-2 所示。

弹出“新建项目”对话框，如图2-3 所示。

左侧菜单栏默认选择的是uni-app，自定义项目名称为uniappdemo，在“选择模板”处选择“默认模板”，“Vue版本选择”默认为2，本书的教程基于Vue3，所以选择3。此处Vue版本选择的作用只是让HBuilder X生成对应Vue版本的初始化代码，在开发过程中HBuilder X同时支持Vue2 和Vue3 版本的代码，所以即便选错了，后期也可以在编码时更改。单击“创建”按钮，完成初始化项目的创建。

项目创建完成后即可运行，运行项目时必须打开项目中的任意一个文件，如打开App.vue，如图2-4所示。

以“浏览器运行模式”为例，选择“运行”→“运行到浏览器”命令，选择浏览器，按照软件的提示操作即可，待运行成功后可以在浏览器中体验uni-app的H5版。

注意：在HBuilder X提示编译工具安装成功后，重新运行。如果重新运行后没有反应，则重启HBuilder X软件，再次运行即可。

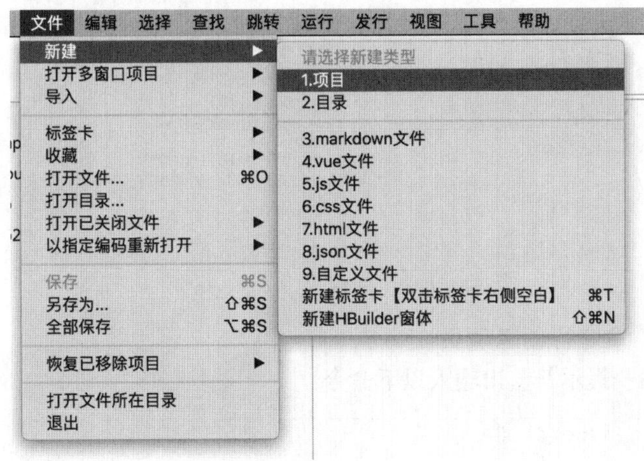

图2-2　选择“1.项目”命令

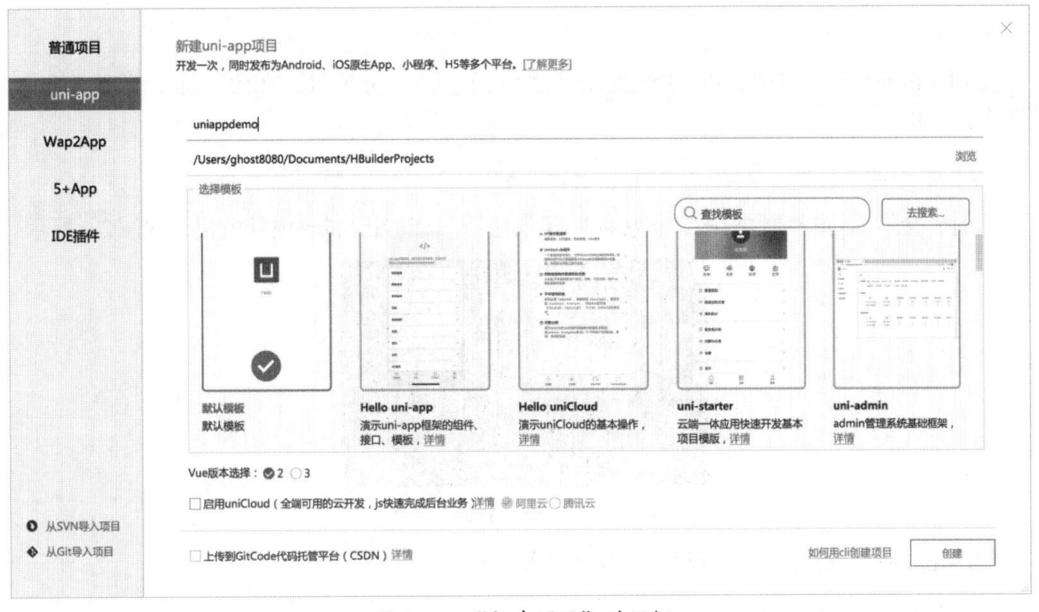

图2-3　“新建项目”对话框

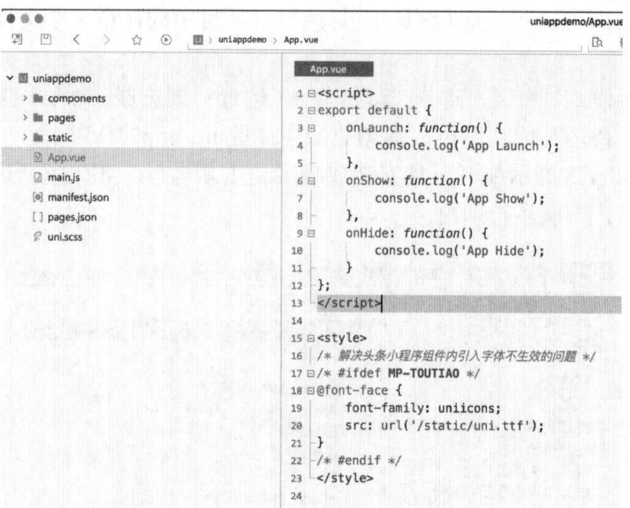

图 2-4 打开 App.vue 文件

2. 通过 vue-cli 命令行创建项目

安装环境，在命令提示符窗口输入以下命令：

```
npm install -g @vue/cli@4
```

全局安装vue-cli，如果安装过vue-cli可省略此步骤。

使用正式版（对应HBuilder X最新正式版）创建uni-app项目，在命令提示符窗口输入以下命令：

```
vue create -p dcloudio/uni-preset-vue 项目名称
```

使用Alpha版（对应HBuilder X最新Alpha版）创建uni-app项目，在命令提示符窗口输入以下命令：

```
vue create -p dcloudio/uni-preset-vue#alpha 项目名称
```

此时，会提示选择项目模板，初次体验建议选择 Hello uni-app 项目模板，如图2-5所示。

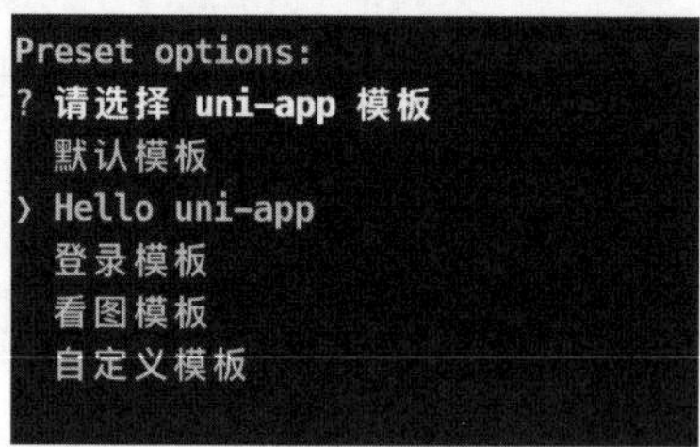

图 2-5 选择 uni-app 模板

创建完成后，使用以下命令运行和发布项目：

```
npm run dev:%PLATFORM%   //运行项目
npm run build:%PLATFORM% //发布项目
```

%PLATFORM% 的取值见表2-1。

表 2-1　%PLATFORM% 的取值

值	平　台
app-plus	App 平台生成打包资源（支持 npm run build:app-plus，可用于持续集成。不支持 run，运行调试仍需在 HBuilder X 中操作）
h5	H5 Web 版，在浏览器中运行
mp-alipay	支付宝小程序
mp-baidu	百度小程序
mp-weixin	微信小程序
mp-toutiao	字节跳动小程序
mp-qq	QQ 小程序
mp-360	360 小程序
mp-kuaishou	快手小程序
mp-jd	京东小程序
mp-xhs	小红书小程序
quickapp-webview	快应用 (webview)
quickapp-webview-union	快应用联盟
quickapp-webview-huawei	快应用华为

以运行、发布微信小程序为例，输入以下命令：

```
npm run dev:mp-weixin //运行微信小程序
npm run build:mp-weixin //发布微信小程序
```

这样就可以运行、发布微信小程序了。当然，运行微信小程序必须要安装“微信小程序开发者工具”，uni-app微信小程序调试会在2.3节讲解。

uni-app标准的项目结构如图2-6所示。使用HBuilder X创建的项目，有些文件夹是没有的，可以根据业务需求自行创建文件夹。

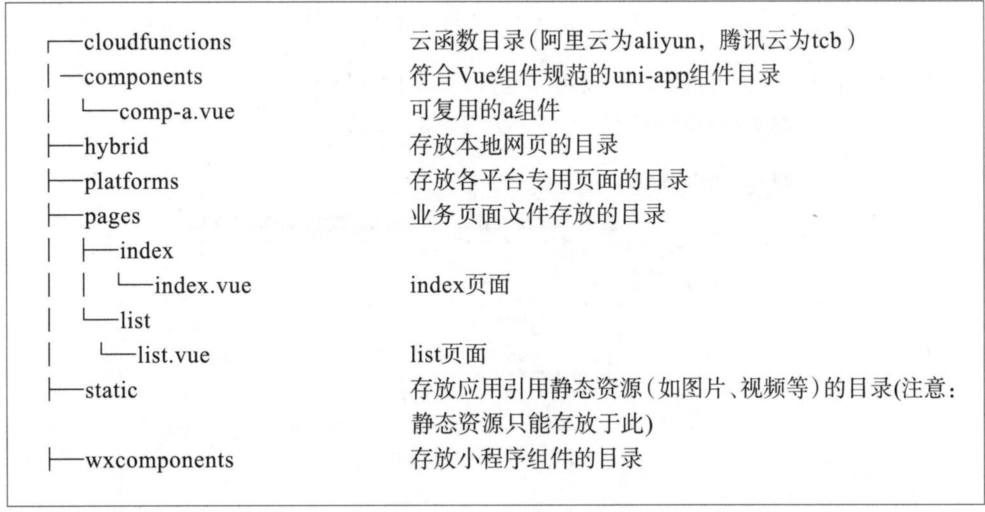

图 2-6　uni-app 标准的项目结构

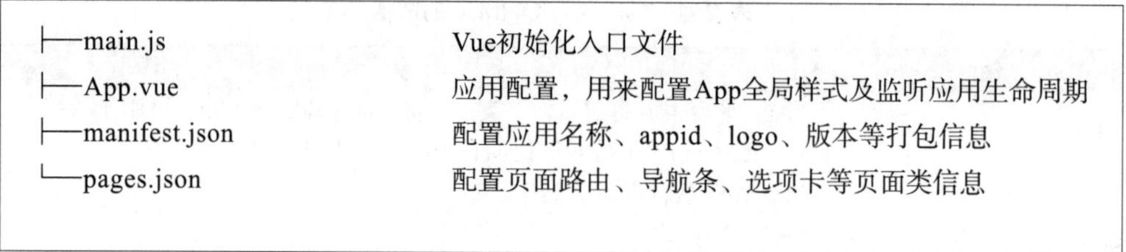

```
├──main.js            Vue初始化入口文件
├──App.vue            应用配置，用来配置App全局样式及监听应用生命周期
├──manifest.json      配置应用名称、appid、logo、版本等打包信息
└──pages.json         配置页面路由、导航条、选项卡等页面类信息
```

图 2-6　uni-app 标准的项目结构（续）

2.2　App 真机运行

开发Android手机App通常使用Windows操作系统，使用uni-app开发App在运行调试时可以使用模拟器或真机。

2.2.1　真机运行

真机运行需要连接手机，开启USB调试，进入uniappdemo项目，选择“运行”→“运行到手机或模拟器”命令，在其下拉菜单中选择运行的设备，即可在该设备中体验uni-app，如图2-7所示。

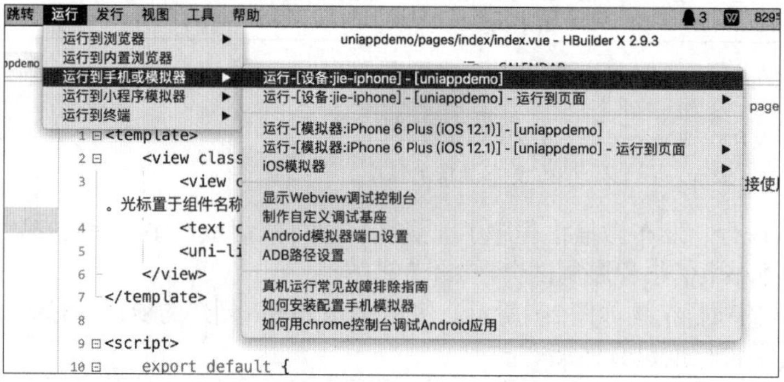

图 2-7　真机运行

如果不用真机运行，也可以使用模拟器运行，如图2-8所示。

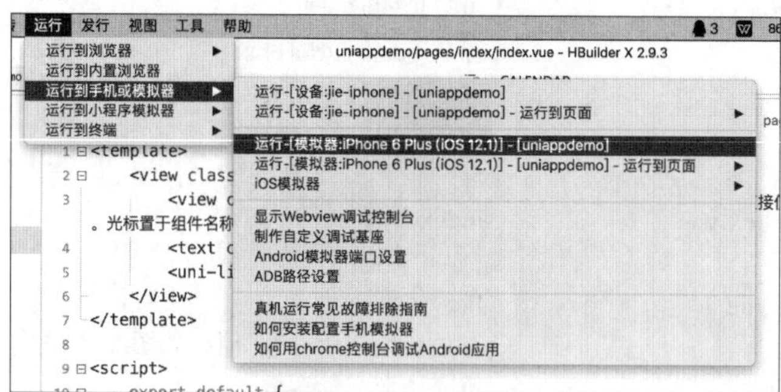

图 2-8　模拟器运行

2.2.2 打包发行

将App打包为原生App有两种模式——云端和离线。

1. 云端

打开HBuilder X，选择“发行”→“原生App-云打包”命令，如图2-9所示，打开图2-10所示的云端打包界面。

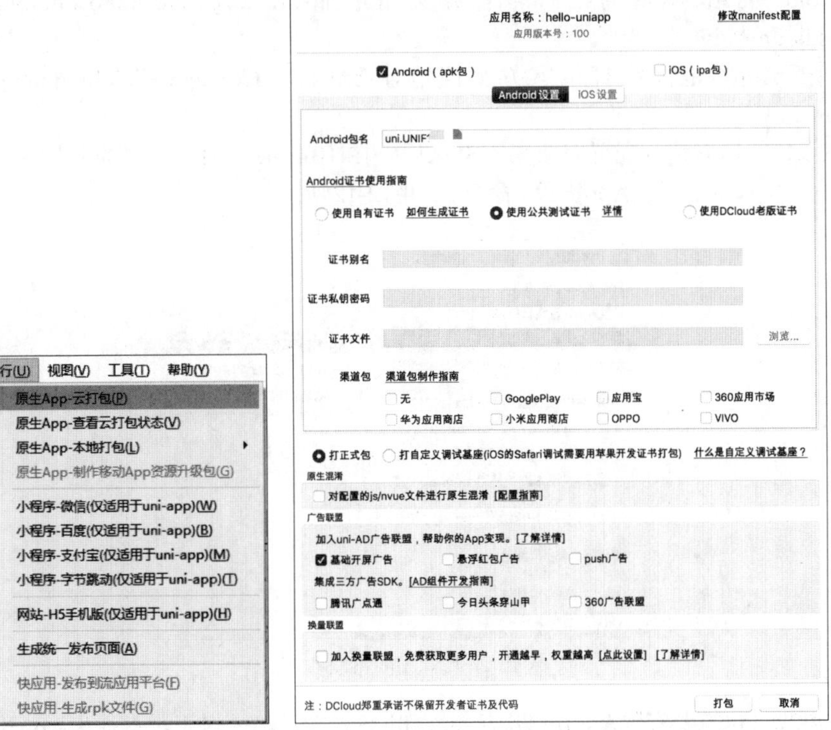

图 2-9 选择“原生 App- 云打包”命令　　　　图 2-10 App 云端打包界面

在App正式运营时要选择自有证书，如果不知道如何生成证书，可单击“如何生成证书”超链接按照教程自己生成。如果不生成iOS版的App包，则取消勾选iOS(ipa包)复选框。单击“打包”按钮，进入打包状态。注意，必须在https://www.dcloud.io官网注册成为会员并登录才能使用云端打包功能，如果没有注册或登录会员，HBuilder X会给出提示和注册地址，按照提示操作即可。

打包成功后，在HBuilder X软件的控制台会给出下载apk包的地址，如图2-11所示。

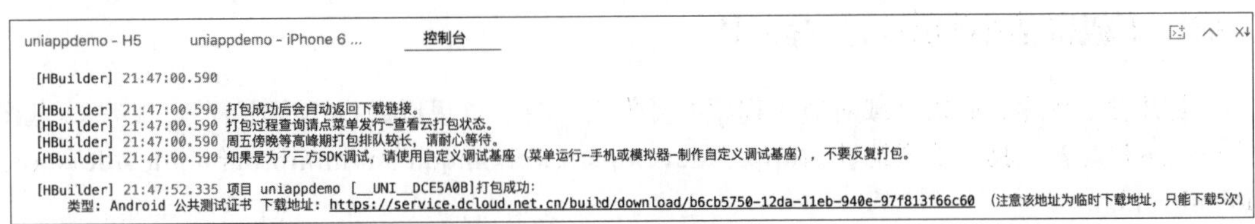

图 2-11 apk 包下载地址

2. 离线

离线打包配置比较复杂，需要使用App离线开发者工具包，即App离线SDK。把App运行环境（runtime）封装为原生开发调用接口，开发者可以在自己的Android及iOS原生开发环境配置工程中使用，包括Android离线开发SDK和iOS离线开发SDK。

App离线SDK主要用于App本地离线打包及扩展原生能力，对应HBuilder X的云端打包功能。当uni-app、5+ App等项目发行为原生App时，无须将App资源及打包要使用的签名证书等提交到云端打包服务器，在开发者本地配置的原生开发环境中即可生成安装apk/ipa包。

Android平台App本地离线打包官方文档地址：https://nativesupport.dcloud.net.cn/AppDocs/usesdk/android。

iOS平台App本地离线打包官方文档地址：https://nativesupport.dcloud.net.cn/AppDocs/usesdk/ios。

按照官方文档安装及配置完成离线SDK后，在HBuilder X中选择“发行”→“原生App-本地打包”→“生成本地打包App资源”命令，如图2-12所示。

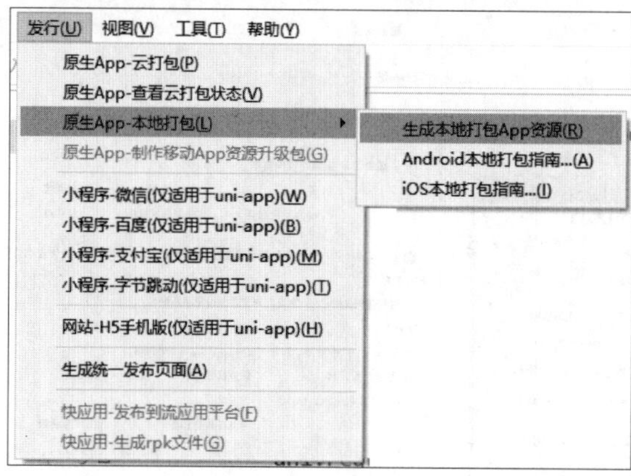

图 2-12　App 离线打包

由于离线打包需要安装Android开发环境和iOS开发环境，对于没有原生App开发基础的人员来说，安装及配置难度较高，因此推荐使用云端打包。

2.3　uni-app 微信小程序调试

使用uni-app开发小程序非常方便，其开发效率远远高于原生小程序开发。本书的实战项目也是小程序项目。

2.3.1　下载微信小程序开发者工具

要开发小程序，必须下载对应小程序的开发者工具。这里以微信小程序为例，首先下载微信小程序开发者工具，下载地址为https://developers.weixin.qq.com/miniprogram/dev/devtools/download.html。

打开网址，进入如图2-13所示页面。

图 2-13　下载微信小程序开发者工具页面

这里选择稳定版，下载的是macOS版本。下载完成后，进行安装，根据提示操作即可。安装完成后，要使用uni-app开发，需要打开微信小程序开发者工具的服务端口，进入微信小程序开发者工具，选择“设置”→“安全设置”命令，打开如图2-14所示的窗口，将“服务端口”改为打开状态。

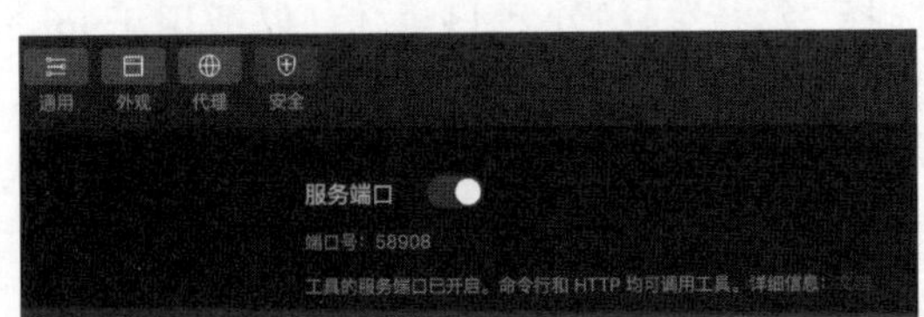

图 2-14　微信小程序开发者工具服务端口

2.3.2　运行项目

安装并配置完成微信小程序开发者工具后，打开HBuilder X，进入项目，选择“运行”→“运行到小程序模拟器”→“微信开发者工具”命令，即可在微信小程序开发者工具中体验uni-app，如图2-15所示。

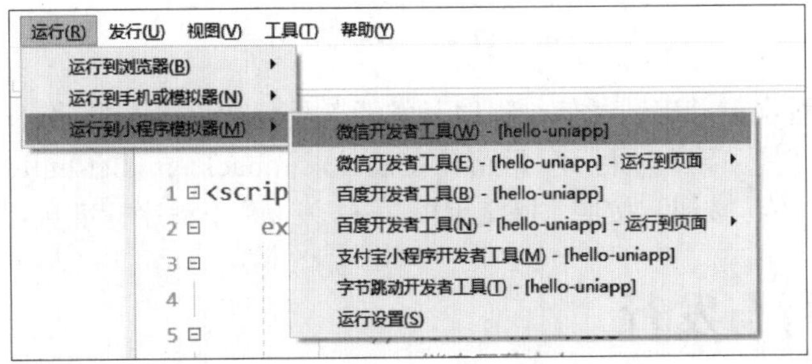

图 2-15　使用 HBuilder X 运行微信小程序

注意：

如果是第一次使用，需要先配置小程序IDE的相关路径才能运行成功。选择“运行”→运行到小程序模拟器→“运行设置”命令，在图2-16所示的位置输入微信小程序开发者工具的安装路径即可。

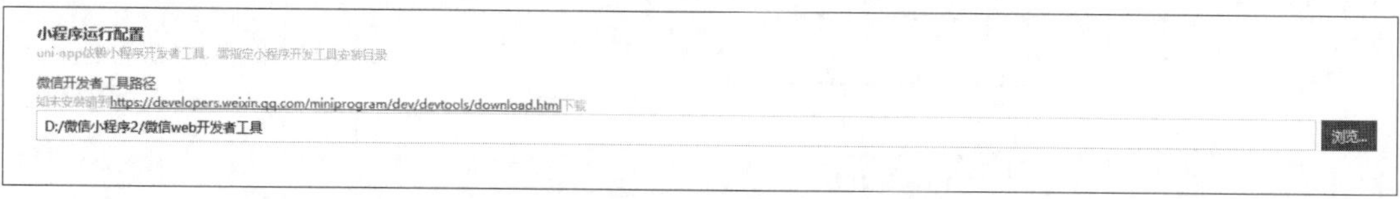

图 2-16　配置微信小程序开发者工具路径

uni-app默认把项目编译到项目根目录的unpackage目录中。

2.3.3　发行小程序

发行微信小程序必须有微信小程序AppID。在微信公众号平台（网址为https://mp.weixin.qq.com）注册账号，进入小程序管理后台，选择“开发”→“开发设置”命令，即可在打开的页面中获取微信小程序AppID。

在HBuilder X中，选择“发行”→“小程序-微信（仅适用于uni-app）”命令，弹出“微信小程序发行”对话框，输入小程序名称和AppID，单击“发行”按钮，即可在 unpackage/dist/build/mp-weixin中生成微信小程序项目代码，如图2-17所示。

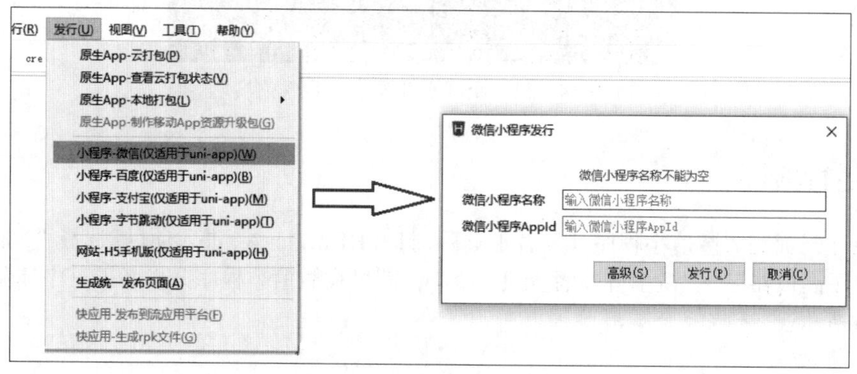

图 2-17　发行微信小程序

稍等片刻，系统会自动启动微信小程序开发者工具。如果没有自动启动，则需要手动打开微信小程序开发者工具。导入项目，项目路径在根目录unpackage/dist/build/mp-weixin 文件中，项目导入后单击“上传”按钮，按照“提交审核”→“发布”小程序标准流程逐步操作即可。

2.4　H5 运行与发行

H5一般指HTML 5，是构建Web内容的一种语言描述方式，是指Web端在浏览器中运行，

HBuilder X提供了Web运行环境，用户只需要选择运行的浏览器。

2.4.1 运行项目

打开项目，选择“运行”→“运行到浏览器”命令，在其子菜单中选择合适的浏览器，如Chrome，即可在浏览器中体验uni-app的H5版，如图2-18所示。

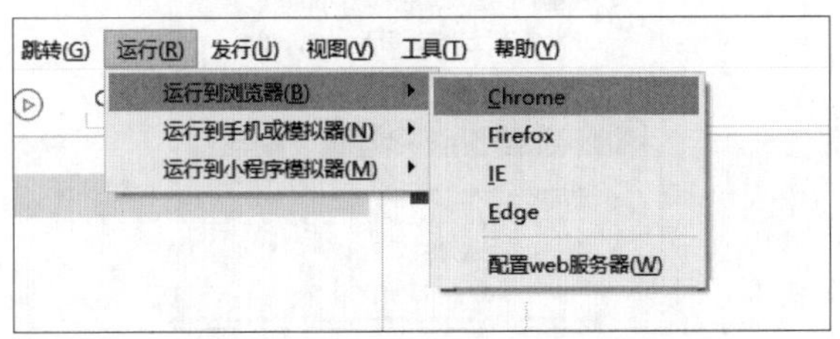

图 2-18 运行 H5 项目到浏览器

2.4.2 发行项目

使用HBuilder X打开项目根目录下的manifest.json文件，进入可视化界面，如图2-19所示。

图 2-19 manifest.json 文件 H5 配置可视化界面

uni-app中的路由模式和Vue一样，具有hash和history两种模式。这里选择hash模式。应用的基础路径相当于vue.config.js配置文件中的publicPath选项，如发行在网站根目录，可不配置应用基本路径。

在HBuilder X中，选择“发行”→“网站-H5手机版(仅适用于uni-app)”命令，即可生成H5的相关资源文件，保存于unpackage目录，如图2-20所示。

图 2-20 发布 H5 手机版

生成的文件保存在unpackage/dist/build/h5文件夹中，将h5文件夹中的所有文件上传到服务器即可。

2.5 小结

"工欲善其事，必先利其器"，这一章主要讲解了uni-app的环境搭建，如何下载与使用HBuilder X创建项目，如何真机运行和打包发行App以及如何下载、运行及发行微信小程序开发者工具等内容。需要注意的是，使用HBuilder X运行微信小程序时，需要在微信开发者工具中打开服务端口。从本章开始，我们就正式学习uni-app了，建议大家把环境搭建起来，跟着我们一起学习uni-app。

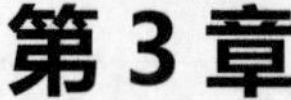

第3章 生命周期

uni-app生命周期是以小程序的生命周期为基础实现的，分为应用生命周期、页面生命周期和组件生命周期，其中组件生命周期就是Vue的生命周期。

扫一扫，看视频

生命周期官网文档地址:https://uniapp.dcloud.io/frame?id=生命周期。

3.1 应用生命周期

uni-app支持的应用生命周期函数见表3-1。

表 3-1 应用生命周期函数

函数名	说 明
onLaunch	当 uni-app 初始化完成时触发（全局只触发一次）
onShow	当 uni-app 启动或从后台进入前台时触发（监听用户进入小程序）
onHide	当 uni-app 从前台进入后台时触发（监听用户离开小程序）
onError	当 uni-app 报错时触发
onUniNViewMessage	对 nvue 页面发送的数据进行监听
onUnhandledRejection	对未处理的 Promise 拒绝事件进行监听
onPageNotFound	页面不存在监听函数
onThemeChange	监听系统主题变化

其中，onLaunch、onShow、onHide这3个生命周期钩子函数在开发中经常使用。注意，应用生命周期仅可在App.vue文件中监听，在其他页面监听无效。尽量不要在onLaunch钩子函数中进行页面跳转，如果进行页面跳转，可能会出现白屏报错问题，其原因可能会和pages.json内配置的第一个页面跳转时出现冲突，可以使用延迟进行跳转处理，如setTimeout。

3.2 页面生命周期

页面生命周期写在pages文件夹下的文件中。页面生命周期函数见表3-2。

表 3-2 页面生命周期函数

函数名	说 明	平台差异说明	最低版本
onLoad	监听页面加载，其参数为上一个页面传递的数据，参数类型为 Object（用于页面传参）		
onShow	监听页面显示。页面每次出现在屏幕上都触发，包括从下级页面返回当前页面		
onReady	监听页面初次渲染完成。注意，如果渲染速度快，会在页面进入动画完成前触发		
onHide	监听页面隐藏		
onUnload	监听页面卸载		

续表

函数名	说　明	平台差异说明	最低版本
onResize	监听窗口尺寸变化	App、微信小程序	
onPullDownRefresh	监听用户下拉动作，一般用于下拉刷新		
onReachBottom	页面上拉触底事件的处理函数		
onTabItemTap	点击 TabBar 时触发，参数为 Object 类型	微信小程序、百度小程序、H5、App（自定义组件模式）	
onShareAppMessage	用户点击右上角分享（可以在分享时设置分享标题、路径等）		
onPageScroll	监听页面滚动，参数为 Object 类型	微信小程序、百度小程序、字节跳动小程序、支付宝小程序	
onNavigationBarButtonTap	监听原生标题栏按钮点击事件，参数为 Object 类型	5+ App、H5	
onBackPress	监听页面返回，返回 event = {from:backbutton| navigateBack}，其中，backbutton 表示来源是左上角返回按钮或 Android 返回键；navigateBack 表示来源是 uni.navigateBack	App、H5	
onNavigationBarSearchInputChanged	监听原生标题栏搜索输入框输入内容变化事件	App、H5	1.6.0
onNavigationBarSearchInputConfirmed	监听原生标题栏搜索输入框搜索事件，用户点击软键盘上的“搜索”按钮时触发	App、H5	1.6.0
onNavigationBarSearchInputClicked	监听原生标题栏搜索输入框点击事件	App、H5	1.6.0
onShareTimeline	监听用户点击右上角转发到朋友圈	微信小程序	2.8.1+
onAddToFavorites	监听用户点击右上角收藏	微信小程序	2.8.1+

注：“平台差异说明”一列中有内容的代表函数（属性或方法等）只在所列平台中可用，无内容的代表全平台通用。其余表中与此相同。

表3-2中字体加粗的页面生命周期函数是开发项目常用的函数，其中，onPullDownRefresh可以实现下拉刷新，onReachBottom可以实现上拉加载数据。这些常用的页面生命周期函数会在后面的实战项目中使用并详细讲解，在这里只需简单了解。

3.3 组件生命周期

组件生命周期写在components文件夹下的文件中，和Vue标准组件生命周期相同，其函数见表3-3。

表 3-3 组件生命周期函数

函数名	说 明
before Create	在实例初始化之后被调用
created	在实例创建完成后被立即调用
beforeMount	在挂载开始之前被调用
mounted	挂载到实例上之后被调用（该函数在服务器端渲染期间不被调用）
beforeUpdate	数据更新时被调用，发生在虚拟 DOM 打补丁之前
updated	由于数据更改导致的虚拟 DOM 重新渲染和打补丁，在这之后会调用此函数
beforeDestroy	在实例销毁之前被调用
destroyed	在 Vue 实例销毁后被调用

3.4 小结

本章主要介绍了uni-app的生命周期，主要分为应用生命周期、页面生命周期和组件生命周期。其中，应用生命周期仅可在App.vue中监听，在其他页面监听无效。另外，尽量不要在onLaunch钩子函数中进行页面跳转，如果进行页面跳转，可能会出现白屏报错问题。页面生命周期函数中，onPullDownRefresh可以实现下拉刷新，onReachBottom可以实现上拉加载数据。组件生命周期就是Vue标准组件的生命周期。

第4章

尺寸单位、Flex布局与背景图片

uni-app支持的通用CSS单位包括 px、rpx。为支持跨平台，在搭建框架时，建议使用Flex布局。uni-app支持在CSS中设置背景图片，使用方式与普通Web项目大体相同。

扫一扫，看视频

4.1 尺寸单位

uni-app支持的通用CSS单位包括px、rpx，其中px为屏幕像素，rpx为响应式px。

rpx是一种根据屏幕宽度自适应的动态单位。以750宽的屏幕为基准，750rpx恰好为屏幕宽度。屏幕变宽，rpx 实际显示效果会等比放大，但在App端和H5端，屏幕宽度达到960px时，默认将按照375px的屏幕宽度进行计算。我们在开发移动端项目时选择rpx作为尺寸单位。

设计师一般只提供一个分辨率的设计图，如果严格按照设计图的px进行开发，在不同宽度的手机上，界面很容易变形，并且主要是宽度变形，因为有滚动条，所以在高度上一般不容易出问题。由此，引发了较强的动态宽度单位需求，微信小程序设计了rpx，以解决该问题。uni-app在App端、H5端、小程序都支持rpx，并且可以配置不同屏幕宽度的计算方式。rpx是相对于基准宽度的单位，可以根据屏幕宽度进行自适应。uni-app规定屏幕基准宽度为750rpx。

页面元素在uni-app中的宽度计算公式如下：

```
750 × 元素在设计稿中的宽度 / 设计稿基准宽度
```

举例说明如下：

（1）若设计稿宽度为750px，元素 A 在设计稿上的宽度为 100px，那么元素 A 在 uni-app 中的宽度应该设计为 750 × 100 / 750，即 100rpx。

（2）若设计稿宽度为 640px，元素 B 在设计稿上的宽度为 100px，那么元素 B 在 uni-app 中的宽度应该设计为 750 × 100 / 640，即约 117rpx。

（3）若设计稿宽度为 375px，元素 C 在设计稿上的宽度为 200px，那么元素 C 在uni-app 中的宽度应该设计为 750 × 200 / 375，即 400rpx。

4.2 Flex 布局

为了支持跨平台，uni-app建议使用Flex布局（Flexible Box，弹性布局）。

传统的布局基于盒状模型，依赖 display 属性、position属性和float属性。如果使用Flex布局，则不建议使用float属性。

4.2.1 Flex 布局概述

Flex用来为盒状模型提供最大的灵活性。

任何一个容器都可以指定为 Flex 布局。代码示例如下：

```
.box{
    display: flex;
}
```

行内元素也可以使用 Flex 布局。代码示例如下：

```
.box{
    display: inline-flex;
}
```

采用 Flex 布局的元素称为 Flex 容器（flex container），简称容器，如图4-1所示。它的所有子元素自动成为容器成员，称为 Flex 项目（flex item），简称项目。

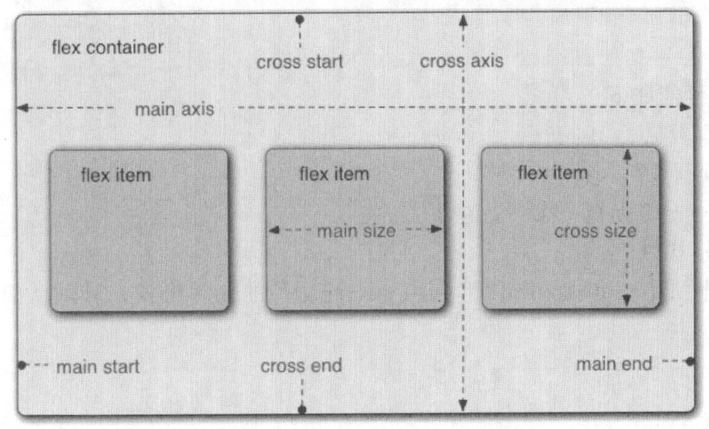

图 4-1　flex container 示意图

容器默认存在两条轴：水平的主轴（main axis）和垂直的交叉轴（cross axis）。主轴的开始位置（与边框的交叉点）称为main start，结束位置称为main end；交叉轴的开始位置称为cross start，结束位置称为cross end。

项目默认沿主轴排列。单个项目占据的主轴空间称为main size，占据的交叉轴空间称为cross size。

4.2.2　容器的属性

Flex布局支持6个容器属性:flex-direction、flex-wrap、flex-flow、justify-content、align-items、align-content。

1. flex-direction 属性

flex-direction属性决定主轴的方向（项目的排列方向）。代码示例如下：

```
.box {
    display:flex;
    flex-direction: row;
}
```

flex-direction属性值如下：

（1）row（默认值）：主轴为水平方向，起点在左端。

（2）row-reverse：主轴为水平方向，起点在右端。

（3）column：主轴为垂直方向，起点在上沿。

（4）column-reverse：主轴为垂直方向，起点在下沿。

2. flex-wrap 属性

在默认情况下，项目都排在一条线（又称轴线）上。如果一条轴线无法排列所有项目，可以换行（换到下一条线上）。代码示例如下：

```
.box {
    display:flex;
    flex-wrap: wrap;
}
```

flex-wrap属性值如下：

（1）nowrap（默认）：不换行。

（2）wrap：换行，第一行在上方。

（3）wrap-reverse：换行，第一行在下方。

3. flex-flow 属性

flex-flow属性是flex-direction属性和flex-wrap属性的简写形式，其默认值为row nowrap。代码示例如下：

```
.box {
    display:flex;
    flex-flow: row nowrap;
}
```

4. justify-content 属性

justify-content属性定义了项目在主轴上的对齐方式。代码示例如下：

```
.box {
    display:flex;
    justify-content: flex-start;
}
```

justify-content属性值如下：

（1）flex-start（默认值）：左对齐。

（2）flex-end：右对齐。

（3）center：居中。

（4）space-between：两端对齐，项目之间的间隔都相等。

（5）space-around：每个项目两侧的间隔相等。因此，项目之间的间隔比项目与边框的间隔大一倍。

5. align-items 属性

align-items属性定义了项目在交叉轴上如何对齐。代码示例如下：

```
.box {
    display:flex;
    align-items: flex-start;
}
```

align-items属性值如下：

（1）flex-start：交叉轴的起点对齐。

（2）flex-end：交叉轴的终点对齐。

（3）center：交叉轴的中点对齐。

（4）baseline：项目的第一行文字的基线对齐。

（5）stretch（默认值）：如果项目未设置高度或设为auto，则将占满整个容器的高度。

6. align-content 属性

align-content属性定义了多条轴线的对齐方式。如果项目只有一条轴线，则该属性不起作用。代码示例如下：

```
.box {
    display:flex;
    align-content: flex-start;
}
```

align-content属性值如下：

（1）flex-start：与交叉轴的起点对齐。

（2）flex-end：与交叉轴的终点对齐。

（3）center：与交叉轴的中点对齐。

（4）space-between：与交叉轴两端对齐，轴线之间的间隔相等。

（5）space-around：每条轴线两侧的间隔都相等。因此，轴线之间的间隔比轴线与边框的间隔大一倍。

（6）stretch（默认值）：轴线占满整个交叉轴。

4.2.3　项目的属性

Flex布局支持6个项目属性：order、flex-grow、flex-shrink、flex-basis、flex、align-self。

1.order 属性

order属性定义项目的排列顺序。order数值越小，排列越靠前，默认为0。代码示例如下：

```
.item {
    order: 1;
}
```

order属性示例如图4-2所示。

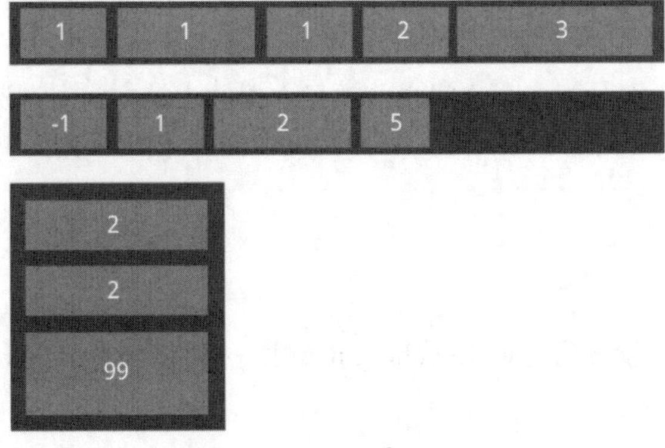

图 4-2　order 属性示例

2. flex-grow 属性

flex-grow属性定义项目的放大比例，默认为0，即如果存在剩余空间，也不放大。代码示例如下：

```
.item {
    flex-grow: 0;
}
```

flex-grow属性示例如图4-3所示。

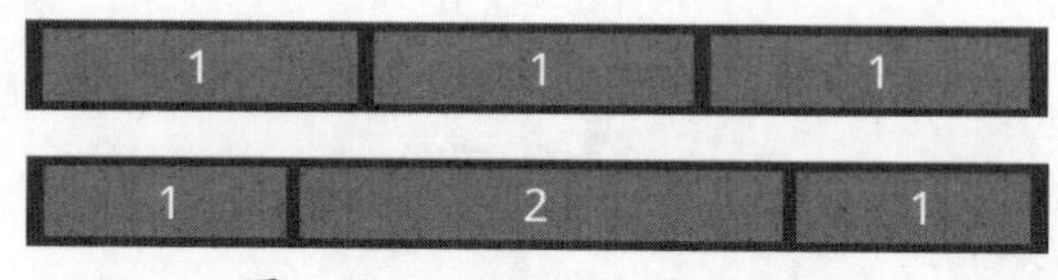

图 4-3 flex-grow 属性示例

如果所有项目的flex-grow属性值都为1，则它们将等分剩余空间（如果有剩余空间）；如果一个项目的flex-grow属性值为2，其他项目都为1，则前者占据的剩余空间将比其他项目多一倍。

3. flex-shrink 属性

flex-shrink属性定义项目的缩小比例，默认为1，即如果空间不足，该项目将缩小。代码示例如下：

```
.item {
    flex-shrink: 1;
}
```

flex-shrink属性示例如图4-4所示。

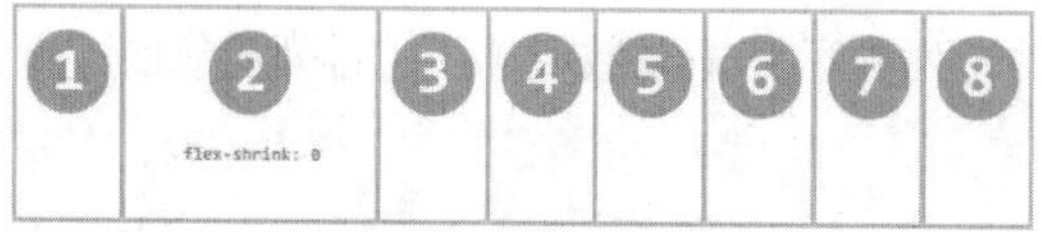

图 4-4 flex-shrink 属性示例

如果所有项目的flex-shrink属性都为1，则当空间不足时，都将等比例缩小；如果一个项目的flex-shrink属性值为0，其他项目都为1，则当空间不足时，前者不缩小。负值对该属性无效。

4. flex-basis 属性

flex-basis属性定义在分配多余空间之前项目占据的主轴空间。浏览器根据该属性计算主轴是否有多余空间。其默认值为auto，即项目的本来大小。代码示例如下：

```
.item {
    flex-basis: auto
}
```

flex-basis属性可以设置为与width或height属性一样的值（如350px），则项目将占据固定空间。

5. flex 属性

flex属性是flex-grow、flex-shrink 和 flex-basis的简写，默认值为0 1 auto，其中后两个属性为可选项。代码示例如下：

```
.item {
    flex: 0 1 auto
}
```

6. align-self 属性

align-self属性允许单个项目有与其他项目不一样的对齐方式，可覆盖align-items属性。其默认值为auto,表示继承父元素的align-items属性,如果没有父元素,则等同于stretch。代码示例如下：

```
.item {
    align-self: auto;
}
```

align-self属性示例如图4-5所示。

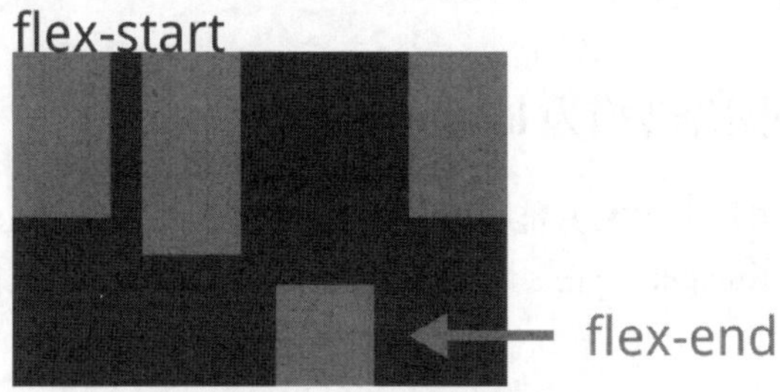

图 4-5　align-self 属性示例

该属性可以取6个值，除了auto外，其他都与align-items属性值完全一致。

4.3　背景图片

uni-app支持在CSS里设置背景图片，设置方式与普通Web项目大体相同，但是也有一些不同，下面介绍注意事项。

4.3.1　使用本地背景图片的问题

在CSS里设置背景图片时，为了多端兼容，需要注意以下几点：

（1）支持 base64 格式图片。

（2）支持网络路径图片。

（3）小程序不支持在CSS中使用本地文件，包括本地的背景图片和字体文件，需要是base64方式才可使用。App端在v3 模式以前也有相同限制，从v3 编译模式起支持直接使用本地背景图片和字体。

使用本地路径背景图片需要注意以下几点：

（1）为方便开发者，当背景图片小于 40KB且uni-app 编译到不支持本地背景图片的平台时，会自动将其转换为 base64 格式。

（2）当背景图片不小于 40KB时，会有性能问题，故不建议使用太大的背景图片。如果开发者必须使用太大的背景图片，则需要自己将其转换为 base64 格式，或将其复制到服务器上，从网络地址引用。

（3）本地背景图片的引用路径推荐使用以 ~@ 开头的绝对路径。

代码示例如下：

```
.test2 {
    background-image: url('~@/static/logo.png');
}
```

注意：

微信小程序真机不支持相对路径，但开发者工具支持。因此，以真机为主的开发不要使用相对路径。

4.3.2 代码演示及把图片转换为 base64 格式

在项目中引入背景图片，观察其能否转换为base64格式。在static文件夹下创建images文件夹，将背景图片复制到images文件夹下，在pages/index/index.vue文件中的代码如下：

```
<template>
<view>
    <view class="bg"></view>
</view>
</template>

<script>
</script>

<style>
    .bg{
        width:200rpx;
        height:200rpx;
        background-image: url("~@/static/images/1.jpg");
        background-size:100%
    }
</style>
```

以上加粗代码为背景图片相关代码。<view>元素是uni-app的组件，uni-app组件名称以小程序命名为基准。有开发微信小程序经验的读者应该对该组件不陌生，在9.1 节会详细讲解<view>组件。接下来使用HBuilder X运行微信小程序，如果运行后在微信开发者工具控制台中报以下错误:Cannot read property 'forceUpdate' of undefined，则需要配置微信小程序的AppID。使用HBuilder X打开mainifest.json文件，选择微信小程序配置，输入自己的微信小程序AppID，勾选“ES6转ES5”“上传代码时自动压缩”“检查安全域名和TLS版本”复选框，如图4-6所示。

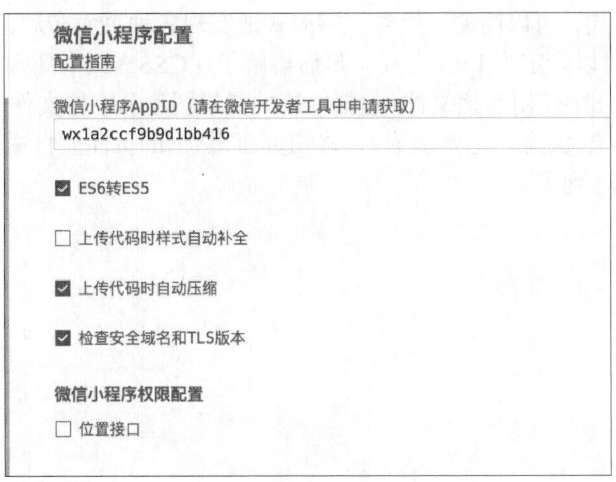

图4-6　微信小程序配置

在微信开发者工具中运行，效果如图4-7所示。

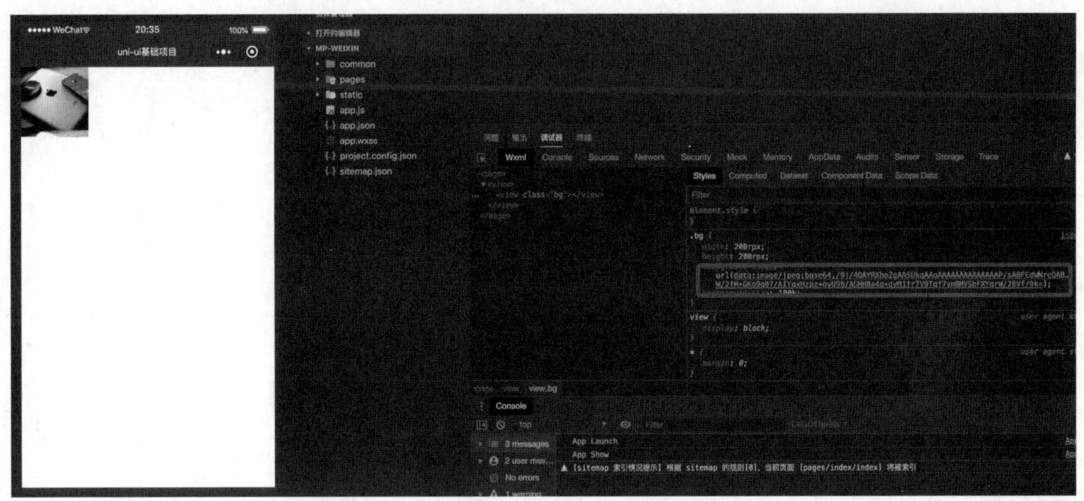

图4-7　背景图片转换为base64格式

由图4-7可以看出，当背景图片小于40KB时，其会自动转换为base64格式；如果背景图片大于40KB，则应使用网络图片。

4.4 小结

本章主要讲解了uni-app中尺寸单位的类型，在开发移动端时，选择rpx作为尺寸单位，同时讲解了rpx换算规则和换算公式。另外，还讲解了Flex布局（弹性布局），这也是微信小程序官方主推的页面布局方式，可以简便、完整、响应式地实现各种页面布局，如垂直居中。任何一个容器、一个项目都可以指定为 Flex 布局。最后讲解了在CSS样式中插入背景图片的注意事项，如小程序不支持在CSS中使用本地文件（包括本地的背景图片和字体文件）。总之，本章讲解的是切页面最基础的知识，大家一定要多练习，多切页面，做出的页面能自适应绝大部分屏幕尺寸，这样前端的基本功就达到了。

第5章 pages.json 配置

pages.json文件用来对 uni-app 进行全局配置，决定页面文件的路径、窗口样式、原生的导航栏、底部的原生tabBar等。

5.1 配置项

pages.json文件用来对uni-app进行全局配置，可以在配置项设置默认页面的窗口、设置页面路径及窗口、设置组件自动引入规则、设置底部tab以及分包加载配置等。

5.1.1 globalStyle 和 pages

扫一扫，看视频

globalStyle用于全局设置应用的状态栏、导航条、标题、窗口背景色等。

globalStyle的官方文档地址:https://uniapp.dcloud.io/collocation/pages?id=globalstyle。

globalStyle的属性选项有很多，使用起来很简单，这里讲解常用的属性，打开pages.json文件，配置如下：

```
{
    "pages": [
        {
            "path": "pages/index/index",
            "style": {
                "navigationBarTitleText": "uni-ui基础项目"  //导航栏标题文字内容
            }
        }
    ],
    "globalStyle": {
        "navigationBarTextStyle": "white",
        //导航栏标题颜色及状态栏前景颜色，仅支持 black/white
        "navigationBarTitleText": "uni-app",              //导航栏标题文字内容
        "navigationBarBackgroundColor": "#007AFF",  //导航栏背景颜色(同状态栏背景色)
        "backgroundColor": "#FFFFFF"                  //下拉显示出来的窗口背景色
    }
}
```

pages可以配置应用由哪些页面组成。pages 节点接收一个数组，数组中的每一项都是一个对象。pages属性见表5-1。

表5-1 pages属性

属　性	类　型	描　述
path	String	配置页面路径
style	Object	配置页面窗口表现。配置项参考地址：https://uniapp.dcloud.io/collocation/pages?id=style

接下来使用pages属性配置新增页面。新建一个“新闻页面”，在pages文件夹下创建news文件夹，在该文件夹下创建index.vue文件。其开发目录如下：

```
┌─pages
│  ├─index
│  │  └─index.vue
│  └─news
│     └─index.vue
├─static
├─main.js
├─App.vue
├─manifest.json
└─pages.json
```

上述加粗代码为新创建的目录及文件。在pages.json中填写如下内容：

```
{
    "pages": [
        {
            "path": "pages/index/index",
            "style": {
                "navigationBarTitleText": "uni-ui基础项目"   //导航栏标题文字内容
            }
        },
        {
            "path": "pages/news/index",
            "style": {
                "navigationBarTitleText": "新闻页面"
            }
        }
    ],
    ...
}
```

注意：

（1）pages节点的第一项为应用入口页（首页）。
（2）应用中新增/减少页面时都需要对pages数组进行修改。
（3）文件名不需要写扩展名，框架会自动寻找路径下的页面资源。
（4）当style配置项与globalStyle配置项相同时，会覆盖globalStyle配置项。

配置完成后保存文件，即可访问“新闻页面”。至于跳转访问“新闻页面”，会在第7章进行讲解。

5.1.2 自定义导航栏使用注意事项

扫一扫，看视频

如果原生导航栏不能满足需求，可以使用自定义导航栏。使用HBuilder X运行微信小程序项目后，在微信开发者工具中渲染的效果即原生导航栏的效果，如图5-1所示。

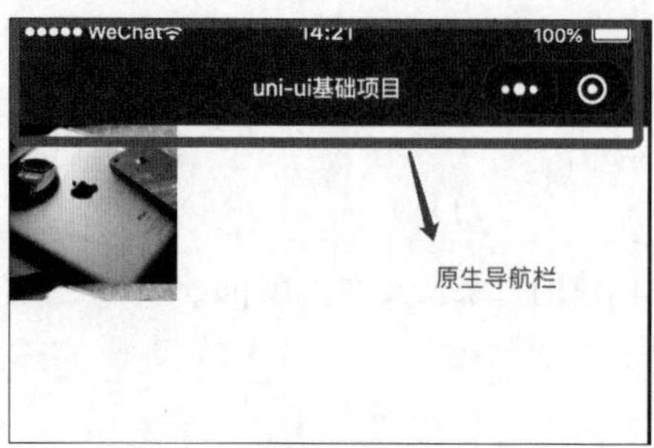

图 5-1 原生导航栏

如果要自定义导航栏，则需要将原生导航栏隐藏。在pages.json文件中配置如下：

```
{
    "pages": [
        {
            "path": "pages/index/index",
            "style": {
                "navigationBarTitleText": "uni-ui基础项目",
                "navigationStyle":"custom"
            }
        },
        ...
    ],
    ...
}
```

当navigationStyle设置为custom时，不再显示原生导航栏，如图5-2所示。

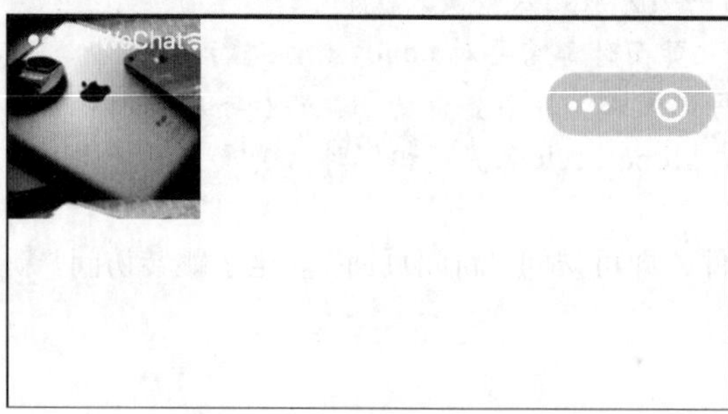

图 5-2 不显示原生导航栏

需要注意的是，如果没有原生导航栏，在非H5端，手机顶部状态栏区域会被页面内容覆盖。这是因为窗体是沉浸式的，即全屏可写内容。uni-app提供了状态栏高度的CSS变量“--status-bar-height”，如果需要把状态栏的位置从前景部分显示出来，可写一个占位div，高度设置为CSS变量。在pages/index/index.vue文件中的代码如下：

```
<template>
    <view>
        <view class="status_bar">
            <!-- 这里是状态栏 -->
        </view>
            <view class="bg"></view>
    </view>
</template>

<script>
</script>

<style>
    .bg{
        width:200rpx;
        height:200rpx;
        background-image: url("~@/static/images/1.jpg");
        background-size:100%
    }
    .status_bar {
        height: var(--status-bar-height);
        width: 100%;
    }
</style>
```

上述加粗代码可以将状态栏显示出来。接下来编写一个导航栏，在pages/index/index.vue文件中的代码如下：

```
<template>
    <view>
        <view class="status_bar">
            <!-- 这里是状态栏 -->
        </view>
        <view class="nav-bar">
            uni-app首页
        </view>
        <view class="bg"></view>
    </view>
```

```
</template>

<script>
</script>

<style>
    .bg{width:200rpx;height:200rpx;background-image: url("~@/static/
    images/1.jpg");background-size:100%}
    .status_bar {height: var(--status-bar-height);width: 100%; background-
    color: #3cc51f;}
    .nav-bar{width:100%;height:80rpx;background-color: #3cc51f;
    color:#FFFFFF;text-align: center;line-height: 80rpx; font-size: 28rpx;}
</style>
```

上述加粗代码为新增代码，在微信开发者工具中渲染的效果如图5-3所示。

图 5-3　自定义导航栏

在后面的实战项目开发中，会根据设计的UI设置不同的自定义导航栏样式。这里需要注意的是，自定义导航栏需要添加CSS变量“--status-bar-height”，这样便可以显示出状态栏。

5.1.3 easycom

扫一扫，看视频

传统Vue组件需要安装、引用、注册后才能使用，而使用easycom组件只需一步。只要组件安装在项目的components目录下，并符合components/组件名称/组件名称.vue目录结构，即可无须引用、注册，直接在页面中使用。代码示例如下：

```
<template>
    <view class="container">
        <uni-list>
            <uni-list-item title="第一行"></uni-list-item>
            <uni-list-item title="第二行"></uni-list-item>
        </uni-list>
    </view>
</template>
```

```
<script>
    // 这里不用import引入，也不需要在components内注册uni-list组件
        export default {
    }
</script>
```

2.1.2节中创建的项目自带UI组件，如uni-list和uni-list-item组件，其存放在components文件夹下，必须符合components/uni-list/uni-list.vue和components/uni-list-item/uni-list-item.vue这样的目录结构和命名规范，才可以支持easycom。打开components文件夹可以看到许多扩展组件，使用方法可以参考官方文档:https://uniapp.dcloud.io/component/README?id=uniui。

easycom是自动开启的，不需要手动开启，有需求时可以在pages.json的easycom节点进行个性化设置，如关闭自动扫描，或自定义扫描匹配组件的策略。pages.json的easycom属性见表5-2。

表5-2　pages.json的easycom属性

属　性	类　型	默认值	说　明
autoscan	Boolean	true	是否开启自动扫描。开启后将会自动扫描符合 components/组件名称 / 组件名称 .vue 目录结构的组件
custom	Object		以正则方式自定义组件匹配规则。如果 autoscan 不能满足需求，则可以使用 custom 自定义匹配规则

自定义easycom配置的示例：如果需要匹配node_modules内的vue文件，需要使用packageName/path/to/vue-file-$1.vue形式的匹配规则，其中packageName为安装的包名，/path/to/vue-file-$1.vue为vue文件在包内的路径。代码示例如下：

```
"easycom": {
    "autoscan": true,
    "custom": {
        "^uni-(.*)": "@/components/uni-$1.vue",
        // 匹配components目录内的vue文件
        "^vue-file-(.*)": "packageName/path/to/vue-file-$1.vue"
        // 匹配node_modules内的vue文件
    }
}
```

注意：

（1）easycom方式引入的组件无须在页面内引入，也不需要在components内声明，即可在任意页面使用。

（2）easycom方式引入的组件不是全局引入，而是局部引入。例如，在H5端只有加载相应页面才会加载使用的组件。

(3)在组件名完全一致的情况下，easycom引入的优先级低于手动引入(区分连字符形式与驼峰形式)。

(4)考虑到编译速度，直接在pages.json内修改easycom不会触发重新编译，需要改动页面内容才会触发。

(5)easycom只处理Vue组件，不处理小程序组件。暂不处理扩展名为.nvue的组件，建议参考uni ui，使用.vue扩展名，同时兼容nvue页面。

(6)nvue页面里的.vue扩展名的组件同样支持easycom。

5.1.4 tabBar

扫一扫，看视频

如果应用是一个多tab应用，可以通过tabBar配置项指定tab栏的表现，以及在tab切换时显示对应页。tabBar的配置选项有很多，可以参考官方文档:https://uniapp.dcloud.io/collocation/pages?id=tabbar。

接下来，在2.1.2节创建好的项目中添加tabBar。将所需图标复制到static/images文件夹下，图标可以在https://www.iconfont.cn网站中获取。先创建与tabBar关联的页面，在pages文件夹下分别创建cart/index.vue文件和my/index.vue文件。接下来在pages.json文件中进行以下配置:

```
{
    "pages": [
        {
            "path": "pages/index/index",
            "style": {
                "navigationBarTitleText": "uni-ui基础项目",
                "navigationStyle":"custom"
            }
        },
        {
            "path": "pages/cart/index",
            "style": {
                "navigationBarTitleText": "购物车"
            }
        },
        {
            "path": "pages/my/index",
            "style": {
                "navigationBarTitleText": "我的"
            }
        },
        {
            "path": "pages/news/index",
            "style": {
```

```
                "navigationBarTitleText": "新闻页面"
            }
        }
    ],
    "globalStyle": {
        "navigationBarTextStyle": "white",
        "navigationBarTitleText": "uni-app",
        "navigationBarBackgroundColor": "#007AFF",
        "backgroundColor": "#FFFFFF"
    },
    "tabBar": {
        "color": "#7A7E83",
        "selectedColor": "#E00023",
        "borderStyle": "black",
        "backgroundColor": "#ffffff",
        "list": [
            {
                "pagePath": "pages/index/index",
                "iconPath": "static/images/home1.png",
                "selectedIconPath": "static/images/home2.png",
                "text": "首页"
            },
            {
                "pagePath": "pages/cart/index",
                "iconPath": "static/images/cart1.png",
                "selectedIconPath": "static/images/cart2.png",
                "text": "购物车"
            },
            {
                "pagePath": "pages/my/index",
                "iconPath": "static/images/my1.png",
                "selectedIconPath": "static/images/my2.png",
                "text": "我的"
            }
        ]
    }
}
```

上述加粗代码用于显示tabBar，配置完成后首页可显示tabBar。tabBar配置项说明和list属性说明分别见表5-3和表5-4。

表5-3 tabBar配置项说明

属 性	类 型	说 明
color	HexColor	tab 上的文字默认颜色
selectedColor	HexColor	tab 上的文字选中时的颜色
borderStyle	String	tab 上边框的颜色，可选值为 black 和 white
backgroundColor	HexColor	tab 的背景色
list	Array	tab 的列表，详见 list 属性说明（表 5-4），最少 2 个、最多 5 个 tab

表5-4 list属性说明

属 性	类 型	说 明
pagePath	String	页面路径，必须在 pages 中先定义
text	String	tab 上的按钮文字，在 App 和 H5 平台为非必填。例如，中间可放一个没有文字的“+”号图标
iconPath	String	图片路径，icon 大小限制为 40KB，建议尺寸为 81px × 81px。当 postion 为 top 时，此参数无效，不支持网络图片，不支持字体图标
selectedIconPath	String	选中时的图片路径，icon 大小限制为 40KB，建议尺寸为 81px × 81px 。当 postion 为 top 时，此参数无效

在微信开发者工具中渲染的效果即为tabBar展示效果，如图 5-4 所示。

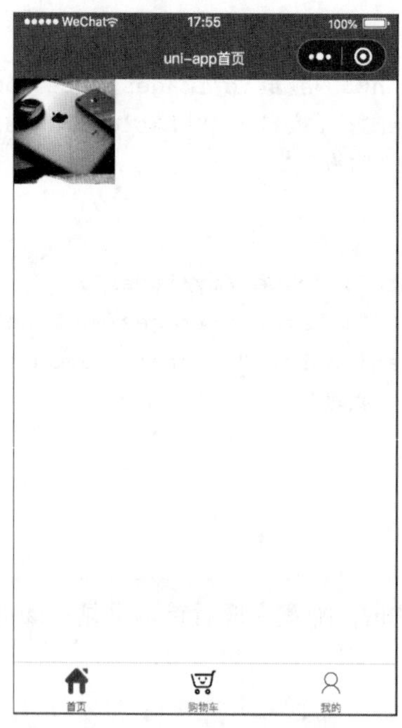

图 5-4 tabBar 展示效果

5.1.5 subPackages 分包加载

分包加载配置是小程序的分包加载机制，因小程序有体积和资源加载限制，各小程序平台提供了分包方式，优化小程序的下载和启动速度。主包主要用于放置默认启动页面（tabBar 页面），以及一些所有分包都需要用到的公共资源（js脚本）；而分包则根据pages.json的配置进行划分。

扫一扫，看视频

假设支持分包的 uni-app 目录结构如下：

```
┌──pages
│   ├──index
│   │   └──index.vue
│   └──login
│       └──login.vue
├──pagesA
│   ├──static
│   └──list
│       └──list.vue
├──pagesB
│   ├──static
│   └──detail
│       └──detail.vue
├──static
├──main.js
├──App.vue
├──manifest.json
└──pages.json
```

则需要在 pages.json 中填写以下内容：

```
{
    "pages": [
        {
            "path": "pages/index/index",
            "style": { ...}
        },
        {
            "path": "pages/login/login",
            "style": { ...}
        }
    ],
    "subPackages": [
        {
```

```
                "root": "pagesA",
                "pages": [
                    {
                        "path": "list/list",
                        "style": { ...}
                    }
                ]
            },
            {
                "root": "pagesB",
                "pages": [
                    {
                        "path": "detail/detail",
                        "style": { ...}
                    }
                ]
            }
        ],
        "preloadRule": {
            "pagesA/list/list": {
                "network": "all",
                "packages": ["__APP__"]
            },
            "pagesB/detail/detail": {
                "network": "all",
                "packages": ["pagesA"]
            }
        }
    }
```

subPackages节点接收一个数组，数组中的每一项都是应用的子包，其属性见表5-5。

表5-5　subPackages的属性

属　性	类　型	是否必填	说　明
root	String	是	子包的根目录
pages	Array	是	子包由哪些页面组成，参考表5-1

注意：

（1）subPackages里的pages的路径是root下的相对路径，不是全路径。

（2）微信小程序每个分包的大小是2MB，总体积不能超过16MB。

（3）百度小程序每个分包的大小是2MB，总体积不能超过8MB。

(4) 支付宝小程序每个分包的大小是2MB，总体积不能超过4MB。
(5) QQ小程序每个分包的大小是2MB，总体积不能超过24MB。
(6) 分包下支持独立的static目录，用来对静态资源进行分包。
(7) uni-app内支持对微信小程序、QQ小程序、百度小程序分包优化，即将静态资源或者js文件放入分包内，不占用主包大小。

preloadRule可以支持分包预载，在进入小程序某个页面时，由框架自动预下载可能需要的分包，提升进入后续分包页面时的启动速度。

preloadRule中，属性（key）是页面路径，值（value）是进入此页面的预下载配置，见表5-6。

表5-6　preloadRule配置属性

字　段	类　型	是否必填	默认值	说　明
packages	StringArray	是	无	进入页面后预下载分包的root或name。__APP__表示主包
network	String	否	wifi	在指定网络下预下载，可选值为all（不限网络）、wifi（仅Wi-Fi下预下载）

App的分包同样支持preloadRule，但网络规则无效。在后面的项目实战开发中会使用分包，读者将会有更深刻的理解。

5.2　小结

pages.json是uni-app的全局配置文件，所有页面文件的路径、窗口样式、原生导航栏和tabBar都存放在pages.json中。本章首先介绍了globalStyle和pages属性，globalStyle配置全局外观，如设置导航栏背景颜色，pages设置页面的路径，与Vue不同，uni-app没有vue-router，也没有主路由和子路由之分，页面的路径需要在pages.json文件中的pages属性中配置；其次介绍了使用自定义导航栏的注意事项；接着介绍了easycom，该组件无须引用、注册，即可直接在页面中使用；最后介绍了subPackages分包加载可以优化小程序的下载和启动速度，需要注意tabBar页面和启动页面不能放在分包中。学习本章时，建议观看视频，同时自己亲手修改一下配置以便查看效果。

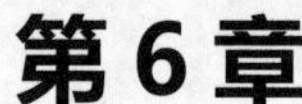

第6章 manifest.json 配置

扫一扫，看视频

manifest.json文件可以配置H5、小程序、App信息，如配置代理，H5模板以及小程序AppID、App的名称、图标、权限等。

6.1　配置项

manifest.json文件配置的选项有很多，这里重点讲解开发项目前必须配置的选项，后面根据开发项目需求进行配置即可。

manifest.json文件官方配置文档地址:https://uniapp.dcloud.io/collocation/manifest。

如果在实际开发中需要配置选项，请仔细阅读manifest.json官方文档。

6.1.1　配置 App 图标

在开发App之前或开发完成之后，应该配上自己的图标，如果没有配置，则默认是HBuilder的图标。使用HBuilder X开发者工具打开manifest.json文件，默认进入基础配置页面，如图6-1所示。

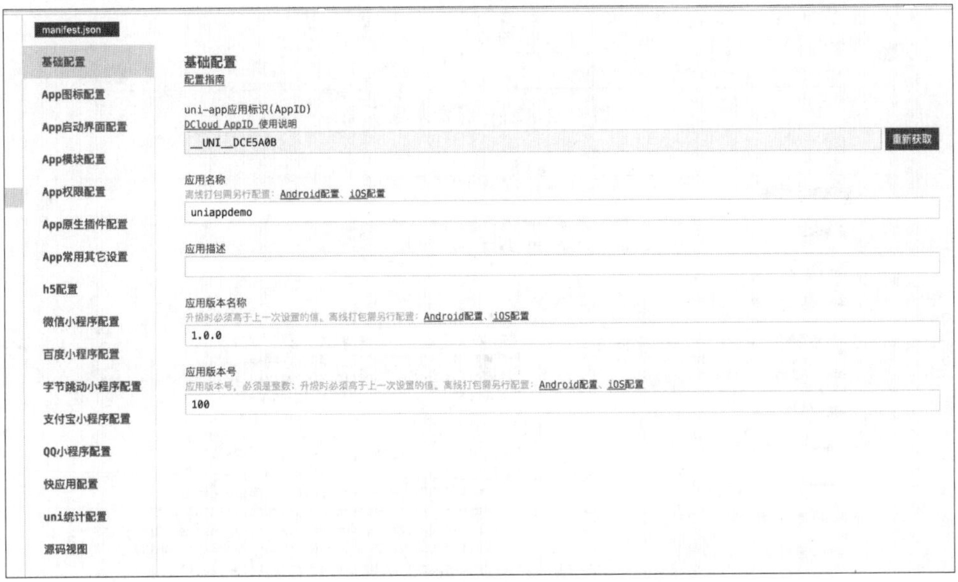

图6-1　基础配置页面

在基础配置页面中可以填写应用的名称和应用的版本号。uni-app的 AppID由DCloud云端分配，主要用于DCloud相关的云服务，请勿自行修改。

选择“App图标配置”选项卡，进入图6-2所示的页面。可以看到图标的尺寸有很多，如果每个都进行制作，则会很浪费时间。在“自动生成图标”选项中单击“浏览”按钮，选择1024 × 1024的图标，单击“自动生成所有图标并替换”按钮，会自动生成Android图标和iPhone图标，生成后图标会自动保存在unpackage/res/icons文件夹下。如果想查看manifest.json文件的源码，可以选择“源码视图”选项卡，如图6-3所示。

manifest.json

基础配置
App图标配置
App启动界面配置
App模块配置
App权限配置
App原生插件配置
App常用其它设置
h5配置
微信小程序配置
百度小程序配置
字节跳动小程序配置
支付宝小程序配置
QQ小程序配置
快应用配置
uni统计配置
源码视图

App图标配置
iOS、Android应用图标配置。云打包后生效。本地打包需另行在原生工程中配置。应用图标配置注意事项

自动生成图标
仅用于自动生成图标，建议使用1024x1024的图标
/Users/ghost8080/Documents/图片/logo1024.png 浏览...
不替换手动设置的图标 自动生成所有图标并替换 浏览生成图标所在目录...

Android图标配置
72x72
hdpi高分屏应用图标
unpackage/res/icons/72x72.png 浏览...
96x96
xhdpi 720P高分屏应用图标
unpackage/res/icons/96x96.png 浏览...
144x144
xxhdpi 1080P高分屏程序图标
unpackage/res/icons/144x144.png 浏览...
192x192
xxxhdpi 4K高分屏程序图标
unpackage/res/icons/192x192.png 浏览...

app store
提交app store使用的图标 1024x1024
unpackage/res/icons/1024x1024.png 浏览...

iPhone图标配置
120x120
iOS7-11程序图标（iPhone4S/5/6/7/8）
unpackage/res/icons/120x120.png 浏览...

图 6-2 App 图标配置页面

manifest.json

基础配置
App图标配置
App启动界面配置
App模块配置
App权限配置
App原生插件配置
App常用其它设置
h5配置
微信小程序配置
百度小程序配置
字节跳动小程序配置
支付宝小程序配置
QQ小程序配置
快应用配置
uni统计配置
源码视图

```
"splashscreen" : {
    "iosStyle" : "common",
    "androidStyle" : "common"
},
"icons" : {
    "android" : {
        "hdpi" : "unpackage/res/icons/72x72.png",
        "xhdpi" : "unpackage/res/icons/96x96.png",
        "xxhdpi" : "unpackage/res/icons/144x144.png",
        "xxxhdpi" : "unpackage/res/icons/192x192.png"
    },
    "ios" : {
        "appstore" : "unpackage/res/icons/1024x1024.png",
        "ipad" : {
            "app" : "unpackage/res/icons/76x76.png",
            "app@2x" : "unpackage/res/icons/152x152.png",
            "notification" : "unpackage/res/icons/20x20.png",
            "notification@2x" : "unpackage/res/icons/40x40.png",
            "proapp@2x" : "unpackage/res/icons/167x167.png",
            "settings" : "unpackage/res/icons/29x29.png",
            "settings@2x" : "unpackage/res/icons/58x58.png",
            "spotlight" : "unpackage/res/icons/40x40.png",
            "spotlight@2x" : "unpackage/res/icons/80x80.png"
        },
        "iphone" : {
            "app@2x" : "unpackage/res/icons/120x120.png",
            "app@3x" : "unpackage/res/icons/180x180.png",
            "notification@2x" : "unpackage/res/icons/40x40.png",
            "notification@3x" : "unpackage/res/icons/60x60.png",
            "settings@2x" : "unpackage/res/icons/58x58.png",
            "settings@3x" : "unpackage/res/icons/87x87.png",
            "spotlight@2x" : "unpackage/res/icons/80x80.png",
            "spotlight@3x" : "unpackage/res/icons/120x120.png"
        }
    }
}
```

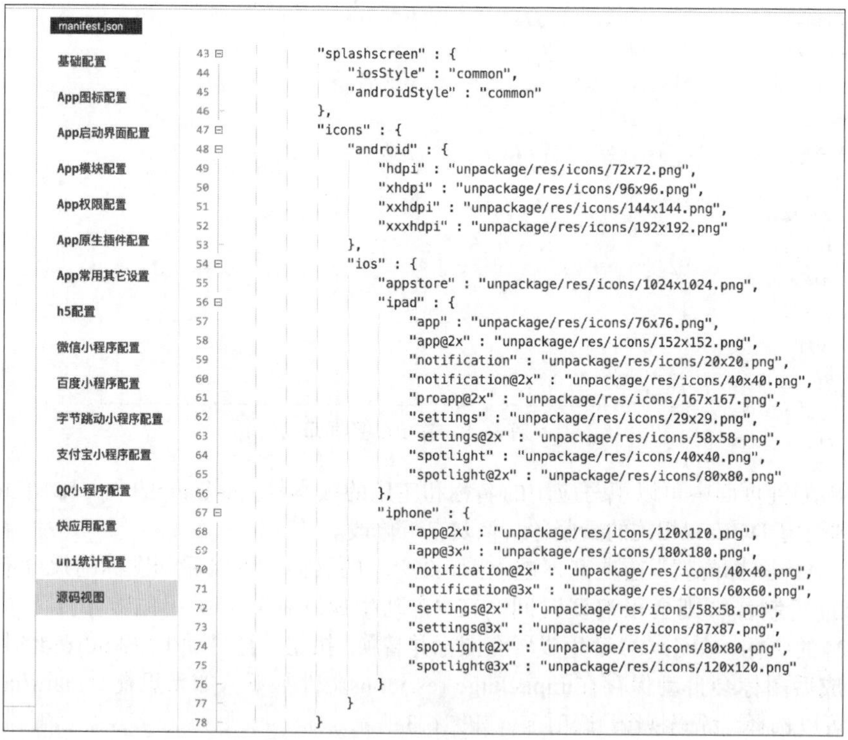

图 6-3 源码视图

图6-2生成的图标会自动生成图6-3中的代码，也可以直接在源码视图中填写配置信息。

6.1.2　配置 AppId

在开发微信小程序时，有些功能必须用自己的AppId，如授权登录、获取手机号、支付等。打开mainifest.json文件，找到以下代码并修改为自己的AppId：

```
"mp-weixin" : {
    "appid" : "wx1a2ccf9b9d1bb416",//填写微信小程序AppID
    ...
}
```

6.1.3　H5 自定义模板

在Vue中，在public文件夹下会有index.html文件，但uni-app的项目中并没有index.html文件，这样在扩展上就会有局限性，如自定义添加meta标签，或者添加第三方统计代码等，但uni-app可以通过manifest.json配置index.html模板文件。

自定义模板官方文档地址：https://uniapp.dcloud.io/collocation/manifest?id=h5。

首先在根目录创建index.html文件，内容如下：

```
<!DOCTYPE html>
<html>
<head>
    <meta charset="UTF-8">
    <meta name="format-detection" content="telephone=no,email=no,date=no,address=no"/>
    <meta http-equiv="X-UA-Compatible" content="IE=edge,chrome=1"/>
   <!--htmlWebpackPlugin.options.title获取HBuilderX配置的标题-->
    <title><%= htmlWebpackPlugin.options.title %></title>
    <script>
        var coverSupport = 'CSS' in window && typeof CSS.supports === 'function'
        && (CSS.supports('top: env(a)') || CSS.supports('top: constant(a)'));
        document.write('<meta name="viewport" content="width=device-width,
        user-scalable=no, initial-scale=1.0, maximum-scale=1.0, minimum-
        scale=1.0' + (coverSupport ? ', viewport-fit=cover' : '') + '" />')
    </script>
    <!--BASE_URL获取manifest.json文件中的运行基础路径，VUE_APP_INDEX_CSS_HASH获取
     自动生成的hash值-->
    <link rel="stylesheet" href="<%= BASE_URL %>static/index.<%= VUE_APP_
     INDEX_CSS_HASH %>.css" />
</head>
<body>
<noscript>
    <strong>Please enable JavaScript to continue.</strong>
</noscript>
```

```
<div id="app"></div>
<!-- built files will be auto injected -->
</body>
</html>
```

上述加粗代码“viewport-fit=cover”可以解决iPhone X两侧边缘出现白边的问题。

接下来在manifest.json文件中配置模板，选择“h5配置”选项卡，如图6-4所示。

图 6-4　h5 配置页面

在“index.html模板路径”中选择index.html，之后使用HBuilder X运行项目，选择“运行到浏览器”，在浏览器中运行测试，如图6-5所示。

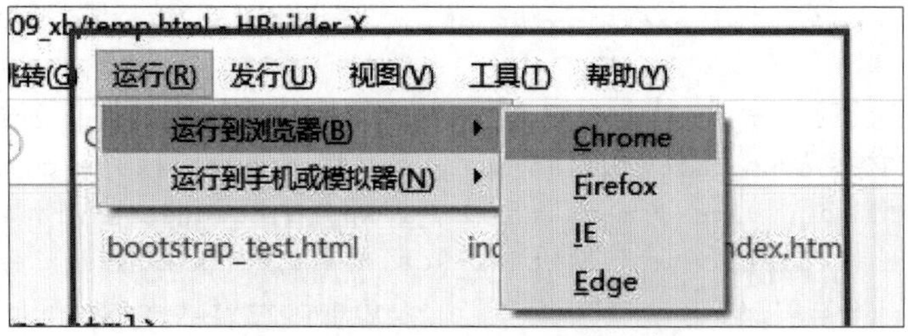

图 6-5　在浏览器中运行测试

6.1.4　配置代理解决跨域问题

在开发H5网站时，在开发环境中需要解决跨域问题，可以在manifest.json文件中配置代理以解决该问题。选择“源码视图”选项卡，在源码视图中配置的代码如下：

```
"h5" : {
        "title" : "hello uniapp",
        "template" : "index.html",
```

```
        "router" : {
            "mode" : "hash",
            "base" : ""
        },
        "sdkConfigs" : {
            "maps" : {}
        },
        "devServer" : {
            "proxy" : {
                "/api" : {
                    "target" : "http://localhost:9191/api",
                    "changeOrigin" : true,
                    "pathRewrite" : {
                        "^/api" : ""
                    }
                }
            }
        }
}
```

6.2 小结

manifest.json文件是应用的配置文件，用于指定应用的名称、图标、权限等。本章主要讲解了manifest.json文件的小程序、H5、App的基础配置，如配置App图标、配置AppID、配置H5 自定义模板以及配置代理解决跨域等问题。当然，第三方地图、支付等功能也可以在manifest.json中进行配置。manifest.json 文件的配置，推荐在 HBuilder X 提供的可视化操作界面完成。

部分配置会在打包时的操作界面自动补全，如证书等信息。Native.js 权限部分会根据配置的模块权限，在打包后会自动填充。部分 modules 是默认的，不需要进行配置。建议各位读者动手配置一下自己项目中的manifest.json文件。

第 7 章

路由

uni-app页面路由由框架统一管理，开发者需要在pages.json文件中配置每个页面路由的路径及页面样式。类似的小程序在app.json中配置页面路由的方式相同。

7.1 组件路由跳转

扫一扫，看视频

uni-app有两种页面路由跳转方式，即使用navigator组件跳转和调用API跳转。本节介绍navigator组件路由跳转。

navigator组件官方文档地址:https://uniapp.dcloud.io/component/navigator? id=navigator。

7.1.1 打开新页面

使用navigator组件将open-type属性值设置为navigate时，会打开新页面。

在pages/index/index.vue文件中的代码如下：

```
<template>
    <view class="page">
        <navigator open-type="navigate" url="/pages/news/index"
         hover-class="navigator-hover">
            <button type="default" class="btn">跳转到新闻页面</button>
        </navigator>
    </view>
</template>

<script>
</script>

<style>
    .page{width:100%;height:100vh;}
    .navigator-hover{width:100%;height:150rpx;background-color:#F56C6C;
     opacity: 1;display:flex;align-items: center}
    .btn{width:50%;height:80rpx;border-radius: 0px;}
    .btn:after{border:0 none;}
</style>
```

如果不在navigator组件中添加open-type属性，则默认为open-type="navigate"。url属性可以设置跳转的路径值为相对路径或绝对路径。hover-class属性指定点击时的样式类，当hover-class="none"时，没有点击效果。注意，在微信小程序中将<button>元素的默认边框隐藏，需要使用伪元素after，在<style>标签内添加.btn:after{border:0 none}可以隐藏<button>元素的默认边框。

7.1.2 页面重定向

使用navigator组件将open-type属性值设置为redirect时，会进行页面重定向，可以理解为Vue路由的replace跳转。

在pages/index/index.vue文件中的代码如下：

```
<navigator open-type="redirect" url="/pages/news/index" hover-class = 
 "navigator-hover">
        <button type="default" class="btn">跳转到新闻页面</button>
  </navigator>
```

使用页面重定向进行跳转，不会进入历史记录，不支持页面返回。可以利用该特点解决微信小程序跳转10层限制的问题，详见7.3.3节。

7.1.3 页面返回

使用组件<navigator open-type="navigate"/>方式跳转页面，在跳转后的页面中可以自定义返回操作。在pages/news/index.vue文件中的代码如下：

```
<template>
    <view>
        <navigator open-type="navigateBack">
            <button type="default">返回</button>
        </navigator>
        新闻页面
    </view>
</template>
```

将open-type属性值设置为navigateBack，单击“返回”按钮可以返回上一页。

7.1.4 Tab 切换

Tab切换主要用于从当前页面跳转到tabBar中的页面，如从新闻页面跳转到购物车页面。在pages/news/index.vue文件中的代码如下：

```
<template>
    <view>
    <navigator url="/pages/cart/index" open-type="switchTab">
            <button type="default">跳转到购物车页面</button>
        </navigator>
    </view>
</template>
```

将open-type属性值设置为switchTab，可以跳转到tabBar中的页面，跳转后页面会全部出栈，只留下新的tabBar中的页面。由于页面全部出栈，因此跳转后的页面不支持页面返回。

7.1.5 reLaunch

使用navigator组件将open-type属性值设置为reLaunch时，会进行重加载，页面全部出栈，只留下新的页面。在pages/news/index.vue文件中的代码如下：

```
<template>
    <view>
        <navigator url="/pages/index/index" open-type="reLaunch">
            <button type="default">跳转到首页</button>
        </navigator>
    </view>
</template>
```

使用重加载可以跳转任何页面，跳转后页面全部出栈，意味着跳转后的页面不支持页面返回。

7.2 API 路由跳转

扫一扫，看视频

本节使用API实现路由跳转页面。

API路由跳转官方文档:https://uniapp.dcloud.io/api/router?id=navigateto。

7.2.1 打开新页面

使用uni.navigateTo()方法跳转页面，可以保留当前页面，跳转到应用内的某个页面；使用uni.navigateBack可以返回到原页面，类似于Vue中的this.$router.push()方法。

在pages/index/index.vue文件中的代码如下：

```
<template>
    <view class="page">
        <button type="default" class="btn" @click="goPage('/pages/news/
         index')">跳转到新闻页面</button>
    </view>
</template>

<script>
    export default {
        name:"index",
        methods:{
            goPage(url){
                uni.navigateTo({
                    url:url
                })
            }
        }
    }
</script>

<style>
    .page{width:100%;height:100vh;}
```

```
    .navigator-hover{width:100%;height:150rpx;background-color:#F56C6C;
     opacity: 1;display:flex;align-items: center}
    .btn{width:50%;height:80rpx;border-radius: 0px;}
    .btn:after{border:0 none;}
</style>
```

上述加粗代码为在<button>组件上添加click事件，使用uni.navigateTo()方法实现页面跳转，该方法传入对象类型的参数url属性为跳转页面的路径。路径后可以带参数，参数与路径之间使用“?”分隔，参数键与参数值用“=”相连，不同参数用“&”分隔，如“path?key=value&key2=value2”。

7.2.2 页面重定向

使用uni.redirectTo()方法关闭当前页面，跳转到应用内的某个页面，类似于Vue的this.$router.replace()方法。

在pages/index/index.vue文件中的代码如下：

```
<template>
    <view class="page">
        <button type="default" class="btn" @click="goPage('/pages/news/
         index')">跳转到新闻页面</button>
    </view>
</template>

<script>
    export default {
        name:"index",
        methods:{
            goPage(url){
                uni.redirectTo({
                    url:url
                })
            }
        }
    }
</script>

<style>
    .page{width:100%;height:100vh;}
    .navigator-hover{width:100%;height:150rpx;background-color:#F56C6C;
     opacity: 1;display:flex;align-items: center}
    .btn{width:50%;height:80rpx;border-radius: 0px;}
    .btn:after{border:0 none;}
</style>
```

注意：

跳转到tabBar页面时只能使用switchTab跳转。

7.2.3 页面返回

使用uni.navigateBack()方法关闭当前页面，返回上一页面或多级页面，并且可以决定需要返回几层。

在pages/news/index.vue文件中的代码如下：

```
<template>
    <view>
        <button type="default" @click="goBack()">返回</button>
    </view>
</template>

<script>
    export default {
        name:"news",
        methods:{
            goBack(){
                uni.navigateBack({
                    delta: 1
                });
            }
        }
    }
</script>
```

uni.navigateBack()方法类似于Vue中的this.$router.go(-1)，其中，delta参数表示返回的页面数，如果delta大于现有页面数，则返回到首页。

7.2.4 Tab 切换

使用uni.switchTab()方法跳转到tabBar页面，并关闭其他所有非tabBar页面。

在pages/news/index.vue文件中的代码如下：

```
<template>
    <view>
        ...
        <button type="default" @click="goTabBar('/pages/cart/index')">跳转到
        购物车页面</button>
    </view>
</template>
```

```
<script>
    export default {
        name:"news",
        methods:{
            ...
            goTabBar(url){
                uni.switchTab({
                    url:url
                });
            }
        }
    }
</script>
```

url参数为需要跳转的tabBar页面的路径(需要在pages.json的tabBar字段中定义的页面),路径后不能带参数。

7.2.5 reLaunch

使用uni.reLaunch()方法可以关闭所有页面,接着跳转到应用内的某个页面。

在pages/news/index.vue文件中的代码如下:

```
<template>
    <view>
        ...
        <button type="default" @click="goReLaunch('/pages/index/index')">跳
         转到首页</button>
    </view>
</template>

<script>
    export default {
        name:"news",
        methods:{
            ...
            goReLaunch(url){
                uni.reLaunch({
                    url: url
                });
            }
        }
    }
</script>
```

注意：

url参数表示需要跳转的应用内页面路径，路径后可以带参数。参数与路径之间使用“?”分隔，参数键与参数值用“=”相连，不同参数用“&”分隔，如“path?key=value&key2=value2”。

如果跳转的页面路径是tabBar页面，则不能带参数。

7.3　接收参数与层级限制

扫一扫，看视频

本节重点讲解如何接收路由跳转过来的动态参数，以及如何解决微信小程序使用navigateTo()方法跳转10层超过页面限制的问题。

7.3.1　接收参数

uni-app中的接收路由参数与Vue不同，uni-app是在onLoad()生命周期钩子函数中接收参数。代码示例如下：

```
//从起始页面跳转到pages/news/index.vue页面并传递参数
uni.navigateTo({
    url:'pages/news/index?id=1&title=新闻动态'
})

//在pages/news/index.vue页面接收参数
onLoad(opts){                    //opts为object类型，会序列化上个页面传递的参数
   console.log(opts.id);         //输出上个页面传递的参数，值为1
   console.log(opts.title);      //输出上个页面传递的参数，值为新闻动态
}
```

上述加粗代码中，onLoad()方法接收的参数为object类型，会序列化上个页面传递的参数。

需要注意的是，url是有长度限制的，太长的字符串会传递失败，这可以使用encodeURIComponent()方法解决。代码示例如下：

```
//从起始页面跳转到pages/news/index.vue页面并传递参数
uni.navigateTo({
   url:"/pages/news/index?id=1&title="+encodeURIComponent('新闻动态')+""
})

//在pages/news/index.vue页面接收参数
onLoad(opts){                    //opts为object类型，会序列化上个页面传递的参数
    console.log(decodeURIComponent(opts.title));
    //输出上个页面传递的参数，值为“新闻动态”
}
```

上述加粗代码中，使用路由传递参数时使用encodeURIComponent()方法将参数的值进行

编码，接收参数时使用decodeURIComponent()方法进行解码。

7.3.2 获取当前页栈

使用getCurrentPages()方法可以获取当前页面栈的实例，并以数组形式按栈的顺序给出，第一个元素为首页，最后一个元素为当前页面。

代码示例如下：

```
onLoad(opts){                              //opts为object类型，会序列化上个页面传递的参数
    //获取页面栈
    let pages = getCurrentPages();
    //第一个页面
    let firstPage= pages[0];
    console.log(firstPage.route);          //结果:pages/index/index
    //获取当前页面
    let curPage=pages[pages.length-1];
    console.log(curPage.route);            //结果:pages/news/index
}
```

使用navigateTo()方法跳转页面会将每一个页面添加到页面栈中。使用getCurrentPages()方法可以获取到所有页面栈，返回的是一个数组，如果获取到第一个页面，索引为0；如果获取当前页面，索引为“总页面栈数量-1”；如果获取到最后一个页面，索引为“总页面栈数量-1”，其实最后一个页面就是当前页面。可以利用该方法解决微信小程序限制10层页面的问题。

7.3.3 解决跳转10层限制的问题

扫一扫，看视频

微信小程序为了解决性能问题，使用navigateTo()方法跳转，其页面限制为10层。因为每次使用navigateTo()方法跳转页面都会将跳转的页面添加到页面栈中，页面栈越多，性能越差。每次使用navigateBack()方法返回页面时，就会减少一层页面栈。例如，从文章列表页面跳转到文章详情页面，文章详情页面中有相关文章，单击相关文章的标题时会使用navigateTo()方法再次跳转到文章详情页面，在这过程中并没有使用navigateBack()方法返回页面，而是无限地使用navigateTo()方法进行跳转，但最多只能跳转10次，超过10次，则跳转无效。

该问题有两个解决方案：

（1）使用getCurrentPages()方法获取栈中的页面数，如果大于等于10层，则使用redirectTo()方法进行跳转。

（2）官方文档给出了完美的解决方案，如图7-1所示。

注意：为了不让用户在使用小程序时造成困扰，[illegible]，请尽量避免多层级的交互方式。

图7-1 官方文档给出的解决跳转10层页面限制的方案

第一种方案的代码示例如下：

```
<template>
    <div>
        <div>相关新闻:</div>
        <div>
            <div class="news-list" @click="pushPage('/pages/news/index?id=3&
             title='+encodeURIComponent('民法典将如何影响婚姻生活')+'')">民法典将如
             何影响婚姻生活</div>
        </div>
    </div>
</template>

<script>
    export default {
        name: "news",
        data(){
            return {
                title:""
            }
        },
        onLoad(opts){
            let pages=getCurrentPages();
            //获取页面栈总页数
            this.pagesCount=pages.length;
        },
        methods:{
            pushPage(url){
                //如果页面栈总数大于10
                if(this.pagesCount>=10){
                   //使用重定向跳转页面
                    uni.redirectTo({
                        url
                    })
                }else{
                    //如果页面栈总数小于10，则打开新页面
                    uni.navigateTo({
                        url: url
                    })
                }
            }
        }
    }
</script>
```

上述代码中的注释很详细，逻辑也比较简单，真实演示效果可以观看视频教程。

7.3.4 解决 tabBar 不能传参的问题

扫一扫，看视频

跳转到tabBar页面时是不能传参的，但是实际开发中需要在跳转tabBar页面时进行传参，这时可以将参数传入本地缓存来实现传参。H5的本地缓存是localStorage，uni-app的本地缓存是uni.setStorageSync。创建pages/shop/index.vue文件，该文件中的代码如下：

```
<template>
    <div>
        <div class="shop-list" @click="goPage('/pages/index/index','1')">北
         京分店</div>
        <div class="shop-list" @click="goPage('/pages/index/index','2')">上
         海分店</div>
        <div class="shop-list" @click="goPage('/pages/index/index','3')">山
         西分店</div>
    </div>
</template>

<script>
    export default {
        name: "shop",
        methods:{
            goPage(url,id){
                //将id传入本地缓存
                uni.setStorageSync("branch_shop_id",id);
                //跳转到tabBar页面
                uni.switchTab({
                    url
                })
            }
        }
    }
</script>

<style scoped>
    .shop-list{height:40px;line-height:40px;}
</style>
```

uni.setStorageSync()方法的第一个参数是属性，第二个参数是值。与H5的localStorage使用方式一样，使用uni.switchTab()方法跳转到tabBar的首页，在pages/index/index.vue文件中接收参数，代码如下：

```
onShow(){
    //获取pages/shop/index.vue文件中uni.setStorageSync()方法属性为
    //branch_shop_id的值
    let branchShopId=uni.getStorageSync("branch_shop_id");
    console.log(branchShopId);              //结果为1
}
```

使用uni.getStorageSync()方法获取本地缓存，这样即可实现路由传参效果。

7.3.5 自定义 tabBar 及封装 tabBar 组件

扫一扫，看视频

5.4节介绍了tabBar的配置，但是系统生成的tabBar有时不能满足我们的需求，而且有局限性，如自定义tabBar的样式、在子页面使用tabBar等这些特殊的需求都无法实现，这时需要自己写一个tabBar。使用Vue开发时，tabBar可以使用嵌套路由的方式实现，而在uni-app中没用嵌套路由的概念，可以使用组件的形式实现。

首先编写tabBar组件，在components文件夹中创建tab-bar/index.vue文件。该文件中的代码如下：

```
<template>
    <view class="tab-bar">
        <view :class="{item:true, home:true, active:homeActive}" @
         click="replacePage('/pages/index/index')">
            <view class="icon"></view>
            <view class="text">首页</view>
        </view>
        <view :class="{item:true, my:true, active:myActive}" @
         click="replacePage('/pages/my/index')">
            <view class="icon"></view>
            <view class="text">我的</view>
        </view>
    </view>
</template>

<script>
    export default {
        name: "tab-bar",
        data(){
          return {
              homeActive:true,
              myActive:false
          }
        },
        props:{
            //当前路由地址
```

```
            curRoute:{
                type:String,
                default:""
            }
        },
        methods:{
            //重定向跳转页面
            replacePage(url){
                uni.redirectTo({
                    url
                })
            }
        },
        mounted(){
            //改变tabBar图标和文字的样式
            switch(this.curRoute){
                case "pages/index/index":    //首页
                    this.homeActive=true;
                    this.myActive=false;
                    break;
                case "pages/my/index":       //我的页面
                    this.homeActive=false;
                    this.myActive=true;
                    break;
                default:                     //默认为首页样式
                    this.homeActive=true;
                    this.myActive=false;
                    break;
            }
        }
    }
</script>

<style scoped>
    .tab-bar{width:100%;height:100rpx;background-color:#FFFFFF;border-
     top:1px solid #EFEFEF;position: fixed;bottom:0px;left:0px;display:flex;
     justify-content: space-between;align-items: center;padding:0px 20%;box-
     sizing: border-box;}
    .tab-bar.item{width:80rpx;height:auto;}
    .tab-bar.item.icon{width:60rpx;height:60rpx;margin:0 auto;}
    .tab-bar.item.text{font-size:24rpx;text-align: center;color:#7F8387;}
    .tab-bar.item.active.text{color:#E00023}
```

```
    .tab-bar.item.home.icon{background-image:url("~@/static/images/home1.png");background-position: center;background-repeat: no-repeat;background-size: 100%;}
    .tab-bar.item.home.active.icon{background-image:url("~@/static/images/home2.png");background-position: center;background-repeat: no-repeat;background-size: 100%;}

    .tab-bar.item.my.icon{background-image:url("~@/static/images/my1.png");background-position: center;background-repeat: no-repeat;background-size: 100%;}
    .tab-bar.item.my.active.icon{background-image:url("~@/static/images/my2.png");background-position: center;background-repeat: no-repeat;background-size: 100%;}
</style>
```

上述代码比较简单，不再赘述。需要注意的是，为了兼容性考虑，在开发uni-app的组件时应避免使用created生命周期，而应使用mounted生命周期。

接下来将tabBar组件分别引入pages/index/index.vue和pages/my/index.vue文件中。pages/index/index.vue文件中的代码如下：

```
<template>
    <view class="page">
        <button type="default" class="btn" @click="goPage('/pages/news/index?id=1&title=新闻动态')">跳转到新闻页面</button>
        <button type="default" class="btn" @click="goPage('/pages/shop/index')">选择商品</button>
        <!--将当前页面的路由地址传递给tabBar组件-->
        <tab-bar :curRoute="curRoute"></tab-bar>
    </view>
</template>

<script>
    import TabBar from "../../components/tab-bar";
    export default {
        name:"index",
        data(){
            return {
                curRoute:""
            }
        },
        onLoad(opts) {
            let pages=getCurrentPages();
            let curPage=pages[pages.length-1];
            //获取当前页面
```

```
                this.curRoute=curPage.route;
            },
            components:{
                TabBar
            },
            methods:{
                goPage(url){
                    uni.navigateTo({
                        url:url
                    })
                }
            }
        }
</script>

<style>
    .page{width:100%;height:100vh;}
    .navigator-hover{width:100%;height:150rpx;background-color:#F56C6C;
     opacity: 1;display:flex;align-items: center}
    .btn{width:50%;height:80rpx;border-radius: 0px;}
    .btn:after{border:0 none;}
</style>
```

上述加粗代码为tabBar相关代码，使用getCurrentPages()函数获取当前路由地址，传递给tabBar组件，在tabBar组件内部利用父组件传过来的路由地址，判断并改变tabBar的图标和文字的样式。

pages/my/index.vue文件中的代码如下：

```
<template>
    <view>
        我的页面
        <tab-bar :curRoute="curRoute"></tab-bar>
    </view>
</template>

<script>
    import TabBar from "../../components/tab-bar";
    export default {
        name: "my",
        data(){
            return {
                curRoute:""
            }
```

```
        },
        onLoad(opts) {
            let pages=getCurrentPages();
            let curPage=pages[pages.length-1];
            this.curRoute=curPage.route;
        },
        components:{
            TabBar
        }
    }
</script>
```

其与pages/index/index.vue文件中引入tabBar组件一样，不作过多解释。在微信开发者工具中渲染的效果，即自定义tabBar预览效果如图7-2所示。

图 7-2　自定义 tabBar 预览效果

点击“我的”或“首页”按钮，即可分别跳转到相应页面。需要注意的是，如果使用微信开发者工具进行测试，跳转页面时会出现滑动效果，而正常情况是不应该出现滑动效果的，所以需要在手机上进行真机测试。打开微信开发者工具，单击“预览”按钮，如图7-3所示。

图 7-3　单击“预览”按钮

这时会将unpackage/dist/dev/mp-weixin文件夹中的所有文件上传到微信小程序服务器上，上传完成后会生成二维码，如图7-4所示。

图 7-4 真机测试二维码

用微信扫一扫功能扫描二维码，可以进行真机测试，在手机上测试自定义的tabBar跳转页面就不会出现滑动效果，可见微信开发者工具的测试和真机测试是有些差异的，最终要以真机测试效果为准。

7.4 小结

本章主要讲解了uni-app的路由，包括组件路由跳转、API路由跳转，同时还讲解了接收参数与层级限制。在组件路由跳转和API路由跳转中，讲解了如何打开新页面、页面重定向、页面返回、Tab切换以及reLaunch（重加载）。另外，讲解了如何接收参数、获取当前栈、解决tabBar跳转无法传参的问题以及使用navigateTo()方法跳转时存在的10层限制的问题。最后还讲解了如何自定义tabBar组件、如何封装tabBar组件以及如何进行真机测试。学习本章时，建议观看视频，效果会一目了然，同时自己多加练习，这样会对路由的知识有更好的理解。

第8章

判断运行环境和平台

由于uni-app的一套代码可多端使用，因此必定存在因不同平台而产生的差异，这时需要根据不同的平台编写不同的代码。

8.1 判断运行环境

扫一扫，看视频

在实际开发中判断运行环境，通常用于调用不同环境的服务端API接口地址。例如，有两个接口，测试接口为http://10.11.29.33:8181，正式接口为http://10.11.29.55，在开发环境下需要调用测试接口，生产环境下需要调用正式接口，这时就需要判断当前的运行环境，从而自动调用不同的接口。

8.1.1 开发环境和生产环境

uni-app可通过process.env.NODE_ENV判断当前运行环境是开发环境还是生产环境，一般用于连接测试服务端API接口或生产服务端API接口的动态切换。

在HBuilder X中，选择“运行”命令编译出的代码是开发环境，选择“发行”命令编译出的代码是生产环境。代码示例如下：

```
if (process.env.NODE_ENV === 'development') {
    console.log('开发环境')
}

if (process.env.NODE_ENV === 'production') {
    console.log('生产环境')
}
```

在实际开发中不可能只有生产环境和开发环境，还需要多环境配置，如测试环境、预览环境等。其配置方式与Vue一样，也需要配置package.json文件。使用HBuilder X创建的项目是没有package.json文件的，需要用vue-cli命令行创建项目。项目创建后即可看到package.json文件，可以根据Vue多环境配置方案自行配置，如果无法配置成功，请看视频教程学习。

8.1.2 配置生产环境和开发环境的 API 接口

在开发项目时一般会给两个服务端API接口，分别是测试接口和正式接口。本节实现不同环境自动切换接口地址的功能。为了在开发项目中使用方便，这里将其封装到一个文件中。在static文件夹中创建js/conf/config.js文件，该文件中的代码如下：

```
let baseApi=process.env.NODE_ENV === 'development'?"http://dev.lucklnk.
com":"http://www.lucklnk.com";
export default {
    baseApi:baseApi
}
```

接下来将config.js文件导入main.js文件中，并挂载到Vue原型中实现全局变量。新增代码如下：

```
...
import config from "./static/js/conf/config";
Vue.prototype.$config=config;
...
```

将config挂载到Vue原型属性$config上，在pages/index/index.vue文件中使用，代码如下：

```
...
onLoad(opts) {
    //如果是开发环境，则结果为http://dev.lucklnk.com
    console.log(this.$config.baseApi);
},
...
```

在开发项目获取服务端数据时就可以使用this.$config.baseApi进行接口拼接。

8.2 判断平台

扫一扫，看视频

uni-app中的平台判断有两种场景，分别是编译期判断和运行期判断，在项目中常用的是编译期判断。

8.2.1 编译期判断

编译期判断，即条件编译，不同平台在编译出包后已经是不同的代码。编译期判断可以写在<script>标签、<template>标签和<style>标签内。

<script></script>标签内的代码示例如下：

```
// #ifdef H5
    alert("只有H5平台才有alert方法")
// #endif
```

判断的格式:#ifdef为开始标签，#endif为结束标签，H5为判断的条件。注意，#ifdef和#endif一定要写在注释中。

在<template>标签内的代码示例如下：

```
<!--#ifdef H5-->
    <section>我是H5标签</section>
<!--#endif-->
```

<style>标签内的代码示例如下：

```
/* #ifdef H5 */
    section{font-size:28rpx;color:#F56C6C}
/* #endif */
```

条件编译的官方文档地址:https://uniapp.dcloud.io/platform?id=条件编译。

条件编译使用特殊的注释作为标记，在编译时根据这些特殊的注释，将注释中的代码编译到不同平台，见表8-1。其写法是以 #ifdef 或 #ifndef 加 %PLATFORM% 开头，以 #endif 结尾。

（1）# ifdef：if defined 仅在某平台存在。

（2）# ifndef：if not defined 除了某平台均存在。

（3）%PLATFORM%：平台名称，其可取值见表8-2。

表8-1 条件编译

条件编译写法	说 明
# ifdef APP-PLUS 需要条件编译的代码 # endif	仅出现在 App 平台下的代码
# ifndef H5 需要条件编译的代码 # endif	除了 H5 平台外，其他平台均存在的代码
# ifdef H5 || MP-WEIXIN 需要条件编译的代码 # endif	在 H5 平台或微信小程序平台存在的代码（这里只有 ||，不可能出现 &&，因为没有交集）

表8-2 %PLATFORM%可取值

值	平 台
APP-PLUS	App
APP-PLUS-NVUE	App nvue
H5	浏览器
MP-WEIXIN	微信小程序
MP-ALIPAY	支付宝小程序
MP-BAIDU	百度小程序
MP-TOUTIAO	字节跳动小程序
MP-QQ	QQ 小程序
MP-360	360 小程序
MP	微信小程序、支付宝小程序、百度小程序、字节跳动小程序、QQ 小程序、360 小程序
QUICKAPP-WEBVIEW	快应用通用 (包含联盟、华为)
QUICKAPP-WEBVIEW-UNION	快应用联盟
QUICKAPP-WEBVIEW-HUAWEI	快应用华为

支持的文件有.vue、.js、.css、pages.json以及各种预编译语言文件，如.scss、.less、.stylus、.ts、.pug。注意，条件编译是利用注释实现的，在不同语法里注释的写法不一样，如.js文件中使用“//

注释”、.css文件中使用“/* 注释 */”、.vue文件中使用“<!-- 注释 -->”。

8.2.2　运行期判断

运行期判断是指代码已经打入包中，但仍然需要在运行期判断平台，此时可使用 uni.getSystemInfoSync().platform 判断客户端环境是 Android、iOS 还是小程序开发者工具。注意，在百度小程序开发者工具、微信小程序开发者工具、支付宝小程序开发者工具中使用 uni.getSystemInfoSync().platform时，其返回值均为 devtools。代码示例如下：

```
switch(uni.getSystemInfoSync().platform){
    case 'android':
        console.log('运行在Android上')
        break;
    case 'ios':
        console.log('运行在iOS上')
        break;
    default:
        console.log('运行在开发者工具上')
        break;
}
```

运行期判断在实际开发中并不常用，只需了解即可。

8.3　小结

本章主要讲解了运行环境判断和平台判断。其中，运行环境判断介绍了根据开发环境与生产环境的不同，全自动切换服务端API接口地址；平台判断的重点是编译期判断，在实际开发中经常会使用编译期判断解决平台兼容性的问题。本章内容看起来简单，但在实际的企业开发中特别重要，如API接口的调用，为了安全起见，只有真正发布后才能调用真实地址的数据接口，平常测试时都使用开发环境的接口；而平台判断更多的是用于解决兼容性问题，或者一些特殊功能。读者可以自己动手进行配置，看是否能正确判断运行环境和平台。

第9章

常用组件

uni-app为开发者提供了一系列基础组件，类似HTML里的基础标签元素，但uni-app的组件不是与HTML相同，而是与小程序相同，更适合手机端使用。虽然不推荐使用HTML标签，但实际上如果开发者写了div等标签，在编译到非H5平台时也会被编译器转换为view标签，类似的还有span转text、a转navigator等，包括CSS里的元素选择器也会转换。但为了管理方便、策略统一，编写代码时建议使用view等组件。

组件的官方文档地址:https://uniapp.dcloud.io/component/README。

9.1 view

扫一扫，看视频

view（视图容器）类似于传统HTML中的div，用于包裹各种元素内容，是页面布局最常用的标签。

view官方文档地址:https://uniapp.dcloud.io/component/view。

view和传统HTML的div有一定的区别，div的样式和事件效果都是由CSS和JavaScript（JS）实现的，而view组件支持CSS样式的同时自带事件效果属性。view组件的属性说明见表9-1。

表9-1 view组件的属性说明

属性名	类 型	默认值	说 明
hover-class	String	none	指定按下去的样式类。当 hover-class="none" 时，没有单击态效果
hover-stop-propagation	Boolean	false	指定是否阻止本节点的祖先节点出现单击态
hover-start-time	Number	50	按住后多久出现单击态，单位为 ms
hover-stay-time	Number	400	手指松开后单击态保留时间，单位为 ms

在pages文件夹创建view/index.vue文件，该文件中的代码如下：

```
<template>
    <view class="page">
        <view class="box1" hover-class="hover" hover-start-time="500"
         hover-stay-time="1600">
            <view class="box2" hover-class="hover" hover-start-time="500"
             hover-stay-time="1600" :hover-stop-propagation="true"></view>
        </view>
    </view>
</template>

<script>
    export default {
        name: "view-index"
    }
</script>

<style scoped>
    .page{width:100%;height:100vh;}
    .box1{width:400rpx;height:400rpx;background-color: #007aff; margin-
     bottom: 20rpx;}
    .box2{width:300rpx;height:300rpx;background-color:#FF0000;}
    .hover{background-color:#00FF00}
</style>
```

如果使用 <div>，编译时会被转换为 <view>。hover-stop-propagation属性类似于阻止冒泡事件，当hover-stop-propagation属性值为true时，单击class="box2"的view组件，不会影响class="box1"的view组件样式的改变。

9.2 scroll-view

扫一扫，看视频

scroll-view（可滚动视图区域）用于区域滚动，类似于H5的iScroll效果。注意，在WebView渲染的页面中，区域滚动的性能不如页面滚动。

scroll-view官方文档地址：https://uniapp.dcloud.io/component/scroll-view。

9.2.1 scroll-view的使用

scroll-view使用非常广泛，如内容溢出时滚动显示内容、横向滚动、纵向滚动、支持下拉刷新、支持上拉加载等。scroll-view组件的属性说明见表9-2。

表9-2 scroll-view组件的属性说明

属性名	类　型	默认值	说　明	平台差异说明
scroll-x	Boolean	false	允许横向滚动	
scroll-y	Boolean	false	允许纵向滚动	
upper-threshold	Number	50	距顶部 / 左边多远时（单位为 px）触发 scrolltoupper 事件	
lower-threshold	Number	50	距底部 / 右边多远时（单位为 px）触发 scrolltolower 事件	
scroll-top	Number		设置竖向滚动条位置	
scroll-left	Number		设置横向滚动条位置	
scroll-into-view	String		值应为某子元素 id（id 不能以数字开头）。设置哪个方向可滚动，则在哪个方向滚动到该元素	
scroll-with-animation	Boolean	false	在设置滚动条位置时使用动画过渡	
enable-back-to-top	Boolean	false	iOS 点击顶部状态栏、Android 双击标题栏时，滚动条返回顶部，只支持竖向	微信小程序
show-scrollbar	Boolean	false	控制是否出现滚动条	App-nvue 2.1.5+
refresher-enabled	Boolean	false	开启自定义下拉刷新	app-vue 2.5.12+、微信小程序基础库 2.10.1+

续表

属性名	类　型	默认值	说　明	平台差异说明
refresher-threshold	Number	45	设置自定义下拉刷新阈值	app-vue 2.5.12+、微信小程序基础库 2.10.1+
refresher-default-style	String	"black"	设置自定义下拉刷新默认样式，支持设置 black、white、none，none 表示不使用默认样式	app-vue 2.5.12+、微信小程序基础库 2.10.1+
refresher-background	String	"#FFF"	设置自定义下拉刷新区域背景颜色	app-vue 2.5.12+、微信小程序基础库 2.10.1+
refresher-triggered	Boolean	false	设置当前下拉刷新状态，true 表示下拉刷新已经被触发，false 表示下拉刷新未被触发	app-vue 2.5.12+、微信小程序基础库 2.10.1+
enable-flex	Boolean	false	启用 flexbox 布局。开启后，当前节点声明了 display: flex 后会成为 flex container，并作用于其孩子节点	微信小程序 2.7.3
scroll-anchoring	Boolean	false	开启 scroll anchoring 特性，即控制滚动位置不随内容变化而抖动。该属性仅在 iOS 下生效，Android 下可参考 CSS overflow-anchor 属性	微信小程序 2.8.2
@scrolltoupper	EventHandle		滚动到顶部 / 左边，会触发 scrolltoupper 事件	
@scrolltolower	EventHandle		滚动到底部 / 右边，会触发 scrolltolower 事件	
@scroll	EventHandle		滚动时触发，event.detail = {scrollLeft, scrollTop, scrollHeight, scrollWidth, deltaX, deltaY}	
@refresherpulling	EventHandle		自定义下拉刷新控件被下拉	app-vue 2.5.12+、微信小程序基础库 2.10.1+
@refresherrefresh	EventHandle		自定义下拉刷新被触发	app-vue 2.5.12+、微信小程序基础库 2.10.1+
@refresherrestore	EventHandle		自定义下拉刷新被复位	app-vue 2.5.12+、微信小程序基础库 2.10.1+
@refresherabort	EventHandle		自定义下拉刷新被中止	app-vue 2.5.12+、微信小程序基础库 2.10.1+

scroll-view的属性非常多，其中较常用的有scroll-x、scroll-y、scroll-top、scroll-left、@scroll。

首先实现纵向滚动效果，在pages文件中创建scroll_view/index.vue文件，该文件中的代码如下：

```
<template>
    <view class="page">
        <scroll-view scroll-y="true" class="classify" :scroll-
         top="scrollTop" @scroll="eventScroll" scroll-with-animation="true">
            <view class="item">早餐1</view>
            <view class="item">套餐</view>
            <view class="item">早餐</view>
            <view class="item">套餐</view>
            <view class="item">早餐</view>
            <view class="item">套餐</view>
            <view class="item">早餐</view>
            <view class="item">套餐</view>
            <view class="item">早餐</view>
            <view class="item">套餐</view>
            <view class="item">早餐</view>
            <view class="item">套餐</view>
            <view class="item">早餐</view>
            <view class="item">套餐</view>
            <view class="item">早餐</view>
            <view class="item">套餐</view>
            <view class="item">早餐</view>
            <view class="item">套餐</view>
        </scroll-view>
        <view @click="goTop()">滚动到顶部</view>
    </view>
</template>

<script>
    export default {
        name: "scroll_view",
        data(){
            return {
                //设置竖向滚动条位置
                scrollTop:0
            }
        },
        onLoad(){
```

```
            //记录当前竖向滚动的位置
            this.tmpScrollTop=0;
        },
        methods:{
            //返回顶部
            goTop(){
                //将滚动位置的值赋值给this.scrollTop
                this.scrollTop=this.tmpScrollTop;
                setTimeout(()=>{
                    this.scrollTop=0;        //滚动到顶部
                },30);
            },
            //@scroll监听的方法
            eventScroll(event){
                //event.detail.scrollTop获取距顶部的位置
                this.tmpScrollTop=event.detail.scrollTop;
            }
        }
    }
</script>

<style scoped>
    .page{width:100%;height:100vh;}
    .classify{width:200rpx;height:400rpx;border:1px solid #007aff; overflow:
     hidden;}
    .classify .item{width:100%;height:60rpx;}
</style>
```

将scroll-view组件的属性scroll-y的值设置为true可以纵向滚动。需要注意的是，实现“滚动到顶部”功能需要以下两步：

（1）将监听滚动位置的值赋值给竖向滚动条位置，代码如下：

```
this.scrollTop=this.tmpScrollTop;
```

（2）使用setTimeout或this.$nextTick异步延迟将竖向滚动条位置设置为0，代码如下：

```
setTimeout(()=>{
    this.scrollTop=0;        //滚动到顶部
},30);
```

接下来介绍横向滚动效果，代码如下：

```
<template>
    <view class="page">
        <scroll-view scroll-x="true" class="nav-wrap">
```

```
                <view class="item"></view>
                <view class="item"></view>
                <view class="item"></view>
                <view class="item"></view>
            </scroll-view>
        </view>
</template>

<script>
    export default {
        name: "scroll_view"
    }
</script>

<style scoped>
    .page{width:100%;height:100vh;}
    .nav-wrap{width:600rpx;height:200rpx;border:1px solid #007aff; overflow:
     hidden; white-space: nowrap;}
    .nav-wrap .item{width:200rpx;height:100%;background-color: # FF0000;
     margin-right: 20rpx;display:inline-block;}
</style>
```

将scroll-view组件的属性scroll-x的值设置为true可以横向滚动，横向滚动一定要设置好CSS样式，使class="item"的view组件不能换行。在scroll-view组件上添加class="nav-wrap"的元素样式，nav-wrap的值必须设置width属性值，overflow属性值为hidden，white-space属性值为nowrap，scroll-view组件内部元素class="item"的view组件的样式值需要设置为display:inline-block，这样scroll-view组件内部的元素便不能换行，从而实现横向滚动。

9.2.2 解决 scroll-view 中内嵌 input 真机卡顿问题

扫一扫，看视频

在开发微信小程序时，scroll-view组件中会嵌套多个input组件，在真机测试上会出现页面卡顿问题，这属于微信小程序底层问题，目前没有很好的解决方案。因此，需要从另一个角度解决问题。例如，在开发购物车时数量使用view或text组件包裹，点击数量时将text组件替换成input组件，当输入完成后，再将input组件替换成text组件，即可解决此问题。接下来看一下代码示例，pages/cart/index.vue文件中的代码如下：

```
<template>
    <view class="page">
        <scroll-view class="scroll" scroll-x="true">
            <view class="item" v-for="(item,index) in items" :key="index">
                <view class="title">{{item.title}}</view>
                <view class="amount-wrap">
```

```
                    <view>数量:</view>
                    <view>
                        <!--如果isInput的值为false，则显示text组件，隐藏input组件-->
                        <text v-if="!item.isInput" @click= "showInput (index)">
                         {{item.amount}}</text>
                        <input type="number" v-model="item.amount" v-if="item.
                         isInput" :focus="true" @blur="hideInput(index)" />
                    </view>
                </view>
            </view>
        </scroll-view>
    </view>
</template>

<script>
    export default {
        name: "cart",
        data(){
            return {
                items:[
                    {
                        title:"JavaScript",  //标题
                        amount:1,            //数量
                        isInput:false        //是否显示input组件
                    },
                    {
                        title:"VUE",
                        amount:1,
                        isInput:false
                    },
                    {
                        title:"React",
                        amount:1,
                        isInput:false
                    }
                ]
            }
        },
        methods:{
            //显示input组件
            showInput(index){
                this.items[index].isInput=true;
```

```
                //使用$set解决改变值视图不渲染的问题
                this.$set(this.items,index,this.items[index]);
            },
            //隐藏input组件
            hideInput(index){
                this.items[index].isInput=false;
                this.$set(this.items,index,this.items[index]);
            }
        }
    }
</script>

<style scoped>
    .page{width:100%;height:auto;}
    .item{width:100%;height:auto;margin-top:20rpx;font-size:28rpx;}
    .amount-wrap{width:100%;display: flex;}
    .amount-wrap text,.amount-wrap input{padding:0 20rpx;}
    .scroll{width:100%;height:500rpx;}
</style>
```

注意，一定要使用v-if，而不要使用v-show，因为v-if不会将元素渲染到视图中。由于scroll-view组件内部并没有input组件，因此不会卡顿，点击text组件时显示input组件删除text组件，失去焦点时显示text组件删除input组件。input组件有两个属性，即focus和@blur。其中，focus的值为true，表示显示input组件时自动获取焦点；@blur表示输入框失去焦点时触发。这里的input并非HTML的input元素，而是uni-app的input组件。

9.3 swiper

扫一扫，看视频

swiper（滑块视图容器）的最常用的功能就是制作轮播图效果，其可以兼容多个平台。

swiper官网文档地址：https://uniapp.dcloud.io/component/swiper。

swiper一般用于左右滑动或上下滑动，如制作banner轮播图。注意，滑动切换和滚动切换是有区别的，滑动切换是一屏一屏地切换。swiper下的每个swiper-item都是一个滑动切换区域，不能停留在2个滑动切换区域之间。swiper组件的属性说明见表9-3。

表9-3 swiper组件的属性说明

属性名	类　型	默认值	说　明	平台差异说明
indicator-dots	Boolean	false	是否显示面板指示点	
indicator-color	Color	rgba(0,0,0,.3)	指示点颜色	
indicator-active-color	Color	#000000	当前选中的指示点颜色	
active-class	String		swiper-item 可见时的 class	支付宝小程序

续表

属性名	类　型	默认值	说　明	平台差异说明
changing-class	String		acceleration 设置为 {{true}} 时且处于滑动过程中，中间若干屏处于可见时的 class	支付宝小程序
autoplay	Boolean	false	是否自动切换	
current	Number	0	当前所在滑块的 index	
current-item-id	String		当前所在滑块的 item-id，不能与 current 被同时指定	支付宝小程序不支持
interval	Number	5000	自动切换时间间隔	
duration	Number	500	滑动动画时长	app-nvue 不支持
circular	Boolean	false	是否采用衔接滑动，即播放到末尾后重新回到开头	
vertical	Boolean	false	滑动方向是否为纵向	
previous-margin	String	0px	前边距，可用于露出前一项的一小部分，接受 px 和 rpx 值	app-nvue、字节跳动小程序不支持
next-margin	String	0px	后边距，可用于露出后一项的一小部分，接受 px 和 rpx 值	app-nvue、字节跳动小程序不支持
acceleration	Boolean	false	当开启时，会根据滑动速度连续滑动多屏	支付宝小程序
disable-programmatic-animation	Boolean	false	是否禁用代码变动触发 swiper 切换时使用动画	支付宝小程序
display-multiple-items	Number	1	同时显示的滑块数量	app-nvue、支付宝小程序不支持
skip-hidden-item-layout	Boolean	false	是否跳过未显示的滑块布局。设为 true 可优化复杂情况下的滑动性能，但会丢失隐藏状态滑块的布局信息	App、微信小程序
disable-touch	Boolean	false	是否禁止用户的 touch 操作。只在初始化时有效，不能动态变更	App 2.5.5+、H5 2.5.5+、支付宝小程序、字节跳动小程序
touchable	Boolean	true	是否监听用户的触摸事件。只在初始化时有效，不能动态变更	字节跳动小程序、uni-app 2.5.5+（推荐统一使用 disable-touch）
easing-function	String	default	指定 swiper 切换缓动动画类型，有效值有 default、linear、easeInCubic、easeOutCubic、easeInOutCubic	微信小程序

续表

属性名	类　型	默认值	说　明	平台差异说明
@change	EventHandle		current 改变时会触发 change 事 件，event.detail = {current: current, source: source}	
@transition	EventHandle		swiper-item 的位置发生改变时会触发 transition 事件，event. detail = {dx: dx, dy: dy}，支付宝小程序暂不支持 dx、dy	App、H5、微信小程序、支付宝小程序、字节跳动小程序、QQ 小程序
@animationfinish	EventHandle		动画结束时会触发 animationfinish 事件，event. detail = {current: current, source: source}	字节跳动小程序不支持

在pages文件夹中创建swiper/index.vue文件，该文件中的代码如下：

```
<template>
    <view class="page">
        <swiper class="swiper" :indicator-dots="true" :autoplay="true"
         interval="3000" duration="500" indicator-color="rgba(0,0,0,0.3)"
         indicator-active-color="#FF0000" @transition="eventTransition"
         :vertical="false">
            <swiper-item>
                <image src="//vueshop.glbuys.com/uploadfiles/1524206455.jpg">
                </image>
            </swiper-item>
            <swiper-item>
                <image src="//vueshop.glbuys.com/uploadfiles/1484285302.jpg">
                </image>
            </swiper-item>
            <swiper-item>
                <image src="//vueshop.glbuys.com/uploadfiles/1484285334.jpg">
                </image>
            </swiper-item>
        </swiper>
    </view>
</template>

<script>
    export default {
        name: "swiper-index",
```

```
        methods:{
            //监听swiper-item的位置发生改变函数
            eventTransition(e){
                console.log(e.detail.dx,e.detail.dy);
            }
        }
    }
</script>

<style scoped>
    .page{width:100%;height:100vh;}
    .swiper{width:100%;height:400rpx;}
    .swiper image{width:100%;height:100%;}
</style>
```

swiper组件使用起来很简单，但要注意swiper-item组件内部的image组件并非HTML元素，而是uni-app内置的image组件，在uni-app中不要使用img元素而要使用image组件装载图片。swiper组件在微信开发者工具中渲染的效果如图9-1所示。

图 9-1 swiper 组件渲染效果

9.4 movable-area 和 movable-view

movable-area是可拖动区域组件，movable-view是可移动视图组件，这两个组件搭配使用可以实现在页面中拖动滑动或双指缩放功能。

扫一扫，看视频

movable-area和movable-view官方文档地址:https://uniapp.dcloud.io/component/movable-view?id=movable-area。

movable-area指代可拖动的范围，在其中可内嵌movable-view组件用于指示可拖动的区域，即手指或鼠标按住movable-view拖动或双指缩放，但拖不出movable-area规定的范围。当然，也可以不拖动，而使用代码触发movable-view在movable-area里的移动缩放。

movable-area和movable-view支持的平台有App、H5、微信小程序、支付宝小程序、百度小程序、QQ小程序、快应用、360小程序，不支持的平台有字节跳动小程序。movable-area组件的属性说明见表9-4。

表9-4 movable-area组件的属性说明

属性名	类 型	默认值	说 明
scale-area	Boolean	false	当其中的 movable-view 设置为支持双指缩放时，设置此值可将缩放手势生效区域修改为整个 movable-area

注意：

movable-area 必须设置 width 和 height 属性，如果不设置则默认为 10px。movable-area app-nvue平台暂不支持手势缩放，并且和滚动冲突。

movable-view必须在movable-area组件中，并且必须是直接子节点，否则不能移动。movable-view组件的属性说明见表9-5。

表9-5 movable-view组件的属性说明

属性名	类 型	默认值	说 明	平台差异说明
direction	String	none	movable-view 的移动方向，属性值有 all（所有方向）、vertical（纵向）、horizontal（横向）、none	
inertia	Boolean	false	movable-view 是否带有惯性	微信小程序、支付宝小程序、App、H5、百度小程序
out-of-bounds	Boolean	false	超过可移动区域后，movable-view 是否还可以移动	微信小程序、支付宝小程序、App、H5、百度小程序
x	Number / String		定义 x 轴方向的偏移。如果 x 的值不在可移动范围内，会自动移动到可移动范围。改变 x 的值会触发动画	
y	Number / String		定义 y 轴方向的偏移。如果 y 的值不在可移动范围内，会自动移动到可移动范围。改变 y 的值会触发动画	
damping	Number	20	阻尼系数，用于控制 x 或 y 改变时的动画和过界回弹的动画，值越大移动越快	微信小程序、支付宝小程序、App、H5、百度小程序

续表

属性名	类 型	默认值	说 明	平台差异说明
friction	Number	2	摩擦系数，用于控制惯性滑动的动画，值越大，摩擦力越大，滑动越快停止；必须大于 0，否则会被设置成默认值	微信小程序、支付宝小程序、App、H5、百度小程序
disabled	Boolean	false	是否禁用	
scale	Boolean	false	是否支持双指缩放，默认缩放手势生效区域是在 movable-view 内	微信小程序、支付宝小程序、App、H5
scale-min	Number	0.5	定义缩放倍数最小值	微信小程序、支付宝小程序、App、H5
scale-max	Number	10	定义缩放倍数最大值	微信小程序、支付宝小程序、App、H5
scale-value	Number	1	定义缩放倍数，取值范围为 0.5~10	微信小程序、支付宝小程序、App、H5
animation	Boolean	true	是否使用动画	微信小程序、支付宝小程序、App、H5、百度小程序
@change	EventHandle		拖动过程中触发的事件，event.detail = {x: x, y: y, source: source}，其中 source 表示产生移动的原因，值可为 touch（拖动）、touch-out-of-bounds（超出移动范围）、out-of-bounds（超出移动范围后的回弹）、friction（惯性）和空字符串（setData）	
@scale	EventHandle		缩放过程中触发的事件，event.detail = {x: x, y: y, scale: scale}	微信小程序、App、H5、百度小程序

在pages文件夹下创建movable/index.vue文件，该文件中的代码如下：

```
<template>
    <view class="page">
        <view class="uni-padding-wrap uni-common-mt">
            <view class="uni-title uni-common-mt">
                示例 1:
                <text>movable-view 区域小于 movable-area</text>
            </view>
            <movable-area class="marea1">
                <movable-view class="mview1" :x="x" :y="y" direction="all" @
                 change="onChange">示例1</movable-view>
            </movable-area>
```

```
            <button @click="moveMview" type="default">点击这里移动至 (30px,
             30px)</button>
            <view class="uni-title uni-common-mt">
                示例 2:
                <text>movable-view区域大于movable-area</text>
            </view>
            <movable-area class="marea2">
                <movable-view class="max" direction="all">
                    <image src="//vueshop.glbuys.com/uploadfiles/1484285334.
                     jpg"></image>
                </movable-view>
            </movable-area>
        </view>
    </view>
</template>

<script>
    export default {
        name: "movable",
        data() {
            return {
                x: 0,
                y: 0
            }
        },
        onLoad(){
            this.tmpX=0;
            this.tmpY=0;
        },
        methods: {
       //单击按钮，实现class="mview1"的movable-view组件，移动至x:30px、y:30px的位置
            moveMview: function(e) {
                this.x = this.tmpX
                this.y = this.tmpY
                // this.$nextTick(function() {
                //     this.x = 30
                //     this.y = 30
                // })
                setTimeout(()=>{
                    this.x = 30
                    this.y = 30
                },30)
```

```
            },
            //记录class="mview1"的movable-view组件移动的位置
            onChange: function(e) {
                this.tmpX = e.detail.x
                this.tmpY = e.detail.y
            }
        }
    }
</script>

<style scoped>
    .page{width:100%;height:100vh;}
    .mview1{width:300rpx;height:300rpx;background-color:#007aff;}
    .marea1{width:100%;height:600rpx;}

    .marea2{width:300rpx;height:300rpx;margin:0 auto;overflow:hidden;}
    .marea2.max{width:600rpx;height:600rpx;background-color:#FF0000}
    .marea2.max image{width:100%;height:100%;}
</style>
```

在上述代码的“示例1”中，单击button按钮，将class="mview1"的movable-view组件移动到指定位置，需要在class="mview1"的movable-view组件上绑定@change事件，并调用自定义的onChange()方法。在onChange()方法接收的参数中，使用e.detail.x获取x坐标的值，e.detail.y获取y坐标的值，分别赋值给this.tmpX和this.tmpY属性。单击button按钮，调用moveMview()方法，在该方法内部，先将this.tmpX和this.tmpY的值分别赋值给this.x和this.y，再使用setTimeout()或this.$nextTick()方法异步延迟、在该方法内部将要移动位置的值赋给this.x和this.y，最后将this.x和this.y分别赋值给class="mview1"的movable-view组件的x和y属性。

9.5 text 和 rich-text

在uni-app中，纯文字建议使用text（文本）组件包裹，以实现长按选择文字、连续空格、解码功能。rich-text（富文本）组件可以解析HTML标签，通常由于显示商品介绍、文章内容等应用场景。

9.5.1 text 的使用

text组件用于包裹文本内容。

text官方文档地址：https://uniapp.dcloud.io/component/text。

text组件的属性说明见表9-6。

扫一扫，看视频

表9-6　text组件的属性说明

属性名	类　型	默认值	说　明	平台差异说明
selectable	Boolean	false	文本是否可选	App、H5、微信小程序
space	String		显示连续空格	App、H5、微信小程序
decode	Boolean	false	是否解码	App、H5、微信小程序

space属性的值见表9-7。

表9-7　space属性的值

值	说　明
ensp	中文字符空格一半大小
emsp	中文字符空格大小
nbsp	根据字体设置的空格大小

在pages文件夹下创建text/index.vue文件，该文件中的代码如下：

```
<template>
    <view class="page">
        <text :selectable="true" space="nbsp" :decode="true">{{text}}</text>
    </view>
</template>

<script>
    export default {
        name: "text-index",
        data(){
            return {
                text:'我是&lt;text&gt; 组件',
            }
        }
    }
</script>

<style scoped>
    .page{width:100%;height:100vh;}
</style>
```

在微信开发者工具中渲染的内容为“我是<text> 组件”。

注意：

<text> 组件内只支持嵌套 <text>，不支持其他组件或自定义组件，否则会引发在不同

平台的渲染差异。如果在app-nvue下，则只有<text>才能包裹文本内容，无法在<view>组件包裹文本。decode可以解析的内容有 、<、>、&、'、 、 ，支持“\n”方式换行。如果使用 <span> 组件，编译时会被转换为 <text>。

各个操作系统的空格标准并不一致。

除了文本节点以外的其他节点都无法长按选中。

9.5.2　rich-text 的使用

扫一扫，看视频

rich-text可以支持部分HTML节点及属性。

rich-text官方文档地址:https://uniapp.dcloud.io/component/rich-text。

rich-text组件的属性说明见表9-8。

表9-8　rich-text组件的属性说明

属性名	类　型	默认值	说　明	平台差异说明
nodes	Array、String	[]	节点列表、HTML String	
space	string		显示连续空格	微信基础库 2.4.1+、QQ 小程序
selectable	Boolean	false	rich-text 是否可以长按选中。可用于复制、粘贴等场景	百度 App 11.10+

nodes属性推荐使用 Array 类型。由于rich-text组件会将 String 类型转换为 Array 类型，因此性能会有所下降。注意，App-nvue 平台的nodes 属性只支持使用 Array 类型，支付宝小程序的nodes 属性只支持使用 Array 类型。rich-text组件支持默认事件，包括click、touchstart、touchmove、touchcancel、touchend、longpress。nodes属性支持的HTML节点及属性见表9-9。

表9-9　nodes属性支持的HTML节点及属性

节　点	属　性
a	
abbr	
b	
blockquote	
br	
code	
col	span、width
colgroup	span、width
dd	
del	

续表

节　点	属　性
div	
dl	
dt	
em	
fieldset	
h1	
h2	
h3	
h4	
h5	
h6	
hr	
i	
img	alt、src、height、width
ins	
label	
legend	
li	
ol	start、type
p	
q	
span	
strong	
sub	
sup	
table	width
tbody	
td	colspan、height、rowspan、width

续表

节 点	属 性
tfoot	
th	colspan、height、rowspan、width
thead	
tr	
ul	

注意：

全局支持class和style属性，但不支持id属性。

首先介绍nodes为String类型的代码示例。在pages文件夹下创建rich_text/index.vue文件，该文件中的内容如下：

```
<template>
    <view class="page">
        <rich-text :nodes="content"></rich-text>
    </view>
</template>

<script>
    export default {
        name: "text-index",
        data(){
            return {
                content:'<div>我  是<span style="color:#FF0000;font-size:
                36rpx;">富文本</span>组件</div>'
            }
        }
    }
</script>

<style scoped>
    .page{width:100%;height:100vh;}
</style>
```

rich-text在微信开发者工具中渲染的效果如图9-2所示。

图 9-2　rich-text 渲染的效果

nodes为String类型时使用很简单，但是性能和平台的兼容性略差，推荐使用Array类型。nodes内的节点现支持两种类型，通过type来区分，分别是元素节点(type=node)和文本节点(type=text)，默认是元素节点，即在富文本区域里显示的HTML节点，属性说明见表9-10和表9-11。

表9-10　nodes为node类型的属性说明

属性名	说　明	类　型	是否必填	备　注
name	标签名	String	是	支持部分受信任的 HTML 节点
attrs	属性	Object	否	支持部分受信任的属性，遵循 Pascal（帕斯卡）命名法
children	子节点列表	Array	否	结构和 nodes 一致

表9-11　nodes为text类型的属性说明

属性名	说　明	类　型	是否必填	备　注
text	文本	String	是	支持 entities

nodes为text类型的文本节点代码示例如下：

```
<template>
    <view class="page">
        <rich-text :nodes="content"></rich-text>
    </view>
</template>

<script>
    export default {
        name: "text-index",
        data(){
            return {
                content:[
                    {
                        type:"text",
                        text:"我是富文本组件"
```

```
                    },
                    {
                        type:"text",
                        text:"我是富文本组件2"
                    }
                ]
            }
        }
    }
</script>

<style scoped>
    .page{width:100%;height:100vh;}
</style>
```

data()方法返回的content属性中type属性值为text，表示文本节点，在text属性值中写入内容会将内容渲染到视图上。

nodes为node的元素节点代码示例如下：

```
<template>
    <view class="page">
        <rich-text :nodes="content"></rich-text>
    </view>
</template>

<script>
    export default {
        name: "text-index",
        data(){
            return {
                content:[
                    {
                        type:"node",
                        name:"div",
                        children:[
                            {
                                type:"text",
                                text:"我是"
                            },
                            {
                                type:"node",
                                name:"span",
                                attrs:{
```

```
                                        style:"color:#FF0000;font-size:36rpx;"
                                    },
                                    children:[
                                        {
                                            type:"text",
                                            text:"富文本"
                                        }
                                    ]
                                },
                                {
                                    type:"text",
                                    text:"组件"
                                }
                            ]
                        }
                    ]
                }
            }
        }
</script>

<style scoped>
    .page{width:100%;height:100vh;}
</style>
```

元素节点中最难的就是content属性值的数据结构，建议读者仔细阅读上述代码，如果阅读中有困难，请看视频教程学习。

在实际开发中，rich-text的内容是通过服务端获取的，而服务端的内容都是String类型。为了解决兼容性和性能问题，uni-app推荐rich-text使用Array类型进行渲染。因此，需要将String类型的数据转成Array类型进行渲染，可以使用html-parser 转换。html-parser下载地址：https://github.com/dcloudio/hello-uniapp/blob/master/common/html-parser.js。

下载完成后，将html-parser.js文件复制到static/js/utils文件夹下，pages/rich_text/index.vue文件中的代码如下：

```
<template>
    <view class="page">
        <rich-text :nodes="content"></rich-text>
    </view>
</template>

<script>
    import parseHtml from "../../static/js/utils/html-parser";
```

```
    export default {
        name: "text-index",
        data(){
            return {
                content:parseHtml('<div>我是<span style="color:#FF0000;font-size:36rpx;">富文本</span>组件</div>')
            }
        }
    }
</script>

<style scoped>
    .page{width:100%;height:100vh;}
</style>
```

导入html-parser.js文件，使用parseHtml()函数将String类型转成Array类型进行渲染输出。

9.6 button

uni-app的button（按钮）功能非常强大，可以实现获取微信用户信息、获取手机号、分享等功能。

button官方文档地址:https://uniapp.dcloud.io/component/button。

9.6.1 button的使用

扫一扫，看视频

uni-app的button组件与H5的button部分功能有所不同，uni-app采用的是小程序button的使用方式，其功能非常强大，属性有很多，属性说明见表9-12~表9-16。

表9-12　button组件的属性说明

属性名	类　型	默认值	说　明	生效时机	平台差异说明
size	String	default	按钮的大小		
type	String	default	按钮的样式类型		
plain	Boolean	false	按钮是否镂空，背景色透明		
disabled	Boolean	false	是否禁用		
loading	Boolean	false	名称前是否带 loading 图标		App-nvue 平台,在 iOS 上为雪花，在 Android 上为圆圈

续表

属性名	类 型	默认值	说 明	生效时机	平台差异说明
form-type	String		用于 <form> 组件，点击后会触发 <form> 组件的 submit/reset 事件		
open-type	String		开放能力		
hover-class	String	button-hover	指定按钮按下去的样式类。当 hover-class="none" 时，没有点击态效果		App-nvue 平台暂不支持
hover-start-time	Number	20	按住后多久出现点击态，单位为 ms		
hover-stay-time	Number	70	手指松开后点击态保留时间，单位为 ms		
app-parameter	String		打开 App 时向 App 传递的参数，open-type=launchApp 时有效		微信小程序、QQ 小程序
hover-stop-propagation	Boolean	false	指定是否阻止本节点的祖先节点出现点击态		微信小程序
lang	String	'en'	指定返回用户信息的语言，zh_CN 为简体中文，zh_TW 为繁体中文，en 为英文		微信小程序
session-from	String		会话来源，open-type="contact" 时有效		微信小程序
send-message-title	String	当前标题	会话内消息卡片标题，open-type="contact" 时有效		微信小程序
send-message-path	String	当前分享路径	会话内消息卡片点击跳转小程序路径，open-type="contact" 时有效		微信小程序
send-message-img	String	截图	会话内消息卡片图片，open-type="contact" 时有效		微信小程序
show-message-card	Boolean	false	是否显示会话内消息卡片，设置此参数为 true，用户进入客服会话会在右下角显示“可能要发送的小程序”提示，用户点击后可以快速发送小程序消息，open-type="contact" 时有效		微信小程序
@getphonenumber	Handler		获取用户手机号回调	open-type="getPhoneNumber"	微信小程序

续表

属性名	类 型	默认值	说 明	生效时机	平台差异说明
@getuserinfo	Handler		用户点击该按钮时，会返回获取到的用户信息，从返回参数的 detail 中获取到的值同 uni.getUserInfo	open-type="getUserInfo"	微信小程序
@error	Handler		当使用开放能力时，发生错误的回调	open-type="launchApp"	微信小程序
@opensetting	Handler		在打开授权设置页并关闭后回调	open-type="openSetting"	微信小程序
@launchapp	Handler		打开 App 成功的回调	open-type="launchApp"	微信小程序

表9-13 size属性有效值说明

值	说 明
default	默认大小
mini	小尺寸

表9-14 type属性有效值说明

值	说 明
primary	微信小程序、360 小程序为绿色，App、H5、百度小程序、支付宝小程序、快应用为蓝色，字节跳动小程序为红色，QQ 小程序为浅蓝色。如果想在多端统一颜色，应改用 default，并自行编写样式
default	白色
warn	红色

表9-15 form-type属性有效值说明

值	说 明
submit	提交表单
reset	重置表单

表9-16 open-type属性有效值说明

值	说 明	平台差异说明
feedback	打开“意见反馈”页面，用户可提交反馈内容并上传日志	App、微信小程序、QQ 小程序
share	触发用户转发	微信小程序、百度小程序、支付宝小程序、字节跳动小程序、QQ 小程序

续表

值	说　明	平台差异说明
getUserInfo	获取用户信息，可以从 @getuserinfo 回调中获取到用户信息，包括头像、昵称等信息	微信小程序、百度小程序、QQ 小程序
contact	打开客服会话，如果用户在会话中点击消息卡片后返回应用，可以从 @contact 回调中获得具体信息	微信小程序、百度小程序
getPhoneNumber	获取用户手机号，可以从 @getphonenumber 回调中获取到用户信息	微信小程序、百度小程序、字节跳动小程序、支付宝小程序
launchApp	打开 App，可以通过 app-parameter 属性设定向 App 传递的参数	微信小程序、QQ 小程序
openSetting	打开授权设置页	微信小程序、百度小程序
getAuthorize	支持小程序授权	支付宝小程序
contactShare	分享到通讯录好友	支付宝小程序
lifestyle	关注生活号	支付宝小程序
openGroupProfile	唤起 QQ 群资料卡页面，可以通过 group-id 属性设定需要打开的群资料卡的群号，同时 manifest 中必须配置 groupIdList	QQ 小程序基础库 1.4.7 版本 +

下面介绍button组件基本使用代码示例。在pages文件夹下创建button/index.vue文件，该文件中的代码如下：

```
<template>
    <view class="page">
        <button class="btn" type="default" form-type=” submit” >button
        组件</button>
    </view>
</template>

<script>
    export default {
        name: "button-index"
    }
</script>

<style scoped>
 .page{width:100%;height:100vh;}
 .btn{width:200rpx;height:80rpx;background-color: #007aff; color: #FFFFFF;
  font-size:32rpx;}
```

```
  .btn:after{ border:0 none;}
</style>
```

button组件使用方法很简单，与传统H5的button类似，但不同之处是type属性值默认是default，不再是button，如果想设置submit或reset，则需要在form-type属性中进行设置。注意，如果想清除button组件的边框，则需要在伪元素after中进行设置。下面介绍button组件的开放能力，如获取微信用户信息，代码示例如下：

```
<template>
    <view class="page">
        <button type="primary" open-type="getUserInfo" @getuserinfo=
         "getUserInfo"> 获取用户信息</button>
        <view>头  像: <image :src="head" style="width:200rpx;height:200rp
         x;"></image></view>
        <view>国家:{{country}}</view>
        <view>省份:{{province}}</view>
        <view>城市:{{city}}</view>
        <view>性别:{{gender==1?'男':(gender==2?'女':'')}}</view>
        <view>昵称:{{nickName}}</view>
    </view>
</template>

<script>
    export default {
        name: "button-index",
        data(){
            return {
                head:"",
                country:"",
                province:"",
                city:"",
                gender:"",
                nickName:""
            }
        },
        methods:{
            getUserInfo(e){
                this.head=e.detail.userInfo.avatarUrl;
                this.country=e.detail.userInfo.country;
                this.province=e.detail.userInfo.province;
                this.city=e.detail.userInfo.city;
                this.gender=e.detail.userInfo.gender;
                this.nickName=e.detail.userInfo.nickName;
```

```
            }
        }
    }
</script>

<style scoped>
 .page{width:100%;height:100vh;}
</style>
```

获取用户信息时，需要在button组件中设置open-type属性，将值设置为getUserInfo。绑定@getuserinfo事件，传入自定义方法getUserInfo()，该方法接收的参数可以获取到用户信息。

9.6.2 获取手机号

扫一扫，看视频

如果想获取微信用户绑定的手机号，需要先调用uni.login接口。因为需要用户主动触发才能发起获取手机号接口请求，所以该功能不由API来调用，而需要通过点击button组件来触发。

注意：

目前该接口针对非个人开发者，且只对完成了认证的小程序开放(不包含海外主体)。该接口需要谨慎使用，若用户举报较多或被发现在不必要的场景下使用，微信有权永久回收该小程序的该接口权限。

获取手机号流程：将 button组件open-type的值设置为getPhoneNumber，当用户点击并同意之后，可以通过@getphonenumber事件回调获取到微信服务器返回的加密数据，然后在第三方服务端结合session_key进行解密获取手机号。

获取手机号需要和服务端配合，前端人员不需要了解服务端，只需要让服务端人员开发出两个接口（uni.login会员登录授权接口和encryptedData解密接口）即可。

uni.login会员登录授权接口官方文档地址：https://developers.weixin.qq.com/miniprogram/dev/api-backend/open-api/login/auth.code2Session.html。

encryptedData解密接口官方文档地址：https://developers.weixin.qq.com/miniprogram/dev/framework/open-ability/signature.html#加密数据解密算法。

在实际开发中，将上面两个官方文档地址发给服务端开发人员即可。对该接口文档感兴趣的读者可以关注“html5程序思维”公众号，发送uniapp进行下载。

使用服务端接口前，应先配置微信开发者工具，否则接口无法使用。打开微信开发者工具，单击“详情”按钮，选择“本地设置”选项卡，勾选“不校验合法域名、web-view（业务域名）、TLS版本以及HTTPS证书”复选框，如图9-3所示。

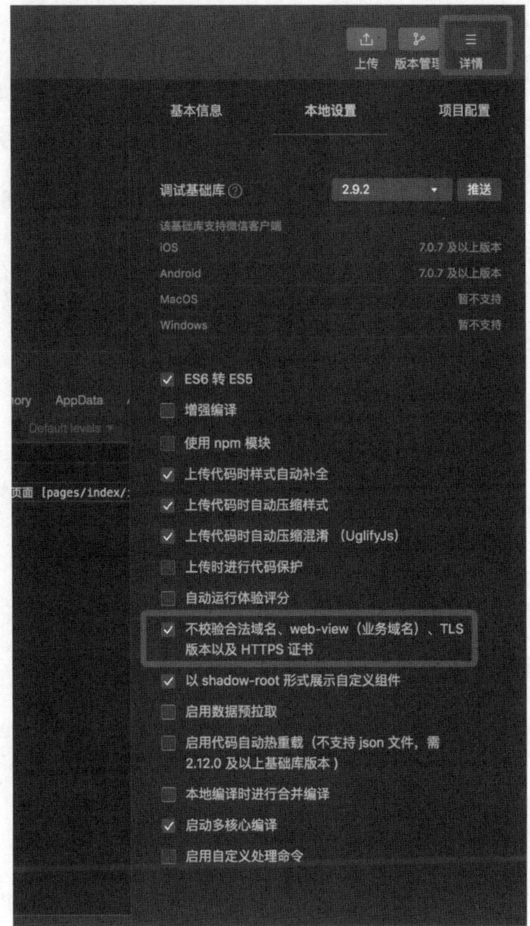

图 9-3 微信开发者工具详情选项

pages/button/index.vue文件中的代码如下：

```
<template>
    <view class="page">
        <button type="warn" open-type="getPhoneNumber" @getphonenumber=
         "getPhoneNumber($event)">获取手机号</button>
    </view>
</template>

<script>
    export default {
        name: "button-index",
        methods:{
            getPhoneNumber(e){
```

```
let iv=e.detail.iv;   //加密算法的初始向量
let enData=e.detail.encryptedData;
//包括敏感数据在内的完整用户信息的加密数据，其中包含手机号
//uni-app的登录API
uni.login({
    provider: 'weixin', //登录服务提供商，使用微信登录
    success: (loginRes) => {
        let code=loginRes.code;//小程序专有，用户登录凭证，通过服
        //务端接口，使用code换取openid和session_key等信息
        //request发起网络请求，类似于ajax
        uni.request({
            method:"POST",
            url: 'https://diancan.glbuys.com/api/v1/wechat_
             openid', //调用服务端接口，获取openid和sessionKey
            data: {
                code: code //用户登录凭证
            },
            header: {
                'Content-Type': 'application/x-www-form-urlencoded'
            },
            success: (res) => {
                let openId=res.data.data.openid;//获取openid
                let sessionKey=res.data.data.session_key;
                //获取sessionKey
                //发起网络请求
                uni.request({
                    method:"POST",
                     url: 'https://diancan.glbuys.com/api/v1/
dewxbizdata',//调用服务端接口解密encryptedData，最终获取手机号
                    data: {
                        session_key: sessionKey,
                        iv:iv,
                        encrypteddata:enData
                    },
                    header: {
                      'Content-Type': 'application/x-www-form-
                      urlencoded'
                    },
                    success: (res) => {
                       //手机号
                       console.log(res.data.data.phoneNumber);
```

```
                                }
                            });
                        }
                    });
                }
            });
        }
    }
}
</script>

<style scoped>
 .page{width:100%;height:100vh;}
</style>
```

上述代码中的注释很清晰，这里不再赘述，请读者仔细阅读。request使用方式很像jQuery，该内容会在10.1节中讲解。

9.6.3 分享小程序

扫一扫，看视频

当button组件open-type属性值为share时，可以自定义分享小程序页面。pages/button/index.vue文件中的代码示例如下：

```
<template>
    <view class="page">
        <button type="primary" open-type="share">分享</button>
    </view>
</template>

<script>
    export default {
        name: "button-index",
        onShareAppMessage(res) {
            console.log(res.from);
            //分享事件来源:button（页面内分享按钮）、menu（右上角分享按钮）
            if (res.from === 'button') {      // 来自页面内分享按钮
                console.log(res.target)
                //如果 from 值是 button，则 target 是触发这次分享事件的 button，
                //否则为 undefined
            }
            return {
                title: 'button组件',           //分享的标题
                path: '/pages/button/index', //页面 path ，必须是以“/”开头的完整路径
```

```
                imageUrl:"http://vueshop.glbuys.com/uploadfiles/1524556409.jpg"
                //分享图标，路径可以是本地文件路径、代码包文件路径或者网络图片路径，支持
                //PNG及JPG。显示图片长宽比是 5 : 4
            }
        }
    }
</script>

<style scoped>
 .page{width:100%;height:100vh;}
</style>
```

分享功能需要配合onShareAppMessage()函数来实现。在小程序中，当用户点击分享后，在JS中定义 onShareAppMessage()处理函数，该函数与onLoad()等生命周期函数同级，设置该页面的分享信息。用户点击分享按钮时会调用该函数。该分享按钮可以是小程序右上角原生菜单自带的分享按钮，也可以是开发者在页面中放置的分享按钮。此事件需要返回一个Object，用于自定义分享内容。

onShareAppMessage官方文档地址:https://uniapp.dcloud.io/api/plugins/share?id=onshareapp-message。

9.7 checkbox

扫一扫，看视频

uni-app的checkbox（复选框）与传统H5的checkbox使用方式不太一样，需要使用checkbox-group组件包裹checkbox才能实现多选。

checkbox官方文档地址:https://uniapp.dcloud.io/component/checkbox。

checkbox-group内部由多个checkbox组成。checkbox组件的属性说明见表9-17和表9-18。

表9-17 checkbox组件的事件属性说明

属性名	类 型	说 明
@change	EventHandle	<checkbox-group> 中的选中项发生改变时触发 change 事件，detail = {value:[选中的 checkbox 的 value 的数组]}

表9-18 checkbox组件的属性说明

属性名	类 型	默认值	说 明
value	String		<checkbox> 标识，选中时触发 <checkbox-group> 的 change 事件，并携带 <checkbox> 的 value
disabled	Boolean	false	是否禁用
checked	Boolean	false	当前是否选中，可用来设置默认选中
color	Color		checkbox 的颜色，同 CSS 的 color

在pages文件夹下创建checkbox/index.vue文件，该文件中的代码如下：

```
<template>
    <view class="page">
        <checkbox-group @change="checkboxChange">
            <label class="uni-list-cell" v-for="(item,index) in items" :key="index">
                <view>
                    <checkbox :value="item.value" :checked="item.checked" />
                </view>
                <view>{{item.name}}</view>
            </label>
        </checkbox-group>
        <button type="button" @click="submit()">提交</button>
    </view>
</template>

<script>
    export default {
        name: "checkbox-index",
        data(){
            return {
                items: [{
                    value: 'USA',
                    name: '美国'
                },
                    {
                        value: 'CHN',
                        name: '中国',
                        checked: 'true'
                    },
                    {
                        value: 'BRA',
                        name: '巴西'
                    },
                    {
                        value: 'JPN',
                        name: '日本'
                    },
                    {
                        value: 'ENG',
                        name: '英国'
                    },
                    {
                        value: 'FRA',
```

```
                    name: '法国'
                }
            ]
        }
    },
    methods: {
        //checkbox-group组件的@change事件的回调方法
        checkboxChange: function (e) {
            var items = this.items,        //data()方法中的items属性
                values = e.detail.value;
                //获取checkbox组件checked属性值为true的所有value，该值为数组
            for (var i = 0; i < items.length; ++i) {
                //如果values包含items[i].value
                if(values.includes(items[i].value)){
                    this.items[i].checked=true;
                    this.$set(this.items,i,this.items[i]);
                }else{
                    this.items[i].checked=false;
                    this.$set(this.items,i,this.items[i]);
                }
            }
        },
        //提交
        submit(){
            let newItems=[];
            for(let i=0;i<this.items.length;i++){
                if(this.items[i].checked){
                    //将已选择的数据添加到数组中
                    newItems.push(this.items[i]);
                }
            }
            //显示选择的数据
            console.log(JSON.stringify(newItems));
        }
    }
}
</script>

<style scoped>
    .page{width:100%;height:100vh;}
    .uni-list-cell{display:flex;}
</style>
```

以上代码中的注释很清晰，不再赘述，在微信开发者工具中渲染的效果，即checkbox组件的预览效果如图9-4所示。

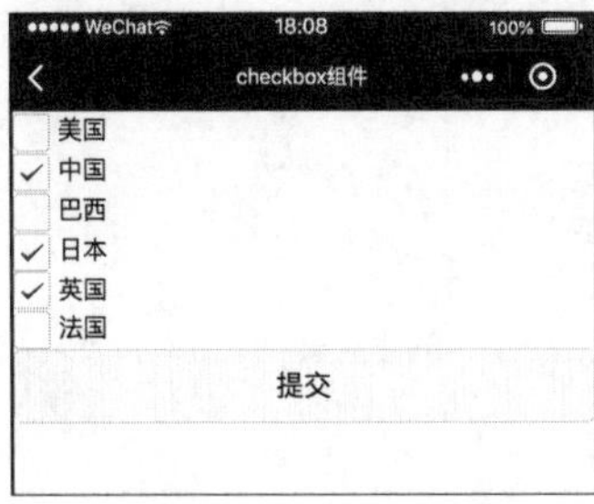

图 9-4 checkbox 组件预览效果

9.8 radio

uni-app的radio（单选按钮）和checkbox一样，需要用radio-group组件包裹radio组件才能使用。

扫一扫，看视频

radio官方文档地址:https://uniapp.dcloud.io/component/radio。

radio-group内部由多个 <radio> 组成。通过把多个radio包裹在一个radio-group下，实现这些radio的单选。radio组件的属性说明见表9-19和表9-20。

表9-19 radio组件的事件属性说明

属性名	类 型	说 明
@change	EventHandle	<radio-group> 中的选中项发生变化时触发 change 事件，event.detail = {value: 选中项 radio 的 value}

表9-20 radio组件的属性说明

属性名	类 型	默认值	说 明
value	String		<radio> 标识。当该 <radio> 选中时，<radio-group> 的 change 事件会携带 <radio> 的 value
checked	Boolean	false	当前是否选中
disabled	Boolean	false	是否禁用
color	Color		radio 的颜色，同 CSS 的 color

在pages文件夹下创建radio/index.vue文件，该文件中的代码如下：

```
<template>
    <view>
        <view class="uni-list">
            <radio-group @change="radioChange">
                <label class="uni-list-cell" v-for="(item, index) in items"
                 :key="item.value">
```

```
                <view>
                    <radio :value="item.value" :checked="item.checked" />
                </view>
                <view>{{item.name}}</view>
            </label>
        </radio-group>
    </view>
    <button type="button" @click="submit">提交</button>
  </view>
</template>

<script>
  export default {
    name: "radio-index",
    data() {
      return {
        items: [{
          value: 'USA',
          name: '美国'
        },
          {
            value: 'CHN',
            name: '中国',
          },
          {
            value: 'BRA',
            name: '巴西'
          },
          {
            value: 'JPN',
            name: '日本'
          },
          {
            value: 'ENG',
            name: '英国'
          },
          {
            value: 'FRA',
            name: '法国'
          },
        ]
      }
    },
```

```
        onLoad(){
            this.newItem={};
        },
        methods: {
          //radio-group组件的@change事件的回调方法
            radioChange: function(evt) {
                for (let i = 0; i < this.items.length; i++) {
                    //如果data方法中的items属性值等于选择数据的值
                    if (this.items[i].value === evt.target.value) {
                        this.items[i].checked=true;
                        this.newItem=this.items[i];
                        //将选择的数据赋值给this.newItem属性
                        break;
                    }
                }
            },
            //提交
            submit(){
                //输出选择的数据
                console.log(this.newItem.name);
            }
        }
    }
</script>

<style scoped>
    .uni-list-cell{display:flex;}
</style>
```

radio组件和checkbox组件使用方式类似，比较简单，在微信开发者工具中渲染的效果，即radio组件预览效果如图9-5所示。

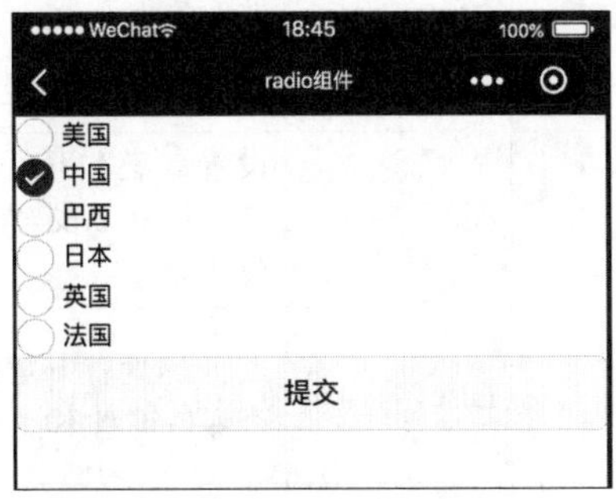

图 9-5　radio 组件预览效果

9.9 input

扫一扫，看视频

uni-app的input（输入框）功能非常强大，使用方式与传统H5的input相似，同样支持v-model。

input官方文档地址：https://uniapp.dcloud.io/component/input。

input组件的属性说明见表9-21~表9~23。

表9-21　input组件的属性说明

属性名	类　型	默认值	说　明	平台差异说明
value	String		input 的初始内容	
type	String	text	input 的类型	H5 暂未支持动态切换，应使用 v-if 进行整体切换
password	Boolean	false	是否为密码类型	
placeholder	String		input 为空时占位符	
placeholder-style	String		指定 placeholder 的样式	
placeholder-class	String	"input-placeholder"	指定 placeholder 的样式类	
disabled	Boolean	false	是否禁用	
maxlength	Number	140	最大输入长度，设置为 -1 时不限制最大长度	
cursor-spacing	Number	0	指定光标与键盘的距离，单位为 px 。取 input 距离底部的距离和 cursor-spacing 指定的距离的最小值作为光标与键盘的距离	App、微信小程序、百度小程序、QQ 小程序
focus	Boolean	false	获取焦点	在 H5 平台能否聚焦及软键盘是否跟随弹出取决于当前浏览器本身的实现
confirm-type	String	done	设置键盘右下角按钮的文字，仅在 type="text" 时生效	
confirm-hold	Boolean	false	点击键盘右下角按钮时是否保持键盘不收起	App、微信小程序、支付宝小程序、百度小程序、QQ 小程序
cursor	Number		指定 focus 时的光标位置	

续表

属性名	类 型	默认值	说 明	平台差异说明
selection-start	Number	–1	光标起始位置，自动聚集时有效，须与 selection-end 搭配使用	
selection-end	Number	–1	光标结束位置，自动聚集时有效，须与 selection-start 搭配使用	
adjust-position	Boolean	true	键盘弹起时是否自动上推页面	App-Android（Vue 页面 softinputMode 为 adjustResize 时无效）、微信小程序、百度小程序、QQ 小程序
hold-keyboard	Boolean	false	聚焦时，点击页面时不收起键盘	微信小程序 2.8.2
@input	EventHandle		当键盘输入时，触发 input 事件，event.detail = {value}	
@focus	EventHandle		input 聚焦时触发，event.detail = { value, height }，height 为键盘高度	仅微信小程序、App（2.2.3+）、QQ 小程序支持 height
@blur	EventHandle		input 失去焦点时触发，event.detail = {value: value}	
@confirm	EventHandle		点击完成按钮时触发，event.detail = {value: value}	
@keyboardheightchange	EventHandle		键盘高度发生变化时触发此事件，event.detail = {height: height, duration: duration}	微信小程序 2.7.0

注意：

@input事件处理函数可以直接返回一个字符串，将替换输入框的内容。仅微信小程序支持，input 组件上有默认的 min-height 样式，如果 min-height 的值大于 height 的值，那么height样式将无效。

表9–22 type属性有效值说明

值	说 明	平台差异说明
text	文本输入键盘	
number	数字输入键盘	均支持，注意 iOS 上 App-vue 弹出的数字键盘并非九宫格方式

续表

值	说 明	平台差异说明
idcard	身份证输入键盘	微信小程序、支付宝小程序、百度小程序、QQ 小程序
digit	带小数点的数字键盘	App 的 nvue 页面、微信小程序、支付宝小程序、百度小程序、头条小程序、QQ 小程序

小程序平台上的number 类型只支持输入整型数字。微信开发者工具上体现不出效果，需要使用真机预览。如果需要在小程序平台输入浮点型数字，请使用 digit 类型。

表9-23 confirm-type属性有效值说明

值	说 明	平台差异说明
send	右下角按钮为“发送”	微信小程序、支付宝小程序、百度小程序、App 的 nvue 页面
search	右下角按钮为“搜索”	
next	右下角按钮为“下一个”	微信小程序、支付宝小程序、百度小程序、App 的 nvue 页面
go	右下角按钮为“前往”	
done	右下角按钮为“完成”	微信小程序、支付宝小程序、百度小程序、App 的 nvue 页面

在pages文件夹下创建input/index.vue文件，该文件中的代码如下：

```
<template>
    <view>
        <view class="uni-common-mt">
            <view class="uni-form-item uni-column">
                <view class="title">可自动聚焦的input</view>
                <input class="uni-input" focus placeholder="自动获得焦点" />
            </view>
            <view class="uni-form-item uni-column">
                <view class="title">键盘右下角按钮显示为搜索</view>
                <input class="uni-input" confirm-type="search" placeholder="
                键盘右下角按钮显示为搜索" />
            </view>
            <view class="uni-form-item uni-column">
                <view class="title">控制最大输入长度的input</view>
                <input class="uni-input" maxlength="10" placeholder="最大输入
                长度为10" />
            </view>
            <view class="uni-form-item uni-column">
                <view class="title">实时获取输入值:{{inputValue}}</view>
```

```
            <input class="uni-input" @input="onKeyInput" placeholder="输
             入同步到view中" />
        </view>
        <view class="uni-form-item uni-column">
            <view class="title">控制输入的input</view>
            <input class="uni-input" @input="replaceInput" v-
             model="changeValue" placeholder="连续的两个1会变成2" />
        </view>
        <!-- #ifndef MP-BAIDU -->
        <view class="uni-form-item uni-column">
            <view class="title">控制键盘的input</view>
            <input class="uni-input" ref="input1" @input="hideKeyboard"
             placeholder="输入123自动收起键盘" />
        </view>
        <!-- #endif -->
        <view class="uni-form-item uni-column">
            <view class="title">数字输入的input</view>
            <input class="uni-input" type="number" placeholder="这是一个
             数字输入框" />
        </view>
        <view class="uni-form-item uni-column">
            <view class="title">密码输入的input</view>
            <input class="uni-input" password type="text" placeholder="
             这是一个密码输入框" />
        </view>
        <view class="uni-form-item uni-column">
            <view class="title">带小数点的input</view>
            <input class="uni-input" type="digit" placeholder="带小数点的
             数字键盘" />
        </view>
        <view class="uni-form-item uni-column">
            <view class="title">身份证输入的input</view>
            <input class="uni-input" type="idcard" placeholder="身份证输
             入键盘" />
        </view>
        <view class="uni-form-item uni-column">
            <view class="title">控制占位符颜色的input</view>
            <input class="uni-input" placeholder-style="color:#F76260"
             placeholder="占位符字体是红色的" />
        </view>
    </view>
  </view>
</template>
```

```
<script>
    export default {
        name: "input-component",
        data() {
            return {
                focus: false,
                inputValue: '',
                changeValue: ''
            }
        },
        methods: {
            //视图中"实时获取输入值"的方法
            onKeyInput: function(event) {
                this.inputValue = event.target.value;
            },
            //视图中"控制输入的input"的方法
            replaceInput: function(event) {
                var value = event.target.value;
                if (value === '11') {
                    this.changeValue = '2';
                }
            },
            //视图中"控制键盘的input"的方法
            hideKeyboard: function(event) {
                if (event.target.value === '123') {
                    uni.hideKeyboard();       //收起键盘，百度小程序不支持
                }
            }
        }
    }
</script>
<style scoped>
//CSS样式省略
</style>
```

请读者仔细阅读上述代码，代码有些多，但是很简单，以上功能请使用真机测试。

9.10 picker

picker（选择器）组件是从底部弹起的滚动选择器。picker支持5种选择器，通过mode来区分，分别是普通选择器、多列选择器、时间选择器、日期选择器、省市区选择器，默认是普通选择器。

picker组件官方文档地址：https://uniapp.dcloud.io/component/picker。

9.10.1　普通选择器

扫一扫，看视频

当mode属性值为selector或没有mode属性时，为普通选择器。普通选择器的属性说明见表9-24。

表9-24　普通选择器的属性说明

属性名	类　型	默认值	说　明
range	Array / Array <Object>	[]	当 mode 为 selector 或 multiSelector 时，range 有效
range-key	String		当 range 是一个 Array <Object> 时，通过 range-key 指定 Object 中 key 的值作为选择器的显示内容
value	Number	0	value 的值表示选择了 range 中的第几个（下标从 0 开始）
@change	EventHandle		value 改变时触发 change 事件，event.detail = {value: value}
disabled	Boolean	false	是否禁用
@cancel	EventHandle		取消选择或点击遮罩层收起 picker 时触发

在pages文件夹下创建picker/selector.vue文件，该文件中的代码如下：

```
<template>
    <view>
        <picker @change="bindPickerChange" :value="index" :range="array"
         range-key="title">
            <view class="picker">{{array[index].title}}</view>
        </picker>
        <button @click="submit">提交</button>
    </view>
</template>

<script>
    export default {
        name: "selector",
        data(){
            return {
                array:[
                    {id:"1",title:'中国'},
                    {id:"2",title:'美国'},
                    {id:"3",title:'巴西'},
                    {id:"4",title:'日本'}
                ],
                index:0 //array数据的索引
            }
        },
        methods:{
```

```
            //视图@change事件回调方法，获取数据索引
            bindPickerChange(e){
                this.index=e.detail.value;    //获取数组array的索引
            },
            //提交数据
            submit(){
                //输出选择的数据
                console.log(this.array[this.index]);
            }
        }
    }
</script>

<style scoped>
    .picker{width:500rpx;height:80rpx;border:1px solid #EFEFEF; line-height:80rpx;}
</style>
```

如果range-key属性值为title，在picker选项中会显示array数组中title的属性值；如果range-key属性值设置为id，则会显示array数组中id的属性值。

注意：

picker在各平台的实现是有UI差异的，有的平台的Android是从中间弹出的，如百度小程序、支付宝小程序；有的平台支持循环滚动，如微信小程序、百度小程序；有的平台没有取消按钮，如App端。以上差异均不影响功能使用。

9.10.2 多列选择器

扫一扫，看视频

当mode属性值为multiSelector时，为多列选择器。多列选择器支持的平台有App、H5、微信小程序、百度小程序、字节跳动小程序、QQ小程序，不支持的平台有支付宝小程序。

多列选择器的属性说明见表9-25。

表9-25 多列选择器的属性说明

属性名	类 型	默认值	说 明
range	二维 Array / 二维 Array <Object>	[]	当 mode 为 selector 或 multiSelector 时，range 有效。二维数组中，长度表示多少列，数组的每项表示每列的数据，如 [["a","b"], ["c","d"]]
range-key	String		当 range 是一个二维 Array <Object> 时，通过 range-key 指定 Object 中 key 的值作为选择器显示内容
value	Array	[]	value 每一项的值表示选择了 range 对应项中的第几个（下标从 0 开始）
@change	EventHandle		value 改变时触发 change 事件，event.detail = {value: value}

续表

属性名	类 型	默认值	说 明
@columnchange	EventHandle		某一列的值改变时触发 columnchange 事件，event.detail = {column: column, value: value}，column 的值表示改变了第几列（下标从 0 开始），value 的值表示变更值的下标
@cancel	EventHandle		取消选择时触发
disabled	Boolean	false	是否禁用

在pages文件夹下创建picker/multiselector.vue文件，该文件中的代码如下：

```
<template>
    <view>
        <picker mode="multiSelector" @change="bindPickerChange" @columnchang
         e="bindPickerColumnChange" :range="array" range-key="title">
            <view class="picker">{{showTitle?showTitle:'请选择'}}</view>
        </picker>
        <button @click="submit">提交</button>
    </view>
</template>

<script>
    export default {
        name: "multiselector",
        data(){
            return {
                array:[
                    [
                        {id:1,title:"主食"},
                        {id:2,title:"饮料"},
                        {id:3,title:"小吃"}
                    ],
                     [
                         {id:1,title:"牛肉汉堡"},
                         {id:2,title:"鸡肉汉堡"},
                         {id:3,title:"培根汉堡"}
                     ]
                ],
                index1:0,       //第一列数据的索引
                index2:0,       //第二列数据的索引
                showTitle:""    //选择后显示的数据
            }
        },
        methods:{
```

```
//视图@columnchange事件回调方法，获取数据索引
bindPickerColumnChange(e){
    let column=e.detail.column;  //获取数组array第几列
    let index=e.detail.value;    //array列中的索引
    if(column==0){//如果是第一列
        switch (index) {
            case 0://如果是主食
                //更新第二列的数据
                this.array[1]=[
                    {id:1,title:"牛肉汉堡"},
                    {id:2,title:"鸡肉汉堡"},
                    {id:3,title:"培根汉堡"}
                ]
                //使用this.$set解决数据更新不渲染视图的问题
                this.$set(this.array,1,this.array[1]);
                break;
            case 1://如果是饮料
                //更新第二列的数据
                this.array[1]=[
                    {id:1,title:"可乐"},
                    {id:2,title:"雪碧"},
                    {id:3,title:"咖啡"}
                ];
                this.$set(this.array,1,this.array[1]);
                break;
            case 2://如果是小吃
                //更新第二列的数据
                this.array[1]=[
                    {id:1,title:"薯条"},
                    {id:2,title:"鸡块"},
                    {id:3,title:"鸡柳"}
                ];
                this.$set(this.array,1,this.array[1]);
                break;
        }
    }
},
//视图@change事件回调方法，获取数据索引
bindPickerChange(e){
    this.index1=e.detail.value[0]//获取第一列数据的索引
    this.index2=e.detail.value[1]//获取第二列数据的索引
    //this.array[0][this.index1]第一列数据的值，this.array[0][this.
    //index1]第二列数据的值
    this.showTitle=this.array[0][this.index1].title+"-"+this.
    array[1][this.index2].title;
```

```
            },
            //提交数据
            submit(){
                //输出选择的数据
                console.log(JSON.stringify(this.array[0][this.index1]),JSON.
                stringify(this.array[1][this.index2]));
            }
        }
    }
</script>

<style scoped>
    .picker{width:500rpx;height:80rpx;border:1px solid #EFEFEF; line-height: 80rpx;}
</style>
```

多列选择器的使用非常灵活，可根据自定义的数组列数显示。以上代码注释很清晰，请读者仔细阅读，如果阅读起来比较困难，推荐配合视频教程学习。

注意：

在微信开发者工具的PC模拟器进行拖动时有可能会出现数据错乱问题，推荐使用真机测试。

在真机测试中渲染的效果如图9-6和图9-7所示。

图 9-6　多列选择器选择“主食”显示的数据

图 9-7　多列选择器选择“饮料”显示的数据

扫一扫，看视频

9.10.3 时间选择器

当mode属性值为time时，为时间选择器。时间选择器的属性说明见表9-26。

表9-26 时间选择器的属性说明

属性名	类 型	默认值	说 明	平台差异说明
value	String		表示选中的时间，格式为 "hh:mm"	
start	String		表示有效时间范围的开始，字符串格式为 "hh:mm"	App 不支持
end	String		表示有效时间范围的结束，字符串格式为 "hh:mm"	App 不支持
@change	EventHandle		value 改变时触发 change 事件，event.detail = {value: value}	
@cancel	EventHandle		取消选择时触发	
disabled	Boolean	false	是否禁用	

在pages文件夹下创建picker/time.vue文件，该文件中的代码如下：

```
<template>
    <view>
        <picker mode="time" :value="time" start="09:01" end="17:30"
         @change="bindTimeChange">
            <view class="picker">{{time}}</view>
        </picker>
        <button @click="submit">提交</button>
    </view>
</template>

<script>
    export default {
        name: "time",
        data(){
            return {
                time:"17:30"
            }
        },
        methods:{
            //视图@change事件的回调函数
            bindTimeChange(e){
                this.time=e.detail.value;              //获取选择时间
            },
            submit(){
```

```
                    console.log(this.time);//输出选择的时间
                }
            }
        }
</script>

<style scoped>
    .picker{width:500rpx;height:80rpx;border:1px solid #EFEFEF; line-height: 80rpx;}
</style>
```

时间选择器在App端调用的是原生时间选择控件，在不同平台有不同的UI表现。

在微信开发者工具中渲染的效果，即时间选择器的预览效果如图9-8所示。

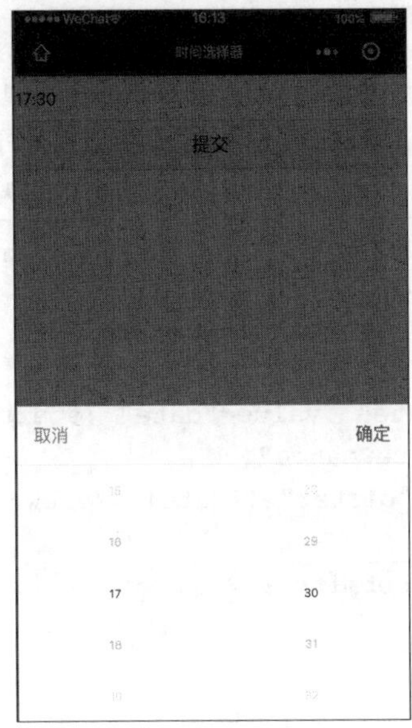

图 9-8　时间选择器的预览效果

9.10.4　日期选择器

扫一扫，看视频

当mode属性值为date时，为日期选择器。日期选择器的属性说明见表9-27和表9-28。

表9-27　日期选择器的属性说明

属性名	类　型	默认值	说　明	平台差异说明
value	String	0	表示选中的日期，格式为 "YYYY-MM-DD"	
start	String		表示有效日期范围的开始，字符串格式为 "YYYY-MM-DD"	App 不支持
end	String		表示有效日期范围的结束，字符串格式为 "YYYY-MM-DD"	App 不支持

续表

属性名	类　型	默认值	说　明	平台差异说明
fields	String	day	有效值为 year、month、day，表示选择器的粒度，默认值为 day。App 端未配置此项时使用系统 UI	H5、App 2.6.3+、微信小程序、百度小程序、字节跳动小程序
@change	EventHandle		value 改变时触发 change 事件，event.detail = {value: value}	
@cancel	EventHandle		取消选择时触发	
disabled	Boolean	false	是否禁用	

表 9–28　fields属性有效值说明

值	说　明
year	选择器粒度为年
month	选择器粒度为月
day	选择器粒度为天

在pages文件夹下创建picker/date.vue文件，该文件中的代码如下：

```
<template>
    <view>
        <picker mode="date" :value="date" :start="startDate" :end="endDate"
         @change="bindDateChange">
            <view class="picker">{{date}}</view>
        </picker>
        <button @click="submit">提交</button>
    </view>
</template>

<script>
    export default {
        name: "time",
        data(){
            return {
                date:this.getDate(),                    //当前日期
                startDate:this.getDate('start'),        //起始日期
                endDate:this.getDate('end')             //结束日期
            }
        },
        methods:{
            //视图@change事件的回调函数
            bindDateChange(e){
                this.date=e.detail.value;               //获取选择的日期
```

```
            },
            submit(){
                console.log(this.date);                  //输出选择的日期
            },
            //获取日期
            getDate(type) {
                const date = new Date();
                let year = date.getFullYear();
                let month = date.getMonth() + 1;
                let day = date.getDate();

                if (type === 'start') {
                    year = year - 60;
                } else if (type === 'end') {
                    year = year + 2;
                }
                month = month > 9 ? month : '0' + month;;
                day = day > 9 ? day : '0' + day;
                return '${year}-${month}-${day}';//将日期格式拼接为"YYYY-MM-DD"
            }
        }
    }
</script>

<style scoped>
    .picker{width:500rpx;height:80rpx;border:1px solid #EFEFEF; line-height:80rpx;}
</style>
```

日期选择器的使用很简单，在App端调用的是原生日期选择控件，在不同平台有不同的UI表现。在微信开发者工具中渲染的效果，即日期选择器的预览效果如图9-9所示。

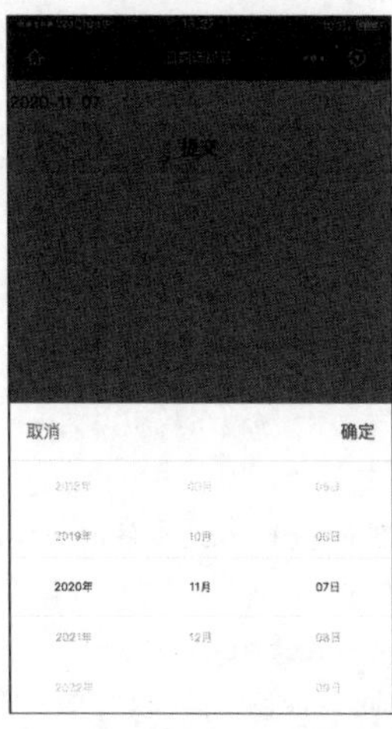

图 9-9　日期选择器的预览效果

9.10.5 省市区选择器

扫一扫，看视频

当mode属性值为region时为省市区选择器。省市区选择器支持的平台有微信小程序、百度小程序、字节跳动小程序、QQ小程序，不支持的平台有App、H5、支付宝小程序。

省市区选择器的属性说明见表9-29。

表9-29 省市区选择器的属性说明

属性名	类 型	默认值	说 明
value	Array	[]	表示选中的省市区，默认选中每一列的第一个值
custom-item	String		可为每一列的顶部添加一个自定义的项
@change	EventHandle		value 改变时触发 change 事件，event.detail = {value: value}
@cancel	EventHandle		取消选择时触发
disabled	Boolean	false	是否禁用

在pages文件夹下创建picker/region.vue文件，该文件中的代码如下：

```
<template>
    <view>
        <picker mode="region" :value="areas" @change="bindAreaChange">
            <view class="picker">{{area?area:'请选择'}}</view>
        </picker>
        <button @click="submit">提交</button>
    </view>
</template>

<script>
    export default {
        name: "region",
        data(){
            return {
                areas:[],        //获取地址数组
                area:"",         //选择的地址
            }
        },
        methods:{
            //视图中picker组件的@change事件的回调函数，获取省市区
            bindAreaChange(e){
                this.areas=e.target.value;    //获取的数组地址
                this.area=this.areas.join(","); //将数组转成字符串并用逗号进行拼接
```

```
            },
            submit(){
                console.log(this.areas,this.area);//输出选择的地址数据
            }
        }
    }
</script>

<style scoped>
    .picker{width:500rpx;height:80rpx;border:1px solid #EFEFEF;line-height:80rpx;}
</style>
```

在微信开发者工具中渲染的效果，即省市区选择器的预览效果如图9-10所示。

图 9-10　省市区选择器的预览效果

由于省市区选择器包含大量数据，占用体积较大，并非所有应用都需要，且很多城市数据有自维护需求，因此在App和H5平台没有内置。可以基于多列picker或picker-view自行填充城市数据，这里推荐使用simpleAddress第三方插件解决平台兼容性问题。

9.10.6　simpleAddress 三级省市区联动

simpleAddress是一款基于uni-app开发的第三方三级省市区联动插件，可以兼容所有平台。

扫一扫，看视频

simpleAddress插件下载地址：https://ext.dcloud.net.cn/plugin?id=1084。

进入simpleAddress插件页面，单击“下载插件ZIP”按钮进行下载，如图9-11所示。

图 9-11　下载 simpleAddress 插件

下载完成后，解压zip包，找到simple-address文件夹，将该文件夹复制到components文件夹下，并将其引入pages/picker/region.vue文件中，该文件中的代码如下：

```
<template>
    <view>
        <view class="picker" @click="openAddres">{{area?area:'请选择'}}</view>
        <button @click="submit">提交</button>
        <simple-address ref="simpleAddress" :pickerValueDefault="cityPickerValue
        Default" @onConfirm="onConfirm" themeColor="#007AFF"></simple-address>
    </view>
</template>

<script>
    import SimpleAddress from "../../components/simple-address/simple-address";
    export default {
        name: "region",
        data(){
            return {
                cityPickerValueDefault: [0, 0, 0],   //地址默认选中值
                area:''                              //选中后的地址
            }
        },
        components:{
            SimpleAddress
        },
        methods:{
            onConfirm(e) {
                this.cityPickerValueDefault=e.value;
                this.area = e.labelArr.join(",");
                //将选择的地址数组转成字符串并用逗号进行拼接
            },
```

```
            //弹出SimpleAddress组件
            openAddress() {
                this.$refs.simpleAddress.open();
            },
            submit(){
                console.log(this.area);                  //输出选择的地址数据
            },
        }
    }
</script>

<style scoped>
    .picker{width:500rpx;height:80rpx;border:1px solid #EFEFEF;line-height:80rpx;}
</style>
```

单击“请选择”按钮，调用自定义openAddress()方法，在该方法内部使用this.$refs.simpleAddress.open()弹出simpleAddress组件。<simple-address>组件属性pickerValueDefault为地址的默认选中值，值为数组类型，数组内的3个值分别代表省、市、区的索引，@onConfirm为监听单击“确定”按钮的事件。

simpleAddress组件的相关说明见表9-30~表9-33。

表9-30 simpleAddress三级省市区联动的属性说明

属性名	类　型	默认值	说　明
animation	Boolean	true	是否打开窗口动画
type	String	bottom	打开弹窗位置，值为 bottom 时从底部弹出
maskClick	Boolean	true	是否开启遮罩层，值为 false 时则不开启
maskBgColor	String	rgba(0, 0, 0, 0.4)	遮罩层背景颜色
themeColor	String	无	按钮主题颜色
cancelColor	String	#1aad19	取消按钮主题颜色
confirmColor	String	themeColor	确认按钮主题颜色
pickerValueDefault	Array	[]	地址默认选中值
btnFontSize	String	uni.scss 里的 $uni-font-size-base	取消、确认按钮字体大小

表9-31 事件说明

事件名称	说　明
onChange	监听 picker-view 的 change 事件
onCancel	监听“取消”按钮单击事件
onConfirm	监听“确定”按钮单击事件

表9-32 事件中返回参数说明

参数名称	说 明
label	省 - 市 - 区
value	获得选中地址区域的 index 索引
cityCode	城市编码
areaCode	区域编码
provinceCode	省份编码

表9-33 方法说明

方法名	类 型	说 明
open	function	弹出地址选择组件，通过 ref 触发
queryIndex	function	自定义信息返回对应的 index，方便设置默认值

9.11 textarea

textarea（多行输入框）的使用方法与传统H5的textarea的使用方法类似，其功能非常强大，实际开发中经常应用于填写备注、个人简介等场景。

textarea官方文档地址:https://uniapp.dcloud.io/component/textarea。

9.11.1 textarea 的使用

textarea的属性说明见表9-34。

表9-34 textarea的属性说明

属性名	类 型	默认值	说 明	平台差异说明
value	String		输入框的内容	
placeholder	String		输入框为空时的占位符	
placeholder-style	String		指定 placeholder 的样式	
placeholder-class	String	textarea-placeholder	指定 placeholder 的样式类	
disabled	Boolean	false	是否禁用	
maxlength	Number	140	最大输入长度，设置为 -1 时不限制最大长度	
focus	Boolean	false	获取焦点	在 H5 平台能否聚焦及软键盘是否跟随弹出取决于当前浏览器本身的实现
auto-height	Boolean	false	是否自动增高，设置为 auto-height 时，style.height 不生效	

续表

属性名	类 型	默认值	说 明	平台差异说明
fixed	Boolean	false	如果 textarea 是在一个 position:fixed 的区域，则需要显式指定属性 fixed 为 true	微信小程序、百度小程序、字节跳动小程序、QQ 小程序
cursor-spacing	Number	0	指定光标与键盘的距离，单位为 px。取 textarea 距离底部的距离和 cursor-spacing 指定的距离的最小值作为光标与键盘的距离	App、微信小程序、百度小程序、字节跳动小程序、QQ 小程序
cursor	Number		指定 focus 时的光标位置	微信小程序、App、H5、百度小程序、字节跳动小程序、QQ 小程序
show-confirm-bar	Boolean	true	是否显示键盘上方带有“完成”按钮一栏	微信小程序、百度小程序、QQ 小程序
selection-start	Number	–1	光标起始位置，自动聚焦时有效，须与 selection-end 搭配使用	微信小程序、App、H5、百度小程序、字节跳动小程序、QQ 小程序
selection-end	Number	–1	光标结束位置，自动聚焦时有效，须与 selection-start 搭配使用	微信小程序、App、H5、百度小程序、字节跳动小程序、QQ 小程序
adjust-position	Boolean	true	键盘弹起时，是否自动上推页面	App-Android（softinputMode 为 adjustResize 时无效）、微信小程序、百度小程序、QQ 小程序
disable-default-padding	Boolean	false	是否去掉 iOS 下的默认内边距	微信小程序 2.10.0
hold-keyboard	Boolean	false	聚焦时，点击页面时不收起键盘	微信小程序 2.8.2
@focus	EventHandle		输入框聚焦时触发，event.detail = { value, height }，height 为键盘高度	仅微信小程序、App（HBuilder X 2.0+ nvue uni-app 模式）、QQ 小程序支持 height

续表

属性名	类　型	默认值	说　明	平台差异说明
@blur	EventHandle		输入框失去焦点时触发，event.detail = {value, cursor}	
@linechange	EventHandle		输入框行数变化时调用，event.detail = {height: 0, heightRpx: 0, lineCount: 0}	字节跳动小程序不支持、nvue iOS 暂不支持
@input	EventHandle		当键盘输入时触发 input 事件，event.detail = {value, cursor}，@input 处理函数的返回值并不会反映到 textarea 上	
@confirm	EventHandle		点击完成时触发 confirm 事件，event.detail = {value: value}	微信小程序、百度小程序、QQ 小程序
@keyboardheigh-tchange	EventHandle		键盘高度发生变化时触发此事件，event.detail = {height: height, duration: duration}	微信小程序 2.7.0

在pages文件夹下创建textarea/index.vue文件，该文件中的代码如下：

```
<template>
    <view>
        <view class="uni-title uni-common-pl">输入区域高度自适应，不会出现滚动条</view>
        <view class="uni-textarea">
            <textarea @blur="bindTextAreaBlur" auto-height />
        </view>
        <view class="uni-title uni-common-pl">占位符字体是红色的textarea</view>
        <view class="uni-textarea">
            <textarea placeholder-style="color:#F76260" placeholder="占位符字
             体是红色的"/>
        </view>
    </view>
</template>

<script>
    export default {
        name: "textarea-component",
        methods: {
            //视图中textarea组件@blur事件的回调函数
```

```
            bindTextAreaBlur: function (e) {
                //获取textarea的值
                console.log(e.detail.value)
            }
        }
    }
</script>

<style scoped>
//CSS样式省略
</style>
```

> **注意：**
> 在微信小程序、百度小程序及字节跳动小程序中，textarea是原生组件，层级高于前端组件。请勿在 scroll-view、swiper、picker-view、movable-view 中使用 textarea 组件，覆盖textarea需要使用cover-view。

9.11.2 textarea 中的换行问题

在实际开发中，textarea组件中的换行符为\r\n，并不是HTML的
标签。因此，在使用rich-text组件输出时并不能显示换行，需要将\r\n转换成
标签。在pages文件夹下创建textarea/br.vue，该文件中的代码如下：

```
<template>
    <view>
        <textarea class="content" v-model="content" placeholder-style="color:#F76260"
         :auto-height="true" placeholder="请输入内容"></textarea>
        <button type="button" @click="submit">提交</button>
    </view>
</template>

<script>
    export default {
        name: "textarea-component",
        data(){
            return {
                content:""
            }
        },
        methods:{
```

```
            submit(){
                let content=this.content.replace(/\n/g,"<br/>");
                content=content.replace(/\r\n/g,"<br/>")
                console.log(content);
            }
        }
    }
</script>
```

使用replace将\n和\r\n替换成
标签即可，\n代表UNIX操作系统中的换行，\r\n代表Windows操作系统中的换行，为了兼容性，须将两者中的换行都替换。

9.12 web-view

web-view（Web浏览器）组件是用来承载网页的容器，其默认自动铺满整个页面，如果使用native vue开发，则需手动指定宽和高。

web-view官方文档地址:https://uniapp.dcloud.io/component/web-view。

在使用真机测试时，web-view加载的URL需要在后台配置域名白名单，包括内部再次使用iframe内嵌的其他URL。通过https://mp.weixin.qq.com 网址登录微信小程序后台，单击左侧导航栏的“开发”按钮，进入开发页面，单击“开发设置”按钮，找到业务域名选项，单击“开始配置”按钮，弹出“配置业务域名”对话框，如图9-12所示。

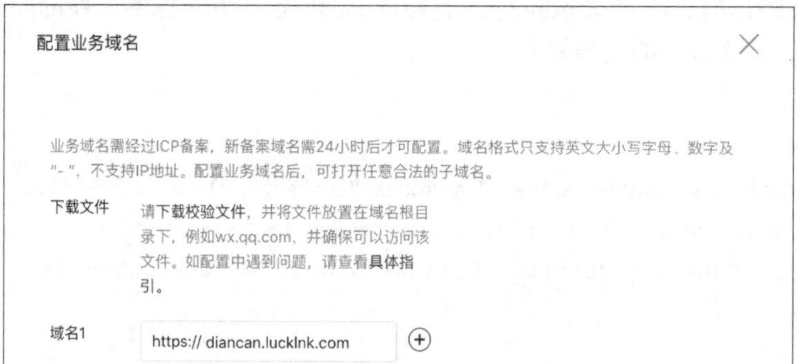

配置业务域名

业务域名需经过ICP备案，新备案域名需24小时后才可配置。域名格式只支持英文大小写字母、数字及“-”，不支持IP地址。配置业务域名后，可打开任意合法的子域名。

下载文件　请下载校验文件，并将文件放置在域名根目录下，例如wx.qq.com，并确保可以访问该文件。如配置中遇到问题，请查看**具体指引**。

域名1　https:// diancan.lucklnk.com

保存　取消

图 9-12　“配置业务域名”对话框

首先下载校验文件，将文件上传到域名根目录下；然后填写域名，必须是HTTPS协议的域名；最后单击“保存”按钮，完成配置。

如果在微信开发者工具中进行测试，则不需要配置业务域名，只需要勾选“不校验合法域名、web-view（业务域名）、TLS版本以及HTTPS证书”复选框即可，如图9-3所示。

web-view的相关属性说明见表9-35~表9-37。

表9-35 web-view的属性说明

属性名	类 型	说 明	平台差异说明
src	String	webview 指向网页的链接	
allow	String	用于为 iframe 指定特征策略	H5
sandbox	String	该属性对呈现在 iframe 框架中的内容启用一些额外的限制条件	H5
webview-styles	Object	webview 的样式	App-vue
@message	EventHandler	网页向应用 postMessage（发布消息）时，会在特定时机（后退、组件销毁、分享）触发并收到消息	H5 暂不支持（可以直接使用 window.postMessage）
@onPostMessage	EventHandler	网页向应用实时 postMessage	App-nvue

表9-36 webview-styles的属性值说明

属性名	类 型	说 明
progress	Object/Boolean	进度条样式。仅加载网络 HTML 时生效，设置为 false 时禁用进度条

表9-37 progress的属性值说明

属性名	类 型	默认值	说 明
color	String	#00FF00	进度条颜色

在pages文件夹下创建webview/index.vue文件，该文件中的代码如下：

```
<template>
    <view>
        <web-view :webview-styles="webviewStyles" src="https://diancan.
         lucklnk.com/"></web-view>
    </view>
</template>

<script>
    export default {
        name: "index",
        data(){
            return {
                //webview的样式
                webviewStyles: {
                    //进度条样式
                    progress: {
```

```
                        color: '#FF3333'
                    }
                }
            }
        }
    }
</script>
```

web-view组件在H5中会转为iframe标签。小程序端仅支持加载网络网页，不支持加载本地网页。小程序端的web-view组件一定要有原生导航栏，并且一定是全屏的 web-view 组件。App平台同时支持网络网页和本地网页，但本地网页及相关资源(.js、.css等文件)必须放在uni-app 项目根目录/hybrid/html 文件夹下。

9.13 image

image（图片）组件在H5中会转为img标签，其功能非常强大，支持图片懒加载、缩放和裁剪。

image官方文档地址：https://uniapp.dcloud.io/component/image。

image组件的属性说明见表9-38。

表9-38 image组件的属性说明

属性名	类 型	默认值	说 明	平台差异说明
src	String		图片资源地址	
mode	String	'scaleToFill'	图片裁剪、缩放的模式	
lazy-load	Boolean	false	图片懒加载。只针对 page 与 scroll-view 下的 image 有效	微信小程序、App、百度小程序、字节跳动小程序
fade-show	Boolean	true	图片显示动画效果	仅 App-nvue 2.3.4+ Android 有效
webp	Boolean	false	默认不解析 webp 格式，只支持网络资源	微信小程序 2.9.0
show-menu-by-longpress	Boolean	false		
@error	HandleEvent		当错误发生时，发布到 AppService 的事件名，事件对象 event.detail = {errMsg: 'something wrong'}	
@load	HandleEvent		当图片载入完毕时，发布到 AppService 的事件名，事件对象 event.detail = {height:' 图片高度 px', width:' 图片宽度 px'}	

mode有14种模式，其中5种是缩放模式，9种是裁剪模式，属性有效值说明见表9-39。

表9-39 image组件的mode属性有效值说明

模 式	值	说 明
缩放	scaleToFill	不保持纵横比缩放图片，使图片的宽高完全拉伸至填满 image 元素
缩放	aspectFit	保持纵横比缩放图片，使图片的长边能完全显示出来，即可以完整地将图片显示出来
缩放	aspectFill	保持纵横比缩放图片，只保证图片的短边能完全显示出来。也就是说，图片通常只在水平或垂直方向是完整的
缩放	widthFix	宽度不变，高度自动变化，保持原图宽高比不变
缩放	heightFix	高度不变，宽度自动变化，保持原图宽高比不变
裁剪	top	不缩放图片，只显示图片的顶部区域
裁剪	bottom	不缩放图片，只显示图片的底部区域
裁剪	center	不缩放图片，只显示图片的中间区域
裁剪	left	不缩放图片，只显示图片的左边区域
裁剪	right	不缩放图片，只显示图片的右边区域
裁剪	top left	不缩放图片，只显示图片的左上边区域
裁剪	top right	不缩放图片，只显示图片的右上边区域
裁剪	bottom left	不缩放图片，只显示图片的左下边区域
裁剪	bottom right	不缩放图片，只显示图片的右下边区域

在pages文件夹下创建image/index.vue文件，该文件中的代码如下：

```
<template>
    <view class="page">
        <view class="image-list">
            <view class="image-item" v-for="(item,index) in array" :key="index">
                <view class="image-content">
                    <image style="width: 200px; height: 200px; background-color:
                     #eeeeee;" :mode="item.mode" :src="src" @error="imageError"
                     :lazy-load="true"></image>
                </view>
                <view class="image-title">{{item.text}}</view>
            </view>
        </view>
    </view>
</template>

<script>
    export default {
```

```
    name: "image",
    data() {
        return {
            array: [{
                mode: 'scaleToFill',
                text: 'scaleToFill：不保持纵横比缩放图片，使图片完全适应'
            }, {
                mode: 'aspectFit',
                text: 'aspectFit：保持纵横比缩放图片，使图片的长边能完全显示出来'
            }, {
                mode: 'aspectFill',
                text: 'aspectFill：保持纵横比缩放图片，只保证图片的短边能完全显示出来'
            }, {
                mode: 'top',
                text: 'top：不缩放图片，只显示图片的顶部区域'
            }, {
                mode: 'bottom',
                text: 'bottom：不缩放图片，只显示图片的底部区域'
            }, {
                mode: 'center',
                text: 'center：不缩放图片，只显示图片的中间区域'
            }, {
                mode: 'left',
                text: 'left：不缩放图片，只显示图片的左边区域'
            }, {
                mode: 'right',
                text: 'right：不缩放图片，只显示图片的右边区域'
            }, {
                mode: 'top left',
                text: 'top left：不缩放图片，只显示图片的左上边区域'
            }, {
                mode: 'top right',
                text: 'top right：不缩放图片，只显示图片的右上边区域'
            }, {
                mode: 'bottom left',
                text: 'bottom left：不缩放图片，只显示图片的左下边区域'
            }, {
                mode: 'bottom right',
                text: 'bottom right：不缩放图片，只显示图片的右下边区域'
            }],
            src: 'https://img-cdn-qiniu.dcloud.net.cn/uniapp/images/shuijiao.jpg'
        }
```

```
        },
        methods: {
            imageError: function(e) {
                console.error('image发生error事件，携带值为' + e.detail.errMsg)
            }
        }
    }
</script>
```

<image> 组件默认的宽度为300px、高度为225px，src属性值支持相对路径、绝对路径，支持base64码。当页面结构比较复杂时，使用 image 可能导致样式生效较慢，出现“闪一下”的情况，此时设置 image{will-change: transform}可优化此问题，will-change会告知浏览器该元素会有哪些变化，这样浏览器可以在元素属性真正发生变化之前做好对应的优化准备工作。

在自定义组件中使用 <image>时，若src使用相对路径，可能出现路径查找失败的情况，因此建议使用绝对路径。

9.14　switch

扫一扫，看视频

switch（开关选择器）在实际开发中经常使用，如是否显示密码、是否打包、是否通知、是否使用余额等场景。

switch官方文档地址:https://uniapp.dcloud.io/component/switch。

switch属性说明见表9-40。

表9-40　switch属性说明

属性名	类　型	默认值	说　明	平台差异说明
checked	Boolean	false	是否选中	
disabled	Boolean	false	是否禁用	字节跳动小程序不支持
type	String	switch	样式，有效值为 switch、checkbox	
@change	EventHandle		checked 改变时触发 change 事件，event.detail={ value:checked}	
color	Color		switch 的颜色，同 CSS 的 color	

在pages文件夹下创建switch/index.vue文件，该文件中的代码如下：

```
<template>
    <view class="form-main">
        密码: <input type="text" :password="isShowPwd" placeholder="请输入密
            码" /> <switch type="switch" @change="showPwd" color="#FF0000"
            style="transform:scale(0.7)" />
    </view>
```

```
</template>

<script>
    export default {
        name: "switch-index",
        data(){
            return {
                isShowPwd:false
            }
        },
        methods: {
            //视图中switch组件@change事件的回调方法
            showPwd: function (e) {
                //获取switch的checked属性，值为true代表选中
                this.isShowPwd=!e.target.value;
            }
        }
    }
</script>

<style scoped>
    .form-main{display:flex;}
</style>
```

在不同平台中，switch的默认颜色不同，如在微信小程序中为绿色，在字节跳动小程序中为红色，在其他平台为蓝色。更改颜色需要使用color属性。如果需要调节switch大小，可通过CSS的scale()方法调节，如缩小到70%，则可设置style="transform:scale(0.7)"。

9.15 audio

audio（音频）组件的使用方法与传统H5的audio的使用方法类似，如果需要带UI样式，也可以自己制作。

audio官方文档地址：https://uniapp.dcloud.io/component/audio。

9.15.1 audio的使用

扫一扫，看视频

audio组件支持的平台有App、H5、微信小程序、百度小程序，不支持的平台有支付宝小程序、字节跳动小程序、QQ小程序。

audio组件的属性说明见表9-41。

表9-41 audio组件的属性说明

属性名	类 型	默认值	说 明
id	String		audio 组件的唯一标识符
src	String		要播放音频的资源地址
loop	Boolean	false	是否循环播放
controls	Boolean	false	是否显示默认控件
poster	String		默认控件上的音频封面的图片资源地址。如果 controls 属性值为 false，则设置 poster 无效
name	String	未知音频	默认控件上的音频名字。如果 controls 属性值为 false，则设置 name 无效
author	String	未知作者	默认控件上的作者名字。如果 controls 属性值为 false，则设置 author 无效
@error	EventHandle		当发生错误时触发 error 事件，detail = {errMsg: MediaError.code}
@play	EventHandle		当开始 / 继续播放时触发 play 事件
@pause	EventHandle		当暂停播放时触发 pause 事件
@timeupdate	EventHandle		当播放进度改变时触发 timeupdate 事件，detail = {currentTime, duration}
@ended	EventHandle		当播放到末尾时触发 ended 事件

在pages文件夹下创建audio/index.vue文件，该文件中的代码如下：

```
<template>
    <view>
        <view class="page-body">
            <view class="page-section page-section-gap" style="text-align: center;">
                <audio style="text-align: left" :src="current.src" :poster="current.
                poster" :name="current.name" :author="current.author" controls>
                </audio>
            </view>
        </view>
    </view>
</template>

<script>
    export default {
        name: "audio-component",
        data() {
            return {
                current: {
                    poster: 'http://www.lucklnk.com/music/jiehun/1.png',  //封面
```

```
                        name: '咱们结婚吧',//音频名字
                        author: '齐晨',    //作者名字
                        src: 'http://www.lucklnk.com/music/jiehun/1.mp3',  //音频地址
                    }
                }
            }
        }
    </script>
```

注意：

微信小程序平台自基础库 1.6.0 版本开始不再维护audio组件，推荐使用API方式而不是组件方式来播放音频，可以使用uni.createInnerAudioContext代替audio组件。

9.15.2 uni.createInnerAudioContext 代替 audio

扫一扫，看视频

uni.createInnerAudioContext能够创建并返回内部audio上下文innerAudioContext对象，从而代替audio组件。innerAudioContext对象的属性列表见表9-42，方法列表见表9-43，errCoder说明见表9-44，支持的音频格式见表9-45。

表9-42 innerAudioContext对象的属性列表

属　性	类　型	说　明	只　读	平台差异说明
src	String	音频的数据链接，用于直接播放	否	
startTime	Number	开始播放的位置（单位为 s），默认为 0	否	
autoplay	Boolean	是否自动开始播放，默认为 false		
loop	Boolean	是否循环播放，默认为 false		
obeyMuteSwitch	Boolean	是否遵循系统静音开关，默认值为 true。当此参数为 false 时，即使用户打开了静音开关，也能继续发出声音	否	微信小程序、百度小程序、字节跳动小程序
duration	Number	当前音频的长度（单位为 s），只有在当前有合法的 src 时返回	是	
currentTime	Number	当前音频的播放位置（单位为 s），只有在当前有合法的 src 时返回，时间不取整，保留到小数点后 6 位	是	
paused	Boolean	当前是否为暂停或停止状态，true 表示暂停或停止，false 表示正在播放	是	
buffered	Number	音频缓冲的时间点，仅保证当前播放时间点到此时间点内容已缓冲	是	
volume	Number	音量，范围为 0~1	否	

表9-43 innerAudioContext对象的方法列表

方 法	参 数	说 明
play		播放（H5 端部分浏览器需要在用户交互时进行）
pause		暂停
stop		停止
seek	position	跳转到指定位置（单位为 s）
destroy		销毁当前实例
onCanplay	callback	音频进入可以播放状态，但不保证后面可以流畅播放
onPlay	callback	音频播放事件
onPause	callback	音频暂停事件
onStop	callback	音频停止事件
onEnded	callback	音频自然播放结束事件
onTimeUpdate	callback	音频播放进度更新事件
onError	callback	音频播放错误事件
onWaiting	callback	音频加载中事件，当音频因为数据不足，需要停下来加载时会触发
onSeeking	callback	音频进行 seek 操作事件
onSeeked	callback	音频完成 seek 操作事件
offCanplay	callback	取消监听 onCanplay 事件
offPlay	callback	取消监听 onPlay 事件
offPause	callback	取消监听 onPause 事件
offStop	callback	取消监听 onStop 事件
offEnded	callback	取消监听 onEnded 事件
offTimeUpdate	callback	取消监听 onTimeUpdate 事件
offError	callback	取消监听 onError 事件
offWaiting	callback	取消监听 onWaiting 事件
offSeeking	callback	取消监听 onSeeking 事件
offSeeked	callback	取消监听 onSeeked 事件

表9-44 errCode说明

errCode	说 明
10001	系统错误
10002	网络错误
10003	文件错误
10004	格式错误
–1	未知错误

表9-45 支持的音频格式

格 式	iOS	Android
flac	×	√
m4a	√	√
ogg	×	√
ape	×	√
amr	×	√
wma	×	√
wav	√	√
mp3	√	√
mp4	×	√
aac	√	√
aiff	√	×
caf	√	×

注：× 表示不支持；√表示支持。其余表中与此相同。

为了兼容性和音质，推荐使用mp3格式。

在pages文件夹下创建audio/audio.vue文件，该文件中的代码如下：

```
<template>
    <view>
        <view class="page-body">
            <view class="page-section page-section-gap" style="text-align: center;">
                <button type="default" @click="doPlay()">播放</button>
                <button type="default" @click="doPause()">暂停</button>
                当前播放时间:{{currentTime}}/{{duration}}
            </view>
        </view>
    </view>
</template>

<script>
    export default {
        name: "audio-component",
        data() {
            return {
                currentTime:"00:00",   //当前音频的播放时间
```

```
            duration:"00:00"        //音频总时长
        }
    },
    onLoad(){
        this.timer=null;
        //创建innerAudioContext实例
        this.innerAudioContext = uni.createInnerAudioContext();
        // 监听音频进入可以播放状态的事件
        this.innerAudioContext.onCanplay(()=>{
            // 必需，可以当作初始化总时长
            this.innerAudioContext.duration;
            // 必需，延迟大约300ms以上才能获取音频总时长
            setTimeout(()=>{
            //获取音频总时长
            this.duration=this.formatSeconds(this.innerAudioContext
            .duration);
            },300)

        });
        //监听音频播放
        this.innerAudioContext.onPlay(()=>{
            //获取当前音频的播放时间
            this.timer=setInterval(()=>{
            this.currentTime=this.formatSeconds(this.innerAudioContext
            .currentTime);
            },1000);
        })

    },
    //离开页面
    onUnload(){
        //清除当前音频播放时间的定时器
        clearInterval(this.timer);
        //销毁innerAudioContext实例
        this.innerAudioContext.destroy();
    },
    methods:{
        //播放
        doPlay(){
            this.innerAudioContext.src="http://www.lucklnk.com/music/
            jiehun/1.mp3";                              //音频地址
```

```
                this.innerAudioContext.play();          //播放音频
            },
            doPause() {
                this.innerAudioContext.pause();         //暂停播放
                //清除获取音频播放时间的定时器
                clearInterval(this.timer);
            },
            //将秒转成03:30格式
            formatSeconds(value){
                let minute=parseInt(value/60);          //分钟
                let second=parseInt(value%60);          //秒
                if(minute<10){                          //如果小于10，补0
                    minute="0"+minute;
                }
                if(second<10){
                    second="0"+second;
                }
                return minute+":"+second;
            }
        }
    }
</script>
```

以上代码注释很清晰，请读者仔细阅读。使用uni.createInnerAudioContext创建audio是没有UI的，UI可以自己定义，也可以去插件市场下载自己喜欢的audio播放器。插件市场地址:https://ext.dcloud.net.cn/search?q=audio。需要注意，获取音频总时长不能直接使用this.innerAudioContext.duration，而是需要进行以下操作：

（1）定义onCanplay回调方法。

（2）在onCanplay回调方法中使用this.innerAudioContext.duration初始化总时长。

（3）在setTimeout回调函数内部使用this.innerAudioContext.duration获取总时长。

9.16 video

扫一扫，看视频

video（视频）组件在H5平台支持mp4、webm 和 ogg格式的视频。如果在H5平台自行开发播放器或使用第三方视频播放器插件，可以自动判断环境兼容性，以决定使用传统H5的video还是uni-app的video。App平台支持本地视频（mp4/flv）、网络视频（mp4/flv/m3u8）及流媒体（rtmp/hls/rtsp）。

video官方文档地址:https://uniapp.dcloud.io/component/video。

uni-app的video组件功能非常强大，其属性说明见表9-46，direction属性的合法值见表9-47。

表9–46 video组件的属性说明

属性名	类 型	默认值	说 明	平台差异说明
src	String		要播放视频的资源地址	
autoplay	Boolean	false	是否自动播放	
loop	Boolean	false	是否循环播放	
muted	Boolean	false	是否静音播放	字节跳动小程序不支持
initial-time	Number		指定视频初始播放位置，单位为 s	字节跳动小程序不支持
duration	Number		指定视频时长，单位为 s	字节跳动小程序不支持
controls	Boolean	true	是否显示默认播放控件（播放 / 暂停按钮、播放进度、时间）	
danmu-list	Object Array		弹幕列表	字节跳动小程序不支持
danmu-btn	Boolean	false	是否显示弹幕按钮。只在初始化时有效，不能动态变更	字节跳动小程序不支持
enable-danmu	Boolean	false	是否展示弹幕。只在初始化时有效，不能动态变更	字节跳动小程序不支持
page-gesture	Boolean	false		
direction	Number		设置全屏时视频的方向，不指定则根据宽高比自动判断。其有效值为 0（正常竖向）、90（屏幕逆时针 90°）、–90（屏幕顺时针 90°）	H5 和字节跳动小程序不支持
show-progress	Boolean	true	是否显示进度条，若不设置，则宽度大于 240 时才会显示	字节跳动小程序不支持
show-fullscreen-btn	Boolean	true	是否显示全屏按钮	
show-play-btn	Boolean	true	是否显示视频底部控制栏的播放按钮	
show-center-play-btn	Boolean	true	是否显示视频中间的播放按钮	字节跳动小程序不支持
show-loading	Boolean	true	是否显示 loading 控件	仅 App 2.8.12+
enable-progress-gesture	Boolean	true	是否开启控制进度的手势	字节跳动小程序不支持
object-fit	String	contain	当视频大小与 video 容器大小不一致时，视频的表现形式。其有效值为 contain（包含）、fill（填充）、cover（覆盖）	微信小程序、字节跳动小程序、H5
poster	String		视频封面的图片网络资源地址。如果 controls 属性值为 false，则设置 poster 无效	
show-mute-btn	Boolean	false	是否显示静音按钮	微信小程序

续表

属性名	类　型	默认值	说　明	平台差异说明
title	String		视频的标题，全屏时在顶部展示	微信小程序
play-btn-position	String	bottom	播放按钮的位置。其有效值为 bottom（controls bar 上）center（视频中间）	微信小程序、字节跳动小程序
enable-play-gesture	Boolean	false	是否开启播放手势，即双击切换播放 / 暂停	微信小程序
auto-pause-if-navigate	Boolean	true	当跳转到其他小程序页面时，是否自动暂停本页面的视频	微信小程序
auto-pause-if-open-native	Boolean	true	当跳转到其他微信原生页面时，是否自动暂停本页面的视频	微信小程序
vslide-gesture	Boolean	false	在非全屏模式下，是否开启亮度与音量调节手势（同 page-gesture）	微信小程序
vslide-gesture-in-fullscreen	Boolean	true	在全屏模式下，是否开启亮度与音量调节手势	微信小程序
ad-unit-id	String		在视频前粘贴广告单元 ID	微信小程序
poster-for-crawler	String		用于在搜索等场景作为视频封面展示，建议使用无播放 icon 的视频封面图，只支持网络地址	微信小程序
@play	EventHandle		当开始 / 继续播放时触发 play 事件	字节跳动小程序不支持
@pause	EventHandle		当暂停播放时触发 pause 事件	字节跳动小程序不支持
@ended	EventHandle		当播放到末尾时触发 ended 事件	字节跳动小程序不支持
@timeupdate	EventHandle		播放进度变化时触发，event.detail = {currentTime, duration} 。触发频率为 250ms 一次	字节跳动小程序不支持
@fullscreen-change	EventHandle		当视频进入和退出全屏时触发，event.detail = {fullScreen, direction}，direction 取为 vertical 或 horizontal	字节跳动小程序不支持
@waiting	EventHandle		视频出现缓冲时触发	字节跳动小程序不支持
@error	EventHandle		视频播放出错时触发	字节跳动小程序不支持
@progress	EventHandle		加载进度变化时触发，只支持一段加载，event.detail = {buffered}	微信小程序、H5
@loadedmeta-data	EventHandle		视频元数据加载完成时触发，event.detail = {width, height, duration}	微信小程序、H5

续表

属性名	类 型	默认值	说 明	平台差异说明
@fullscreen-click	EventHandle		全屏播放时的点击事件，event.detail = { screenX:"Number 类型，点击点相对于屏幕左侧边缘的 X 轴坐标", screenY:"Number 类型，点击点相对于屏幕顶部边缘的 Y 轴坐标", screenWidth:"Number 类型，屏幕总宽度", screenHeight:"Number 类型，屏幕总高度"}	App 2.6.3+
@controlstoggle	EventHandle		切换 controls 显示 / 隐藏时触发，event.detail = {show}	微信小程序 2.9.5

表9–47 direction属性的合法值

值	说 明
0	正常竖向
90	屏幕逆时针 90°
–90	屏幕顺时针 90°

在pages文件夹下创建video/index.vue文件，该文件中的代码如下：

```
<template>
    <view>
        <video :src="videoUrl" controls></video>
    </view>
</template>

<script>
    export default {
        name: "video-component",
        data(){
            return {
                videoUrl:"http://www.lucklnk.com/video/4.mp4"
            }
        }
    }
</script>

<style scoped>
    video{width:100%;height:500rpx;}
</style>
```

uni-app的video组件的使用方法与传统H5的video组件的使用方法类似，各小程序平台支持程度不同，详见以下各个文档。

微信小程序video组件文档：https://developers.weixin.qq.com/miniprogram/dev/component/video.html。

支付宝小程序video组件文档：https://opendocs.alipay.com/mini/component/video。

百度小程序video组件文档：https://smartprogram.baidu.com/docs/develop/component/media_video。

字节跳动小程序video组件文档：https://microapp.bytedance.com/docs/zh-CN/mini-app/develop/component/media-component/video。

QQ小程序video组件文档：https://q.qq.com/wiki/develop/miniprogram/component/media/video.html。

华为快应用video组件文档：https://developer.huawei.com/consumer/cn/doc/development/quickApp-References/webview-component-video。

9.17 小结

本章主要讲解了uni-app的常用组件，包括view组件、scroll-view组件、swiper组件、movable-area和movable-view组件、text组件、rich-text组件、button组件、checkbox组件、radio组件、input组件、picker组件（普通选择器、多列选择器、时间选择器、日期选择器、省市区选择器、simple-Address三级省市区联动）、switch组件、textarea组件、web-view组件、image组件、audio组件、video组件等。为了平台的兼容性，不要随意使用HFML标签，开发时一定要使用uni-app官方文档的组件或是本章讲解的组件。以上组件使用频率很高，读者应多加练习并记住这些组件的用法。

第 10 章

常用 API

uni-app的JS API由标准ECMAScript的JS API 和 uni 扩展 API 两部分组成，标准ECMAScript的API非常多，扩展API命名与小程序相同。在H5端，uni-app的JS代码运行于浏览器中；在非H5端，在Android平台中运行在v8引擎中，在iOS平台中运行在iOS自带的jscore引擎中。在开发时不要把浏览器里的JS等价于标准JS。浏览器基于标准JS扩充了window、document等JS API，Node.js基于标准JS扩充了Fs等模块，小程序也基于标准JS扩展了各种wx.××、my.××、swan.××的API。因此，uni-app的非H5端同样支持标准JS，支持if、for等语法，支持字符串、数组、时间等变量及各种处理方法，但不支持浏览器专用对象。

API官方文档地址：https://uniapp.dcloud.io/api/README。

10.1 request 发起请求

uni.request发起网络请求相当于ajax在实际开发中获取服务端接口数据，其使用方式类似于jQuery的ajax，简单实用。

request官方文档地址:https://uniapp.dcloud.io/api/request/request。

10.1.1 request 的使用

扫一扫，看视频

如果在真机测试下调试，服务端接口地址域名必须是HTTPs协议，并且需要配置域名白名单。可以关注“html5 程序思维”公众号，发送uniapp，即可获取服务端接口文档下载地址，进行下载。

配置域名白名单操作：进入微信小程序官网（https://mp.weixin.qq.com），输入用户名和密码进入管理平台，单击左侧导航“开发”按钮，进入开发页面，单击“开发设置”按钮，进入开发设置页面，找到“服务器域名”选项，配置“request合法域名”。

如果是在微信开发者工具中测试，服务端接口地址域名可以是HTTP或HTTPs协议，无须配置域名白名单，但必须勾选“不校验合法域名、web-view（业务域名）、TLS版本以及HTTPS证书”复选框，如图9-3 所示。

request的参数说明见表10-1。

表 10-1 request的参数说明

参数名	类 型	是否必填	默认值	说 明	平台差异说明
url	String	是		开发者服务器接口地址	
data	Object String ArrayBuffer	否		请求的参数	App（自定义组件编译模式）不支持 ArrayBuffer 类型
header	Object	否		设置请求的 header，header 中不能设置 Referer	H5 端会自动带上 cookie，不可手动覆盖
method	String	否	GET		
timeout	Number	否	60000	超时时间，单位为 ms	微信小程序（2.10.0）、支付宝小程序
dataType	String	否	json	如果设为 json，会尝试对返回的数据进行一次 JSON.parse	
responseType	String	否	text	设置响应的数据类型。其合法值为 text、arraybuffer	App、支付宝小程序不支持
sslVerify	Boolean	否	true	验证 SSL 证书	仅 App Android 端支持（HBuilder X 2.3.3+）
withCredentials	Boolean	否	false	跨域请求时是否携带凭证（cookies）	仅 H5 支持（HBuilder X 2.6.15+）

续表

参数名	类 型	是否必填	默认值	说 明	平台差异说明
firstIpv4	Boolean	否	false	DNS 解析时优先使用 IPv4	仅 App-Android 支持（HBuilder X 2.8.0+）
success	Function	否		收到开发者服务器成功返回的回调函数	
fail	Function	否		接口调用失败的回调函数	
complete	Function	否		接口调用结束的回调函数（调用成功、失败都会执行）	

data数据说明：最终发送给服务器的数据是 String 类型，如果传入的 data 不是 String 类型，会被转换成 String。其转换规则如下：

（1）GET 方法会将数据转换为 query string。例如，{ name: 'name', age: 18 } 转换后的结果是 name=name&age=18。

（2）POST 方法且 header['content-type'] 为 application/json 的数据，会进行 JSON 序列化。

（3）POST 方法且 header['content-type'] 为 application/x-www-form-urlencoded 的数据，会将数据转换为 query string。

method属性有效值说明见表10-2。

表10-2 method属性有效值说明

method	App	H5	微信小程序	支付宝小程序	百度小程序	字节跳动小程序
GET	√	√	√	√	√	√
POST	√	√	√	√	√	√
PUT	√	√	√	×	√	√
DELETE	√	√	√	×	√	×
CONNECT	√	√	√	×	×	×
HEAD	√	√	√	×	√	×
OPTIONS	√	√	√	×	√	×
TRACE	√	√	√	×	×	×

最常用的并且兼容性最好的method属性值为GET、POST，method属性值在服务端接口文档中获取，并不是自己随意编写的。

注意：

有效值必须大写。

success返回参数说明见表10-3。

表10-3 success返回参数说明

参 数	类 型	说 明
data	Object/String/ArrayBuffer	开发者服务器返回的数据
statusCode	Number	开发者服务器返回的 HTTP 状态码
header	Object	开发者服务器返回的 HTTP Response Header
cookies	Array.<string>	开发者服务器返回的 cookies，格式为字符串数组

在pages文件夹下创建request/get.vue文件，该文件中的代码如下（GET请求）：

```
<template>
    <view>
        <view>LOGO:<image :src="logo"></image></view>
        <view>店铺名称:{{shopName}}</view>
        <view>公告:{{notice}}</view>
    </view>
</template>

<script>
    export default {
        name: "request",
        data(){
            return {
                logo:"",
                shopName:"",    //店铺名称
                notice:""       //公告
            }
        },
        onLoad(){
            //GET请求数据
            uni.request({
                url:"https://diancan.glbuys.com/api/v1/business/
                    info?branch_shop_id=900874159",
                method:"GET",
                success:(res)=>{
                    //如果服务端返回的数据正确
                    if(res.data.code==200){
                        this.logo=res.data.data.logo;
                        this.shopName=res.data.data.shop_name;    //店铺名称
                        this.notice=res.data.data.notice;         //公告
                    }
```

```
                }
            });
        }
    }
</script>

<style scoped>
    image{width:200rpx;height:200rpx;}
</style>
```

使用GET请求，服务端接口参数不必写在data选项中，直接拼接在接口地址“?”的后面即可。该接口的参数属性为branch_shop_id，该属性的值为90087415。

如果将服务端接口的参数写在data选项中，可以将以上代码改成如下形式：

```
uni.request({
    url:"https://diancan.glbuys.com/api/v1/business/info",
    method:"GET",
    data:{
        branch_shop_id:"900874159"
    },
    success:(res)=>{
        //如果服务端返回的数据正确
        if(res.data.code==200){
            this.logo=res.data.data.logo;
            this.shopName=res.data.data.shop_name;  //店铺名称
            this.notice=res.data.data.notice;       //公告
        }
    }
});
```

以上加粗代码为修改部分。

在pages文件夹下创建request/post.vue文件，该文件中的代码如下（POST请求）：

```
<template>
    <view>
        <view v-for="(item,index) in classifys" :key="index">{{item.title}}</view>
    </view>
</template>

<script>
    export default {
        name: "request",
        data(){
            return {
```

```
                classifys:[]
            }
        },
        onLoad(){
            //POST请求数据
            uni.request({
                url:"https://diancan.glbuys.com/api/v1/goods/classify",
                method:"POST",
                data:{
                    branch_shop_id:"333071944"
                },
                header:{
                    "content-type":"application/x-www-form-urlencoded"
                },
                success:(res)=>{
                    if(res.data.code==200){
                        this.classifys=res.data.data;
                    }
                }
            });
        }
    }
</script>
```

使用POST请求必须配置header选项，在header选项中添加content-type属性，其值为application/x-www-form-urlencoded。content-type属性值取决于后端接口的content-type类型，示例中的接口content-type类型为x-www-form-urlencoded，在工作中服务端接口的content-type类型也可能是application/json。

如果想要中断请求，代码示例如下：

```
let requestTask = uni.request({
    url: 'https://www.example.com/request', //仅为示例，并非真实接口地址
    complete: ()=> {}
});
requestTask.abort();                          //中断请求
```

使用abort()方法可以中断请求。

10.1.2 Promise 方式请求数据

扫一扫，看视频

uni-app内部对request进行了Promise封装，返回的第一个参数是错误对象，第二个参数是返回数据。

在pages文件下创建request/promise.vue文件，该文件中的代码如下：

```
<template>
    <view>
        <view>LOGO:<image :src="logo"></image></view>
        <view>店铺名称:{{shopName}}</view>
        <view>公告:{{notice}}</view>
    </view>
</template>

<script>
    export default {
        name: "promise",
        data(){
            return {
                logo:"",
                shopName:"",            //店铺名称
                notice:""               //公告
            }
        },
        onLoad(){
            uni.request({
                url:"https://diancan.glbuys.com/api/v1/business/info",
                method:"GET",
                data:{
                    branch_shop_id:"900874159"
                }
            }).then((data)=>{     //data为一个数组，第一项为错误信息，第二项为返回数据
                let [err,res]=data;    //使用解构的方式获取数组的值
                if(res.data.code==200) {
                    this.logo=res.data.data.logo;
                    this.shopName=res.data.data.shop_name;          //店铺名称
                    this.notice=res.data.data.notice;               //公告
                }
            })
        }
    }
</script>

<style scoped>
    image{width:200rpx;height:200rpx;}
</style>
```

注意上述加粗代码，在then()方法中，回调函数的data参数返回的是一个数组，第一项为错误信息，第二项为返回数据。使用解构的方式获取data数组中的数据，如果不用解构的方式获取，则第一项为data[0]，第二项为data[1]。

10.2 uploadFile 文件上传

uploadFile API可以将本地资源上传到服务器，客户端发起一个POST请求，其中 content-type为multipart/form-data。

uploadFile官方文档地址:https://uniapp.dcloud.io/api/request/network-file?id=uploadfile。

10.2.1 uploadFile 的使用

扫一扫，看视频

如果在真机测试上传文件，则与request一样需要配置域名白名单；如果在微信开发者工具中上传文件，只需要勾选“不校验合法域名、web-view（业务域名）、TLS版本以及HTTPS证书”复选框即可使用接口。

uni.uploadFile的参数说明见表10-4。

表10-4 uni.uploadFile的参数说明

参数名	类 型	是否必填	说 明	平台差异说明
url	String	是	开发者服务器 URL	
files	Array	否	需要上传的文件列表。使用 files 时，filePath 和 name 不生效	App、H5(2.6.15+)
fileType	String		文件类型，有效值为 image、video、audio	仅支付宝小程序支持，且必填
file	File	否	要上传的文件对象	仅 H5（2.6.15+）支持
filePath	String	是	要上传文件资源的路径	
name	String	是	文件对应的 key。开发者在服务器端通过该 key 可以获取到文件二进制内容	
header	Object	否	HTTP 请求 Header，header 中不能设置 Referer	
formData	Object	否	HTTP 请求中其他的 form data	
success	Function	否	接口调用成功时的回调函数	
fail	Function	否	接口调用失败时的回调函数	
complete	Function	否	接口调用结束时的回调函数（调用成功、失败都会执行）	

files参数是一个 file 对象的数组，其说明见表10-5。

表10-5 files参数说明

参数名	类 型	是否必填	说 明
name	String	否	提交 multipart 时表单的项目名，默认为 file
file	File	否	要上传的文件对象，仅 H5（2.6.15+）支持
uri	String	是	文件的本地地址

注意：

如果不填name或填的值相同，可能导致服务端读取文件时只能读取到一个文件。

success返回参数说明见表10-6。

表10-6 success 返回参数说明

参数名	类 型	说 明
data	String	开发者服务器返回的数据
statusCode	Number	开发者服务器返回的 HTTP 状态码

在pages文件夹下创建uploadfile/index.vue文件，该文件中的代码如下：

```
<template>
    <view>
        图片预览:<image :src="showImage"></image>
        <button type="default" @click="uploadImage">上传图片</button>
        <button type="default" @click="doAbort">终止上传</button>
    </view>
</template>

<script>
    export default {
        name: "uploadfile",
        data(){
            return {
                showImage:""
            }
        },
        onLoad(){
            this.uploadTask=null;              //uploadFile实例
        },
        methods:{
            //上传图片方法
```

```
            uploadImage(){
                //选择图片
                uni.chooseImage({
                    success: (chooseImageRes) => {
                        //图片的本地文件路径列表
                        const tempFilePaths = chooseImageRes.tempFilePaths;
                        //上传图片
                        this.uploadTask=uni.uploadFile({
                            url: 'http://vueshop.glbuys.com/api/user/myinfo/
                                formdatahead?token=1ec949a15fb709370f',
                            //仅为示例，非真实的接口地址
                            filePath: tempFilePaths[0],//图片的本地文件
                            name: 'headfile',//上传图片接口参数的属性名称为headfile
                            success: (uploadFileRes) => {
                                //上传成功后返回的数据为String类型，转成对象类型
                                let data=JSON.parse(uploadFileRes.data);
                                this.showImage="http://vueshop.glbuys.com/
                                userfiles/head/"+data.data.msbox;//图片网络地址
                            }
                        });
                    }
                });
            },
            //终止上传
            doAbort(){
                this.uploadTask.abort();
            }
        }
    }
</script>

<style scoped>
    image{width:200rpx;height:200rpx;}
</style>
```

先使用uni.chooseImage API从本地相册选择图片或使用相机拍照，选择上传的图片后，再使用uni.uploadFile上传图片。使用abort()方法可以终止上传。

10.2.2 多文件上传

扫一扫，看视频

在实际开发中有很多场景需要批量上传文件，微信小程序只支持单文件上传，多文件上传需要反复调用uni.uploadFile() API。

在pages文件夹下创建uploadfile/multiple.vue文件，该文件中的代码如下：

```
<template>
    <view>
        图片预览:
        <view v-for="(item,index) in showImages" :key="index"><image:src=
         "item"></image></view>
        <button type="default" @click="uploadImage">上传图片</button>
    </view>
</template>

<script>
    export default {
        name: "uploadfile",
        data(){
            return {
                showImages:[]
            }
        },
        methods:{
            //上传图片方法
            uploadImage(){
                //选择图片
                uni.chooseImage({
                    success: (chooseImageRes) => {
                        //图片的本地文件路径列表
                        const tempFilePaths = chooseImageRes.tempFilePaths;
                        //如果选择多个图片
                        if(tempFilePaths.length>0){
                            for(let i=0;i<tempFilePaths.length;i++){
                                //上传图片
                                uni.uploadFile({
                                url: 'http://vueshop.glbuys.com/api/user/
                                myinfo/formdatahead?token=1ec949a15fb709370f',
                                //仅为示例，非真实的接口地址
                                filePath: tempFilePaths[i],//图片的本地文件
                                name: 'headfile',
                                //上传图片接口参数的属性名称为headfile
                                success: (uploadFileRes) => {
                                //上传成功后返回的数据为String类型，转成对象类型
                                let data=JSON.parse(uploadFileRes.data);
                                //将返回的图片网络地址添加到showImages数组中，并在
                                //视图中显示上传的图片
                                this.showImages.push("http://vueshop.glbuys.
                                com/userfiles/head/"+data.data.msbox);
```

```
                                }
                            });
                        }
                    }

                }
            });
        }
    }
}
</script>

<style scoped>
    image{width:200rpx;height:200rpx;}
</style>
```

使用uni.chooseImage() API选择多个图片，循环调用uni.uploadFile() API实现批量上传功能。

10.3　数据缓存

扫一扫，看视频

数据缓存可以将数据存储在本地缓存中，实现长期保存数据的功能，在uni-app中分为两种存储方式：异步存储和同步存储，在实际开发中经常用于保存会员登录状态信息、购物车、历史记录等场景。

数据缓存官方文档地址:https://uniapp.dcloud.io/api/storage/ storage?id=setstorage。

1.setStorage

setStorage可以异步将数据存储到本地缓存中。setStorage的参数说明见表10-7。

表10-7　setStorage的参数说明

参数名	类　型	是否必填	说　明
key	String	是	本地缓存中的指定的 key
data	Any	是	需要存储的内容，支持原生类型及能够通过 JSON.stringify 序列化的对象
success	Function	否	接口调用成功的回调函数
fail	Function	否	接口调用失败的回调函数
complete	Function	否	接口调用结束的回调函数（调用成功、失败都会执行）

代码示例如下：

```
uni.setStorage({
        key: 'name',
        data: '张三',
```

```
        success: ()=> {
            console.log('success');
        }
});
```

上述代码中，key为属性，data为存储的数据。接下来即可使用getStorage()读取本地缓存的数据。

2.getStorage

getStorage可以从本地缓存中异步获取指定 key 对应的内容。getStorage的参数说明见表10-8。

表10-8 getStorage的参数说明

参数名	类 型	是否必填	说 明
key	String	是	本地缓存中的指定的 key
success	Function	是	接口调用的回调函数，res = {data: key 对应的内容 }
fail	Function	否	接口调用失败的回调函数
complete	Function	否	接口调用结束的回调函数（调用成功、失败都会执行）

success返回参数说明见表10-9。

表10-9 success返回参数说明

参数名	类 型	说 明
data	Any	key 对应的内容

代码示例如下：

```
uni.getStorage({
    key:"name",
    success:(res)=>{
        console.log(res.data);//结果：张三
    }
})
```

3.setStorageSync

setStorageSync可以同步将数据存储到本地缓存中。setStorageSync参数说明见表10-10。

表 10-10 setStorageSync 的参数说明

参数名	类 型	是否必填	说 明
key	String	是	本地缓存中的指定的 key
data	Any	是	需要存储的内容，只支持原生类型及能够通过 JSON.stringify 序列化的对象

代码示例如下：

```
uni.setStorageSync('name', '张三');
```

setStorageSync()和H5的localstorage.setItem()使用方式类似，第一个参数为key，第二个参数为值。接下来即可使用getStorageSync()读取本地缓存的数据。

4.getStorageSync

getStorageSync可以从本地缓存中同步获取指定 key 对应的内容。getStorageSync的参数说明见表10-11。

表10-11 getStorageSync的参数说明

参数名	类 型	是否必填	说 明
key	String	是	本地缓存中的指定的 key

代码示例如下：

```
let name=uni.getStorageSync('name');
console.log(name);                //结果：张三
```

getStorageSync()和H5的localstorage.getItem()使用方式类似，传入key可以同步获取key对应的内容。

5.removeStorage

removeStorage可以从本地缓存中异步移除指定的key。removeStorage的参数说明见表10-12。

表10-12 removeStorage的参数说明

参数名	类 型	是否必填	说 明
key	String	是	本地缓存中的指定的 key
success	Function	是	接口调用的回调函数
fail	Function	否	接口调用失败的回调函数
complete	Function	否	接口调用结束的回调函数（调用成功、失败都会执行）

代码示例如下：

```
uni.removeStorage({
    key: 'name',
    success: function (res) {
        console.log('success');
    }
});
```

6.removeStorageSync

removeStorageSync可以从本地缓存中同步移除指定的key。removeStorageSync的参数说明见表10-13。

表10-13 removeStorageSync的参数说明

参数名	类 型	是否必填	说 明
key	String	是	本地缓存中的指定的 key

代码示例如下：

```
uni.removeStorageSync('name');
```

removeStorageSync和H5的localstorage.removeItem()使用方式类似，可以删除指定的key。

7.clearStorage

clearStorage可以异步清除本地所有的数据缓存。代码示例如下：

```
uni.clearStorage();
```

8.clearStorageSync

clearStorageSync可以同步清除本地所有的数据缓存。代码示例如下：

```
uni.clearStorageSync();
```

clearStorageSync()和H5的localstorage.clear()使用方式类似，可以清除所有的本地缓存数据。

10.4 获取位置

获取当前位置，可以实现查找附近的人、附近的商铺、地图定位、详细地址、所在城市等功能。获取位置官方文档地址：https://uniapp.dcloud.io/api/location/location。

10.4.1 getLocation

getLocation可以获取当前的地理位置。注意，在微信小程序中，当用户离开应用后，此接口无法调用，除非申请后台持续定位权限。

扫一扫，看视频

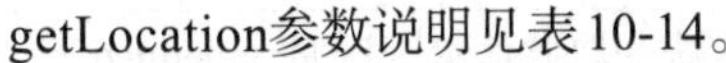

getLocation参数说明见表10-14。

表10-14 getLocation的参数说明

参数名	类 型	是否必填	说 明	平台差异说明
type	String	否	默认为 WGS-84 返回 GPS 坐标，GCJ-02 返回国家测绘局坐标，可用于 uni.openLocation 的坐标，App 平台高德 SDK 仅支持返回 GCJ-02	
altitude	Boolean	否	传入 true 会返回高度信息。由于获取高度需要较高精确度，因此会降低接口返回速度	App 和字节跳动小程序不支持
geocode	Boolean	否	是否解析地址信息，默认为 false	仅 App 平台支持
success	Function	是	接口调用成功的回调函数，返回内容详见返回参数说明	
fail	Function	否	接口调用失败的回调函数	
complete	Function	否	接口调用结束的回调函数（调用成功、失败都会执行）	

success返回参数说明见表10-15。

表10-15 success返回参数说明

参数	说明
latitude	纬度，浮点数，范围为 -90~+90，负数表示南纬
longitude	经度，浮点数，范围为 -180~+180，负数表示西经
speed	速度，浮点数，单位为 m/s
accuracy	位置的精确度
altitude	高度，单位为 m
verticalAccuracy	垂直精度，单位为 m（Android 无法获取，返回 0）
horizontalAccuracy	水平精度，单位为 m
address	地址信息（仅 App 端支持，需要配置 geocode 为 true）

address地址信息说明见表10-16。

表10-16 address地址信息说明

属性	类型	描述	说明
country	String	国家	如“中国”，如果无法获取此信息则返回 undefined
province	String	省份名称	如“北京市”，如果无法获取此信息则返回 undefined
city	String	城市名称	如“北京市”，如果无法获取此信息则返回 undefined
district	String	区（县）名称	如“朝阳区”，如果无法获取此信息则返回 undefined
street	String	街道信息	如“酒仙桥路”，如果无法获取此信息则返回 undefined
streetNum	String	获取街道门牌号信息	如“3 号”，如果无法获取此信息则返回 undefined
poiName	String	POI 信息	如“电子城国际电子总部”，如果无法获取此信息则返回 undefined
postalCode	String	邮政编码	如“100016”，如果无法获取此信息则返回 undefined
cityCode	String	城市代码	如“010”，如果无法获取此信息则返回 undefined

address地址信息仅支持App端，其他平台获取地址信息请参阅10.4.3节。

在pages文件夹下创建map/location.vue文件，该文件中的代码如下：

```
<script>
    export default {
        name: "location",
        onLoad(){
            uni.getLocation({
                type: 'gcj02',
                success: function (res) {
                    console.log('当前位置的经度:' + res.longitude);
                    console.log('当前位置的纬度:' + res.latitude);
                }
            });
```

```
            }
        }
</script>
```

uni.getLocation()的参数type属性值“gcj02”为国家测绘局坐标，定位更加准确，不推荐使用WGS-84(GPS坐标)，因为其误差比较大。但是，H5端不支持GCJ-02，需要在manifest.json文件中配置腾讯地图申请的key，才能在H5端使用GCJ-02获取坐标。

进入腾讯地图开发平台https://lbs.qq.com，注册并登录，单击网站右上角的“控制台”按钮，进入控制台页面。单击顶部导航栏中的“开发文档”选项下的“微信小程序JavaScript SDK”，进入图10-1所示的页面。在图10-1中单击“添加key”按钮，打开图10-2所示的页面。

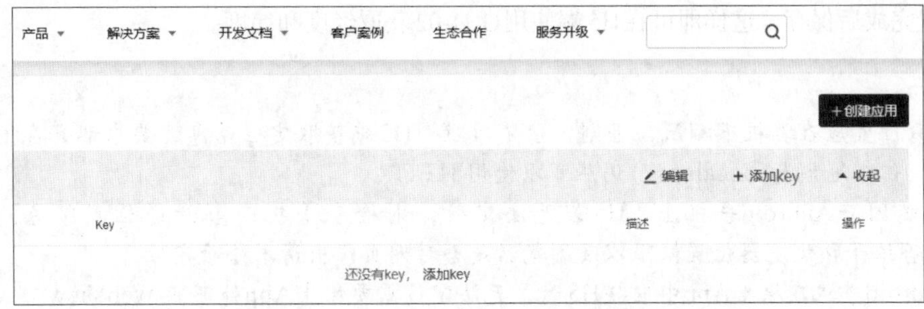

图 10-1 微信小程序 Java Script SDK 页面

添加key到「默认应用」应用

Key名称* 请您填写Key名称
请输入Key名称

描述 例：路线规划、获取用户位置

启用产品 Javascript API、Javascript API GL、地图组件、Android定位SDK、iOS定位SDK
在成功创建key后，即可使用

WebServiceAPI

地图SDK（Android/iOS）

微信小程序

授权 APP ID*

验证码* 看不清，换一张

阅读并同意《腾讯位置服务开放API服务协议》和《腾讯位置服务隐私协议》

添加 取消

图 10-2 添加 key

在图10-2中填写Key名称、描述和验证码，单击“添加”按钮，完成key的申请。在manifest.json配置腾讯地图key，在manifest.json文件中添加以下代码：

```
"h5" : {
        ...
        "sdkConfigs" : {
            "maps" : {
                "qqmap" : {
                    "key" : "你的key"
                }
            }
        },
        ...
}
```

配置完成后保存，这样即可在H5端使用GCJ-02获取经度和纬度。

注意：

获取位置应在真机下测试，否则定位不准确。H5端获取定位信息，要求部署在HTTPs服务上，但本地预览(localhost)仍然可以使用HTTP。

H5端国产Android手机上，H5若无法定位，则检查手机是否开通位置服务、GPS，ROM是否给予该浏览器位置权限以及浏览器是否对网页弹出请求给予定位。

Android手机在原生App中内嵌H5时，无法定位需要原生App处理的WebView。

移动端浏览器普遍仅支持GPS定位，在GPS信号弱的地方可能定位失败。

PC设备使用Chrome浏览器时，位置信息是通过连接谷歌服务器获取的，国内用户获取位置信息时可能会失败。

10.4.2 配合map组件定位“我的位置”显示到地图上

扫一扫，看视频

getLocation配合map组件，可以将“我的位置”显示在地图上。map组件的官网文档地址:https://uniapp.dcloud.io/component/map。

注意：自2022年7月14日起，在微信小程序中使用uni.getLocation获取地理位置必须先在manifest.json文件中进行配置，配置代码如下：

```
/* 小程序特有相关 */
"mp-weixin" : {
    ...
    "requiredPrivateInfos": [
        "getLocation",
        "chooseAddress",
        "onLocationChange",
        "chooseLocation",
        "startLocationUpdateBackground"
    ],
    "permission": {
          "scope.userLocation": {
             "desc": "获取我的位置方便查看附近商铺"
```

```
            }
        }
}
```

requiredPrivateInfos属性值说明如下：

- getFuzzyLocation: 获取模糊地理位置。
- getLocation: 获取精确地理位置。
- onLocationChange: 监听试试地理位置变化事件。
- startLocationUpdate: 接收位置消息（前台）。
- startLocationUpdateBackground: 接收位置消息（前后台）。
- chooseLocation: 打开地图选择位置。
- choosePoi: 打开 POI 列表选择位置。
- chooseAddress: 获取用户地址信息。

permission字段下面的scope.userLocation内部的desc为小程序获取权限时展示的接口用途说明，最长 30 个字符。配置完成后，使用HBuilder X重启项目，即可使用uni.getLocation。

在pages文件下创建map/map.vue文件，文件中的代码如下：

```
<template>
    <view>
        <map class="map" :latitude="latitude" :longitude="longitude":
         markers="markers"></map>
    </view>
</template>

<script>
    export default {
        name: "location",
        data(){
            return {
                latitude:"",            //纬度
                longitude:"",           //经度
                markers:[               //标记点用于在地图上显示标记的位置
                        id:1,
                        latitude:"",  //纬度
                        longitude:"", //经度
                        iconPath:"../../static/images/my.png",    //图标路径
                        width:"30",    //图标的宽
                        height:"30"    //图标的高
                ]
            }
        },
        onLoad(){
            uni.getLocation({
                type: 'gcj02',
                success: (res)=> {
                    console.log('当前位置的经度:' + res.longitude);
```

```
                        console.log('当前位置的纬度:' + res.latitude);
                        this.latitude=res.latitude;
                        this.longitude=res.longitude;
                        //将获取的longitude和latitude赋值给markers
                        //中的longitude、latitude
                        this.markers[0].longitude=this.longitude;
                        this.markers[0].latitude=this.latitude;
                    }
                });
            }
        }
    </script>

    <style scoped>
        .map{width:100%;height:500rpx;}
    </style>
```

map组件的属性latitude代表纬度，longitude代表经度，markers代表标记点用于在地图上显示标记的位置，值是一个数组类型，可以有多个标记。要了解更多map组件的属性，可以查看官网文档。

在微信开发者工具中渲染的效果如图 10-3 所示。

图 10-3　在 map 组件中显示“我的位置”

10.4.3　微信小程序获取地址详情

扫一扫，看视频

如果想在微信小程序中通过getLocation获取位置，显示地址详情（如省份、城市、街道等信息），必须调用腾讯地图开放平台中的微信小程序JavaScript SDK地图版本。

请按以下步骤完成此SDK的调用，建议大家观看视频完成此操作。

（1）注册并登录腾讯地图开放平台，官方地址：https://lbs.qq.com/。

（2）申请开发者密钥（key），申请地址：https://lbs.qq.com/dev/console/application/mine，如果没有应用，需要先创建应用，然后在应用中添加key（按照10.4.1节中的步骤添加即可），在添加时要勾选“WebServiceAPI”复选框（微信小程序SDK需要用到WebServiceAPI的部分服务，所以使用该功能的key需要具备相应的权限）。

（3）下载微信小程序JavaScript SDK，下载地址：https://mapapi.qq.com/web/miniprogram/JSSDK/qqmap-wx-jssdk1.2.zip，下载完成后解压，找到qqmap-wx-jssdk1.2文件夹，将其重命名为qqmap-wx-jssdk，复制到static/js/utils文件夹下。

（4）设置安全域名，登录微信小程序管理后台，设置地址：https://mp.weixin.qq.com/，在左侧导航栏中单击“开发”按钮，进入开发主页面，单击“开发管理”按钮，然后单击“开发设置”按钮，在此页面中找到服务器域名，在服务器域名下有request合法域名，在该域名中添加https://apis.map.qq.com。

（5）在代码中调用该API，pages/map/location.vue文件中的代码如下：

```
<template>
    <view class="page">
        <map style="width: 100%; height: 300px;" :latitude="lat":
         longitude="lng" :markers="markers">
        </map>
        <view>城市:{{city}}</view>
        <view>省份:{{province}}</view>
        <view>区(县):{{district}}</view>
        <view>街道:{{street}}</view>
        <view>门牌号:{{streetNum}}</view>
        <view>城市代码:{{cityCode}}</view>
        <view>地址详情:{{address}}</view>
    </view>
</template>

<script>
    //导入微信小程序JavaScript SDK
    import QQMapWX from "../../static/js/utils/qqmap-wx-jssdk/qqmap-wx-jssdk.js";
    export default {
        name: "location",
        data(){
            return {
                lng:0,                          //经度
                lat:0,                          //纬度
                markers: [{
                    latitude: 0,
                    longitude: 0,
                    iconPath: '../../static/images/location.png',
                    width:40,
                    height:40,
```

```
                callout:{                   //自定义标记点上方的气泡窗口
                    content:"",             //文本内容
                    display:"ALWAYS",       //常显示气泡
                    padding:5,              //文本边缘留白
                    borderRadius:8,         //callout边框圆角
                    fontSize:14             //文字大小
                }
            }],
            city:"",                        //城市
            province:"",                    //省份
            district:"",                    //区(县)
            street:"",                      //街道
            streetNum:"",                   //门牌号
            cityCode:"",                    //城市代码
            address:""                      //地址详情
        }
    },
    onLoad(){
        // 实例化微信小程序JavaScript SDK的API核心类
        this.qqmapsdk = new QQMapWX({
            key: 'OG4BZ-DEPHF-36FJE-NLIJP-H3JV3-2EFWR' //腾讯地图的key
        });
        uni.getLocation({
            type: 'gcj02',
            success: (res)=> {
                console.log('当前位置的经度:' + res.longitude);
                console.log('当前位置的纬度:' + res.latitude);
                this.lng=res.longitude;
                this.lat=res.latitude;
                this.markers[0].latitude=this.lat;
                this.markers[0].longitude=this.lng;
                //获取地理位置信息和附近POI列表
                this.qqmapsdk.reverseGeocoder({
                    location: {
                        latitude: this.lat,
                        longitude: this.lng
                    },
                    //返回成功
                    success:(res)=>{
                        console.log(res);
                        this.city=res.result.ad_info.city;
                        this.province=res.result.ad_info.province;
                        this.district=res.result.ad_info.district;
                        this.street=res.result.address_component.street;
```

```
                    this.streetNum=res.result.address_component.
                    street_number;
                    this.cityCode=res.result.ad_info.city_code;
                    this.address=res.result.address;
                    //将获取的城市赋值到气泡content中，在map组件中显示
                    this.markers[0].callout.content=this.city;
                },
                //返回失败
                fail:(res)=>{
                    console.log(res);
                }
            })
        }
    });
    }
  }
</script>

<style scoped>
    .page{width:100%;}
</style>
```

上述代码注释很清晰，请读者仔细阅读。reverseGeocoder为微信小程序JavaScript SDK的接口，可以获取地理位置信息和附近POI列表，该接口文档地址:https://lbs.qq.com/miniProgram/jsSdk/jsSdkGuide/methodReverseGeocoder。

在微信开发者工具中渲染的效果如图10-4所示。

图 10-4 获取到的地理位置信息

10.5 获取系统信息

扫一扫，看视频

获取系统信息便于用户查看手机品牌、手机型号、屏幕宽度、屏幕高度、操作系统版本、允许微信通知的开关、地理位置的系统开关等信息。在实际开发中，可以根据不同的手机型号解决兼容性问题，如自定义导航时iPhone X "刘海"挡住导航栏的问题。

获取系统信息官方文档地址:https://uniapp.dcloud.io/api/system/info。

10.5.1 getSystemInfo 的使用

获取系统信息分为uni.getSystemInfo（异步获取）和uni.getSystemInfoSync（同步获取）。getSystemInfo和getSystemInfoSync参数说明见表10-17。

表10-17 getSystemInfo和getSystemInfoSync的参数说明

参数名	类 型	是否必填	说 明
success	Function	是	接口调用成功的回调函数
fail	Function	否	接口调用失败的回调函数
complete	Function	否	接口调用结束的回调函数（调用成功、失败都会执行）

success返回参数说明见表10-18。

表10-18 success返回参数说明

参数名	说 明	平台差异说明
brand	手机品牌	App、微信小程序、百度小程序、字节跳动小程序、QQ 小程序
model	手机型号	
pixelRatio	设备像素比	
screenWidth	屏幕宽度	
screenHeight	屏幕高度	
windowWidth	可使用窗口宽度	
windowHeight	可使用窗口高度	
windowTop	可使用窗口的顶部位置	App、H5
windowBottom	可使用窗口的底部位置	App、H5
statusBarHeight	状态栏高度	字节跳动小程序不支持
navigationBarHeight	导航栏高度	百度小程序

续表

参数名	说　明	平台差异说明
titleBarHeight	标题栏高度	支付宝小程序
language	应用设置的语言	字节跳动小程序不支持
version	引擎版本号	H5 不支持
storage	设备磁盘容量	支付宝小程序
currentBattery	当前电量百分比	支付宝小程序
appName	宿主 App 名称	字节跳动小程序
AppPlatform	App 平台	QQ 小程序
host	宿主平台	百度小程序
app	当前运行的客户端	支付宝小程序
cacheLocation	上一次缓存的位置信息	百度小程序
system	操作系统版本	
platform	客户端平台，值域为 iOS、Android	
fontSizeSetting	用户字体大小设置，单位为 px	微信小程序、支付宝小程序、百度小程序、QQ 小程序
SDKVersion	客户端基础库版本	支付宝小程序和 H5 不支持
swanNativeVersion	宿主平台版本号	百度小程序
albumAuthorized	允许微信使用相册的开关（仅 iOS 有效）	微信小程序
cameraAuthorized	允许微信使用摄像头的开关	微信小程序
locationAuthorized	允许微信使用定位的开关	微信小程序
microphoneAuthorized	允许微信使用麦克风的开关	微信小程序
notificationAuthorized	允许微信通知的开关	微信小程序
notificationAlertAuthorized	允许微信通知带有提醒的开关（仅 iOS 有效）	微信小程序
notificationBadgeAuthorized	允许微信通知带有标记的开关（仅 iOS 有效）	微信小程序
notificationSoundAuthorized	允许微信通知带有声音的开关（仅 iOS 有效）	微信小程序
bluetoothEnabled	蓝牙的系统开关	微信小程序
locationEnabled	地理位置的系统开关	微信小程序
wifiEnabled	Wi-Fi 的系统开关	微信小程序
safeArea	在竖屏正方向下的安全区域	App、H5、微信小程序
safeAreaInsets	在竖屏正方向下的安全区域插入位置（2.5.3+）	App、H5、微信小程序

> **提示：**
>
> 屏幕高度＝原生navigationBar高度（含状态栏高度）＋可使用窗口高度＋原生tabBar高度。
>
> windowHeight不包含navigationBar(导航栏)和tabBar的高度。
>
> H5端，windowTop等于navigationBar高度，windowBottom等于tabBar高度。
>
> App端，windowTop等于透明状态navigationBar高度，windowBottom等于透明状态tabBar高度。

safeArea参数说明见表10-19。

表10-19 safeArea参数说明

参数名	类 型	说 明
left	Number	安全区域左上角横坐标
right	Number	安全区域右下角横坐标
top	Number	安全区域左上角纵坐标
bottom	Number	安全区域右下角纵坐标
width	Number	安全区域的宽度，单位为 px
height	Number	安全区域的高度，单位为 px

safeAreaInsets参数说明见表10-20。

表10-20 safeAreaInsets参数说明

参数名	类 型	说 明
left	Number	安全区域左侧插入位置
right	Number	安全区域右侧插入位置
top	Number	安全区域顶部插入位置
bottom	Number	安全区域底部插入位置

使用uni.getSystemInfo异步获取系统信息的代码示例如下：

```
uni.getSystemInfo({
    success: (res)=> {
    console.log(res.model);            //手机型号
        console.log(res.pixelRatio);  //设备像素比
        console.log(res.windowWidth); //可使用窗口宽度
        console.log(res.windowHeight); //可使用窗口高度
        console.log(res.language);    //应用设置的语言
        console.log(res.version);     //引擎版本号
        console.log(res.platform);    //客户端平台，在微信开发者工具中值为devtools
    }
});
```

使用uni.getSystemInfoSync同步获取系统信息的代码示例如下：

```
let res = uni.getSystemInfoSync();
console.log(res.model);
console.log(res.pixelRatio);
console.log(res.windowWidth);
console.log(res.windowHeight);
console.log(res.language);
console.log(res.version);
console.log(res.platform);
```

10.5.2 解决 iPhone X“刘海”兼容性问题

开发微信小程序时设置为自定义导航，在iPhone X中预览会发现自定义导航与“刘海”之间的距离对比原生导航有些差异，用户体验不是很好，如图10-5和图10-6所示。

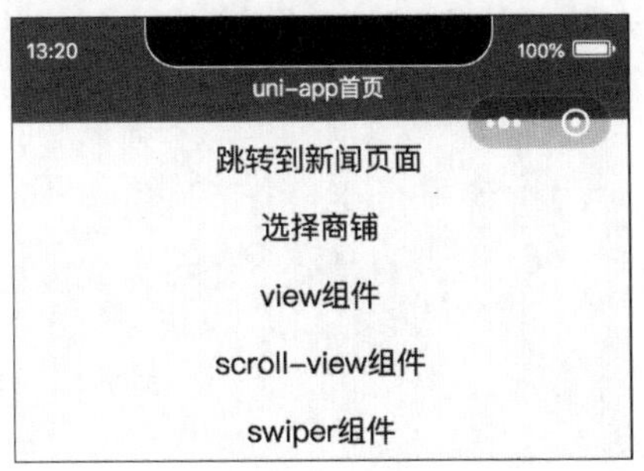

图 10-5 自定义导航在 iPhone X 中显示的效果

图 10-6 原生导航在 iPhone X 中显示的效果

从图10-5中可以看到，自定义导航与“刘海”的距离比较近，如果想要达到原生导航的效果，可以使用uni.getSystemInfoSync()的model属性获取手机型号，判断是否在iPhone X中运行，如果是，则重新设定padding-top的值即可。

pages/index/index.vue文件中的代码如下：

```
<template>
    <view class="page">
        <view class="status_bar">
            <!-- 这里是状态栏 -->
        </view>
        <view :class="{'nav-bar':true,iPX:isIpx}">
            uni-app首页
        </view>
        ...
    </view>
</template>

<script>
    export default {
        name:"index",
```

```
        data(){
            return {
                isIpx:false            //值为true表示在iPhone X中运行
            }
        },
        onLoad(opts) {
            uni.getSystemInfo({
                success: (res)=> {
                    var name = 'iPhone X';
                    //如果在iPhone X中运行
                    if(res.model.indexOf(name) > -1){
                        this.isIpx=true;
                    }
                }
            });
        },
        ...
    }
</script>

<style scoped>
    .page{width:100%;}
    ...
    .nav-bar{width:100%;height:80rpx;background-color:#3cc51f;color:#FFFFFF;text-
     align: center; line-height: 80rpx; font-size: 28rpx;}
    /*在iPhone X中的导航样式*/
    .iPX{padding-top:45rpx;}
    .status_bar {height: var(--status-bar-height);width: 100%; background-
     color: #3cc51f;}
</style>
```

上述代码注释很清晰，请读者仔细阅读。在微信开发者工具中预览的效果如图10-7所示，可以看到基本和原生导航显示效果一样。

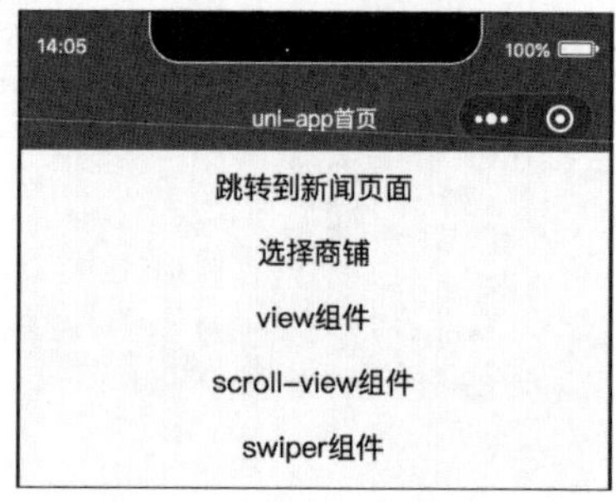

图 10-7　自定义导航在 iPhone X 中显示的效果

10.6 交互反馈

交互反馈包括消息提示框、loading 提示框、模态弹窗(alert、confirm)、操作菜单，这些都是在开发中常用的功能。

交互反馈官方文档地址:https://uniapp.dcloud.io/api/ui/prompt。

10.6.1 消息提示框

扫一扫，看视频

uni.showToast显示消息提示框，其参数说明见表10-21。

表10-21 showToast的参数说明

参数名	类 型	是否必填	说 明	平台差异说明
title	String	是	提示的内容，长度与 icon 取值有关	
icon	String	否	图标	
image	String	否	自定义图标的本地路径	App、H5、微信小程序、百度小程序
mask	String	否	是否显示透明蒙层，防止触摸穿透，默认为 false	App、微信小程序
duration	Number	否	提示的延迟时间，单位为 ms，默认为 1500	
position	String	否	纯文本轻提示显示位置，填写有效值后只有 title 属性生效	App
success	Function	否	接口调用成功的回调函数	
fail	Function	否	接口调用失败的回调函数	
complete	Function	否	接口调用结束的回调函数（调用成功、失败都会执行）	

icon值的说明见表10-22。

表10-22 icon值的说明

值	说 明	平台差异说明
success	显示成功图标，此时 title 文本最多显示 7 个汉字长度。此为默认值	
loading	显示加载图标，此时 title 文本最多显示 7 个汉字长度	支付宝小程序不支持
none	不显示图标，此时 title 文本在小程序最多可显示两行，App 仅支持单行显示	

position值的说明（仅App生效）见表10-23。

表10-23 position值的说明（仅App生效）

值	说 明
top	居上显示
center	居中显示
bottom	居下显示

代码示例如下：

```
uni.showToast({
    title: '标题',
    duration: 2000
});
```

如果想隐藏消息提示框，可以使用uni.hideToast()方法。代码示例如下：

```
uni.hideToast();
```

可以用showToast开发一个表单验证功能。在pages文件夹下创建toast/index.vue文件，该文件中的代码如下：

```
<template>
    <view>
        <view>用户名:<input type="text" placeholder="请输入用户名"
         v-model="userName"></view>
        <view>密码:<input type="password" placeholder="请输入密码"
         v-model="password"></view>
        <button type="default" @click="doLogin()">登录</button>
    </view>
</template>

<script>
    export default {
        name: "toast",
        data(){
            return {
                userName:"",
                password:""
            }
        },
        methods:{
            doLogin(){
                if(this.userName.trim()==''){
```

```
                uni.showToast({
                    title:"请输入用户名",
                    duration:2000,
                    icon:"none"
                });
                return;
            }
            if(this.password.trim()==''){
                uni.showToast({
                    title:"请输入密码",
                    duration:2000,
                    icon:"none"
                });
                return;
            }
        }
    }
}
</script>
```

在微信开发者工具中预览的效果如图10-8示。

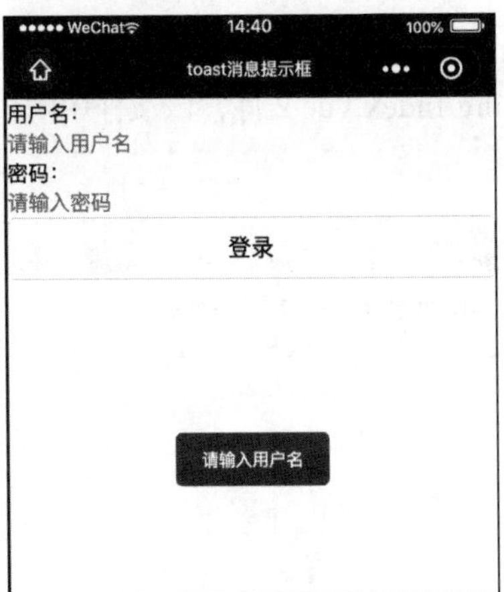

图 10-8 toast 消息提示框

10.6.2 loading 提示框

uni.showLoading显示loading提示框，一般用于请求服务端数据时使用。showLoading的参数说明见表10-24。

表10-24 showLoading的参数说明

参数名	类 型	是否必填	说 明	平台差异说明
title	String	是	提示的内容	
mask	Boolean	否	是否显示透明蒙层，防止触摸穿透，默认为 false	App、微信小程序、百度小程序
success	Function	否	接口调用成功的回调函数	
fail	Function	否	接口调用失败的回调函数	
complete	Function	否	接口调用结束的回调函数（调用成功、失败都会执行）	

代码示例如下：

```
uni.showLoading({
    title: '加载中'
});
```

如果想隐藏loading提示框，可以使用uni.hideLoading()方法。代码示例如下：

```
uni.showLoading({
    title: '加载中'
});

setTimeout(function () {
    uni.hideLoading();                    //隐藏
}, 2000);
```

在pages文件夹下创建loading/index.vue文件，该文件中的代码如下：

```
<template>
    <view>
        <view>LOGO:<image :src="logo"></image></view>
        <view>店铺名称:{{shopName}}</view>
        <view>公告:{{notice}}</view>
    </view>
</template>

<script>
    export default {
        name: "loading",
        data(){
            return {
                logo:"",
                shopName:"",            //店铺名称
                notice:""               //公告
            }
        },
        onLoad(){
```

```
            //显示loading
            uni.showLoading({
                title:"加载中..."
            });
            uni.request({
                url:"https://diancan.glbuys.com/api/v1/business/info",
                method:"GET",
                data:{
                    branch_shop_id:"900874159"
                }
            }).then((data)=>{
                //data为一个数组，数组第一项为错误信息，第二项为返回数据
                //请求完成后，隐藏loading
                uni.hideLoading();

                let [err,res]=data;
                if(res.data.code==200) {
                    this.logo=res.data.data.logo;
                    this.shopName=res.data.data.shop_name; //店铺名称
                    this.notice=res.data.data.notice;      //公告
                }
            })
        }
    }
</script>

<style scoped>
    image{width:200rpx;height:200rpx;}
</style>
```

上述代码实现了请求完成数据之前显示loading，请求完成之后隐藏loading。在微信开发者工具中预览的效果如图10-9所示。

图 10-9　loading 演示效果

10.6.3 模态弹窗

扫一扫，看视频

uni.showModal显示模态弹窗（alert、confirm），类似于标准HTML的消息框。showModal的参数说明见表10-25。

表10-25 showModal的参数说明

参数名	类 型	是否必填	说 明	平台差异说明
title	String	否	提示的标题	
content	String	否	提示的内容	
showCancel	Boolean	否	是否显示取消按钮，默认为 true	
cancelText	String	否	取消按钮的文字，默认为“取消”，最多4个字符	
cancelColor	HexColor	否	取消按钮的文字颜色，默认为 #000000	H5、微信小程序、百度小程序
confirmText	String	否	确定按钮的文字，默认为“确定”，最多4个字符	
confirmColor	HexColor	否	确定按钮的文字颜色，H5 平台默认为 #007aff，微信小程序平台默认为 #3CC51F，百度小程序平台默认为 #3c76ff	H5、微信小程序、百度小程序
success	Function	否	接口调用成功的回调函数	
fail	Function	否	接口调用失败的回调函数	
complete	Function	否	接口调用结束的回调函数（调用成功、失败都会执行）	

注意：

钉钉小程序真机与模拟器表现有差异，真机的title、content均为必填项。

success返回参数说明见表10-26。

表10-26 success返回参数说明

参数名	类 型	说 明
confirm	Boolean	为 true 时，表示用户点击了“确定”按钮
cancel	Boolean	为 true 时，表示用户点击了“取消”按钮（用于 Android 操作系统区分点击蒙层关闭还是点击取消按钮关闭）

1. 实现 alert 效果的代码示例

在pages文件夹下创建modal/alert.vue文件，该文件中的代码如下：

```
<template>
    <view>
```

续表

```
        <button @click="showAlert()">显示alert模态框</button>
    </view>
</template>

<script>
    export default {
        name: "alert",
        methods:{
            showAlert(){
                uni.showModal({
                    title: 'alert模态框',
                    content: '我是alert模态框的内容',
                    showCancel:false,         //隐藏“取消”按钮，实现alert
                    success: function (res) {
                        if (res.confirm) {
                            console.log('用户点击确定');
                        }
                    }
                });
            }
        }
    }
</script>
```

将showCancel属性值设置为false，隐藏“取消”按钮，可以实现alert效果。在微信开发者工具中预览的效果如图10-10所示。

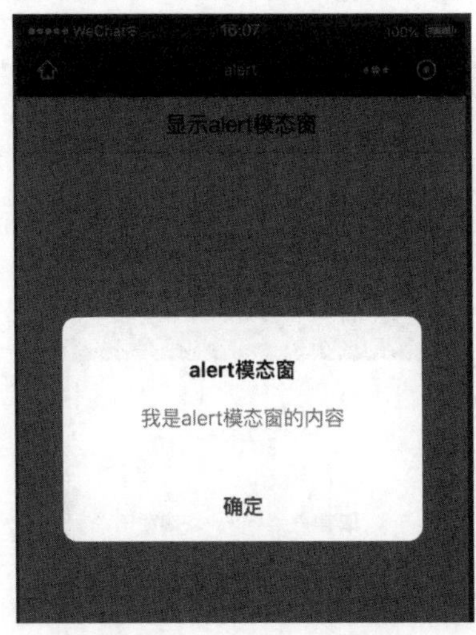

图 10-10 alert 模态窗

2.confirm 效果的代码示例

在pages文件夹下创建modal/confirm.vue文件，该文件中的代码如下：

```
<template>
    <view>
        <button @click="del()">删除</button>
    </view>
</template>

<script>
    export default {
        name: "confirm",
        methods:{
            del(){
                uni.showModal({
                    content: '确认要删除吗？ ',
                    success: function (res) {
                        if (res.confirm) {
                            console.log('确定');
                        }else if (res.cancel) {
                            console.log('取消');
                        }
                    }
                });
            }
        }
    }
</script>
```

将showCancel属性值设置为true或不添加showCancel属性，可以实现confirm效果，常用于删除时提示。在微信开发者工具中预览的效果如图10-11所示。

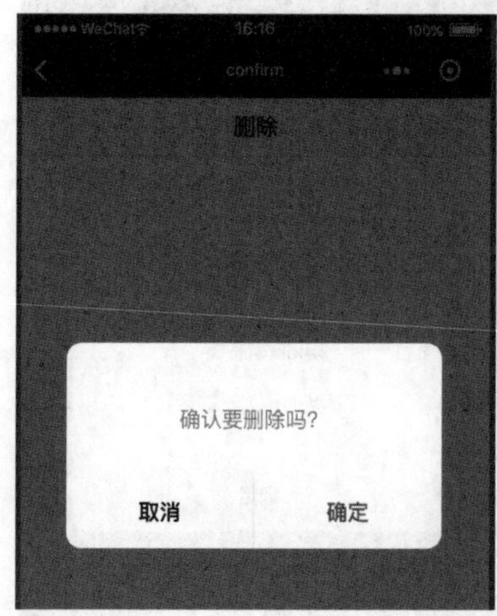

图 10-11 confirm 预览效果

10.6.4 操作菜单

扫一扫，看视频

uni.showActionSheet显示操作菜单，经常用于选择性别以及选择相册还是拍照等功能。showActionSheet的参数说明见表10-27。

表10-27 showActionSheet的参数说明

参 数	类 型	是否必填	说 明	平台差异说明
itemList	Array<String>	是	按钮的文字数组	微信小程序、百度小程序、字节跳动小程序的数组长度最大为6个
itemColor	HexColor	否	按钮的文字颜色，字符串格式，默认为“#000000”	App-iOS、字节跳动小程序不支持
popover	Object	否	大屏设备弹出原生选择按钮框的指示区域，默认居中显示	App-iPad（2.6.6+）、H5（2.9.2）
success	Function	否	接口调用成功的回调函数，详见返回参数说明（表10-29）	
fail	Function	否	接口调用失败的回调函数	
complete	Function	否	接口调用结束的回调函数（调用成功、失败都会执行）	

popover值的说明（仅App生效）见表10-28。

表10-28 popover值的说明（仅App生效）

值	类 型	说 明
top	Number	指示区域坐标，使用原生navigationBar时一般需要加上navigationBar的高度
left	Number	指示区域坐标
width	Number	指示区域宽度
height	Number	指示区域高度

success返回参数说明见表10-29。

表10-29 success返回参数说明

参 数	类 型	说 明
tapIndex	Number	用户点击的按钮，从0开始（顺序为由上至下）

在pages/modal文件夹下创建actionsheet.vue文件，该文件中的代码如下：

```
<template>
    <view>
        <view class="row">
            <view class="col-1">性别:</view>
```

```
            <view class="col-2" @click="selectGender()">{{gender==1?'男':
            (gender==2?'女':"请选择")}}</view>
        </view>
        <button @click="submit()">提交</button>
    </view>
</template>

<script>
    export default {
        name: "actionsheet",
        data(){
            return {
                genders:["男","女"],
                gender:""
            }
        },
        methods:{
            selectGender(){
                uni.showActionSheet({
                    itemList: this.genders,
                    success: (res)=> {
                        console.log('选中了第' + (res.tapIndex + 1) + '个按钮');
                        this.gender=res.tapIndex + 1;
                    }
                });
            },
            submit(){
                console.log(this.gender);            //值:1为男，2为女
            }
        }
    }
</script>

<style scoped>
    .row{width:100%;display: flex;font-size:28rpx;}
    .row.col-2{padding:0px 20rpx;}
</style>
```

res.tapIndex为选择的数据索引，利用索引可以自由地获取值。在微信开发者工具中预览的效果如图10-12所示。

图 10-12 操作菜单预览效果

10.7 动态设置导航条

扫一扫，看视频

导航条中可以动态设置的功能包括当前页面的标题、页面导航条颜色、显示/隐藏导航条加载动画、隐藏返回首页按钮等。

设置导航条官方文档地址：https://uniapp.dcloud.io/api/ui/navigationbar。

10.7.1 动态设置标题

uni.setNavigationBarTitle可以动态设置当前页面的标题，其参数说明见表10-30。

表10-30 setNavigationBarTitle的参数说明

参数名	类 型	是否必填	说 明
title	String	是	页面标题
success	Function	否	接口调用成功的回调函数
fail	Function	否	接口调用失败的回调函数
complete	Function	否	接口调用结束的回调函数（调用成功、失败都会执行）

代码示例如下：

```
onReady(){
    uni.setNavigationBarTitle({
```

```
        title: 'uni-app基础'
    });
}
```

注意：

如果需要在页面进入时设置标题，推荐在onReady内执行，以避免被框架内的修改所覆盖。如果必须在onShow内执行，可以使用setTimeout延迟一小段时间。

10.7.2 设置导航条颜色

uni.setNavigationBarColor可以设置页面导航条颜色。如果需要一进入页面就设置颜色，则应延迟执行，防止被框架内设置的颜色逻辑覆盖。setNavigationBarColor的参数说明见表10-31。

表10-31 setNavigationBarColor的参数说明

参数名	类　型	是否必填	说　明	平台差异说明
frontColor	String	是	前景颜色值，包括按钮、标题、状态栏的颜色，仅支持 #ffffff 和 #000000	App、H5、微信小程序、百度小程序
backgroundColor	String	是	背景颜色值，有效值为十六进制颜色	
animation	Object	否	动画效果，{duration,timingFunc}	微信小程序、百度小程序
success	Function	否	接口调用成功的回调函数	
fail	Function	否	接口调用失败的回调函数	
complete	Function	否	接口调用结束的回调函数（调用成功、失败都会执行）	

animation属性说明见表10-32。

表10-32 animation属性说明

属性名	类　型	默认值	是否必填	说　明
duration	Number	0	否	动画变化时间，单位为 ms
timingFunc	String	'linear'	否	动画变化方式

timingFunc有效值说明见表10-33。

表10-33 timingFunc有效值说明

值	说　明
linear	动画从头到尾的速度是相同的
easeIn	动画以低速开始
easeOut	动画以低速结束
easeInOut	动画以低速开始和结束

success返回参数说明见表10-34。

表10-34 success返回参数说明

参数名	类 型	说 明
errMsg	String	调用结果

代码示例如下：

```
onReady(){
    uni.setNavigationBarColor({
        frontColor: '#ffffff',
        backgroundColor: '#ff0000',
        animation: {
            duration: 400,
            timingFunc: 'easeIn'
        }
    })
}
```

注意：

Android 上的 backgroundColor 参数有限制，黑色大于 GRB(30,30,30)，白色小于 GRB(235, 235,235)。

10.7.3 显示 / 隐藏导航条加载动画

uni.showNavigationBarLoading可以显示导航条加载动画，uni.hideNavigationBarLoading可以隐藏导航条加载动画，通常用于从服务端请求数据之前显示导航条加载动画，请求完成之后隐藏导航条加载动画。其支持的平台有H5、微信小程序、支付宝小程序、百度小程序、QQ小程序，不支持的平台有App、字节跳动小程序。

showNavigationBarLoading和hideNavigationBarLoading参数说明见表10-35。

表10-35 showNavigationBarLoading和hideNavigationBarLoading的参数说明

参数名	类 型	是否必填	说 明
success	Function	否	接口调用成功的回调函数
fail	Function	否	接口调用失败的回调函数
complete	Function	否	接口调用结束的回调函数（调用成功、失败都会执行）

在pages文件夹下创建setnav/index.vue文件，该文件中的代码如下：

```
<template>
    <view>
        <view>LOGO:<image :src="logo"></image></view>
```

```
        <view>店铺名称:{{shopName}}</view>
        <view>公告:{{notice}}</view>
    </view>
</template>

<script>
    export default {
        name: "setnav",
        data(){
            return {
                logo:"",
                shopName:"",            //店铺名称
                notice:""               //公告
            }
        },
        onLoad(){
            //显示导航条加载动画
            uni.showNavigationBarLoading();
            uni.request({
                url:"https://diancan.glbuys.com/api/v1/business/info",
                method:"GET",
                data:{
                    branch_shop_id:"900874159"
                }
            }).then((data)=>{
                //data为一个数组，数组第一项为错误信息，第二项为返回数据
                //请求完成后，导航条加载动画
                uni.hideNavigationBarLoading()

                let [err,res]=data;
                if(res.data.code==200) {
                    this.logo=res.data.data.logo;
                    this.shopName=res.data.data.shop_name;          //店铺名称
                    this.notice=res.data.data.notice;               //公告
                }
            })
        }
    }
</script>

<style scoped>
    image{width:200rpx;height:200rpx;}
</style>
```

在微信开发者工具中预览的效果如图10-13所示。

图 10-13　导航条加载动画的预览效果

10.7.4　隐藏返回首页按钮

uni.hideHomeButton可以隐藏返回首页按钮，支持的平台有微信小程序、字节跳动小程序1.48.0+、QQ小程序1.10.0+，不支持的平台有App、H5、支付宝小程序、百度小程序。

hideHomeButton的参数说明见表10-36。

表10-36　hideHomeButton的参数说明

参数名	类　型	是否必填	说　明
success	Function	否	接口调用成功的回调函数
fail	Function	否	接口调用失败的回调函数
complete	Function	否	接口调用结束的回调函数（调用成功、失败都会执行）

在pages/setnav/index.vue文件中增加的代码如下：

```
onLoad(){
    //隐藏返回首页按钮
    uni.hideHomeButton();
    ...
}
```

在微信开发者工具中预览的效果如图10-14所示。

图 10-14　隐藏返回首页按钮

10.8 动态设置 tabBar

扫一扫，看视频

可以动态地设置tabBar某一项的内容、整体样式、隐藏/显示tabBar以及为某一项右上角添加/删除文本、显示/隐藏某一项右上角的红点等功能。

10.8.1 动态设置 tabBar 某一项的内容

使用uni.setTabBarItem可以动态设置tabBar某一项的内容，支持的平台有App、H5、微信小程序、支付宝小程序、百度小程序、QQ小程序，不支持的平台有钉钉小程序、字节跳动小程序。

setTabBarItem的参数说明见表10-37。

表10-37 setTabBarItem的参数说明

属性名	类 型	是否必填	说 明	平台差异说明
index	Number	是	tabBar 的哪一项，从左边算起	
text	String	否	tab 上的按钮文字	
iconPath	String	否	图片路径，icon 大小限制为 40KB，建议尺寸为 81px × 81px。当 position 为 top 时，此参数无效，不支持网络图片	
selectedIconPath	String	否	选中时的图片路径，icon 大小限制为 40KB，建议尺寸为 81px × 81px。当 position 为 top 时，此参数无效	
pagePath	String	否	页面绝对路径，必须在 pages 中先定义，被替换的 pagePath 不会变成普通页面（仍然需要使用 uni.switchTab 跳转）	App（2.8.4+）、H5（2.8.4+）
success	Funtion	否	接口调用成功的回调函数	
fail	Funtion	否	接口调用失败的回调函数	
complete	Funtion	否	接口调用结束的回调函数（调用成功、失败都会执行）	

在pages/my/index.vue文件中的代码如下：

```
<template>
    <view>
        <button @click="setTabBar()">更改tabBar</button>
    </view>
</template>

<script>
    export default {
        name: "my",
        methods:{
```

```
                setTabBar(){
                    uni.setTabBarItem({
                        index: 0,
                        text: 'Home'
                    })
                }
            }
        }
</script>
```

点击“更改tabBar”按钮，会将tabBar第一项的按钮文字更改为Home。

10.8.2 动态设置 tabBar 的整体样式

使用uni.setTabBarStyle可以动态设置 tabBar 的整体样式，支持的平台有App、H5、微信小程序、支付宝小程序、百度小程序、QQ小程序，不支持的平台有字节跳动小程序。

setTabBarStyle的参数说明见表10-38。

表10-38 setTabBarStyle的参数说明

属性名	类 型	是否必填	说 明
color	String	否	tabBar 上的文字默认颜色
selectedColor	String	否	tabBar 上的文字选中时的颜色
backgroundColor	String	否	tabBar 的背景色
backgroundImage	String	否	图片背景。支持设置本地图片或创建线性渐变，优先级高于 backgroundColor，仅 App 2.7.1+ 支持
backgroundRepeat	String	否	背景图平铺方式。repeat：背景图片在垂直方向和水平方向平铺；repeat-x：背景图片在水平方向平铺，垂直方向拉伸；repeat-y：背景图片在垂直方向平铺，水平方向拉伸；no-repeat：背景图片在垂直方向和水平方向都拉伸。默认使用 no-repeat，仅 App 2.7.1+ 支持
borderStyle	String	否	tabBar 上边框的颜色，仅支持 black、white
success	Funtion	否	接口调用成功的回调函数
fail	Funtion	否	接口调用失败的回调函数
complete	Funtion	否	接口调用结束的回调函数（调用成功、失败都会执行）

backgroundImage创建线性渐变的代码示例如下：

```
backgroundImage: linear-gradient(to top, #a80077, #66ff00);
```

目前暂不支持 radial-gradient（径向渐变），同时只支持两种颜色的渐变，渐变方向如下：

（1）to right：从左向右渐变。

（2）to left：从右向左渐变。
（3）to bottom：从上到下渐变。
（4）to top：从下到上渐变。
（5）to bottom right：从左上角到右下角渐变。
（6）to top left：从右下角到左上角渐变。
在pages/my/index.vue文件中的新增代码如下：

```
<template>
    <view>
        ...
        <button @click="setTabBarStyle()">更改tabBar整体样式</button>
    </view>
</template>

<script>
    export default {
        name: "my",
        methods:{
            ...
            setTabBarStyle(){
                uni.setTabBarStyle({
                    color: '#FF0000',
                    selectedColor: '#00FF00',
                    backgroundColor: '#0000FF',
                    borderStyle: 'white'
                });
            }
        }
    }
</script>
```

单击“更改tabBar整体样式”按钮，在自定义setTabBarStyle()方法中调用uni.setTabBarStyle，可以更新tabBar整体样式。

10.8.3 隐藏 / 显示 tabBar

使用uni.hideTabBar可以隐藏tabBar，使用uni.showTabBar可以显示tabBar。其支持的平台有App、H5、微信小程序、支付宝小程序、百度小程序、QQ小程序，不支持的平台有字节跳动小程序。

showTabBar和hideTabBar的参数说明见表10-39。

表10-39 showTabBar和hideTabBar的参数说明

属性名	类 型	默认值	是否必填	说 明
animation	Boolean	false	否	是否需要动画效果，仅微信小程序和百度小程序支持
success	Funtion		否	接口调用成功的回调函数
fail	Funtion		否	接口调用失败的回调函数
complete	Funtion		否	接口调用结束的回调函数（调用成功、失败都会执行）

在pages/my/index.vue文件中的新增代码如下：

```
<template>
    <view>
        ...
        <button @click="hideTabBar()">隐藏tabBar</button>
        <button @click="showTabBar()">显示tabBar</button>
    </view>
</template>

<script>
    export default {
        name: "my",
        methods:{
            ...
            //隐藏tabBar
            hideTabBar(){
                uni.hideTabBar({
                    animation:true
                })
            },
            //显示tabBar
            showTabBar(){
                uni.showTabBar({
                    animation:true
                })
            }
        }
    }
</script>
```

10.8.4 为 tabBar 某一项的右上角添加 / 删除文本

使用uni.setTabBarBadge可以为tabBar 某一项的右上角添加文本，使用uni.removeTabBarBadge

可以移除tabBar某一项右上角的文本。其支持的平台有App、H5、微信小程序、支付宝小程序、百度小程序、QQ小程序，不支持的平台有字节跳动小程序。

setTabBarBadge和removeTabBarBadge的参数说明见表10-40。

表10-40　setTabBarBadge和removeTabBarBadge的参数说明

参数名	类　型	是否必填	说　明
index	Number	是	tabBar 的哪一项，从左边算起
success	Function	否	接口调用成功的回调函数
fail	Function	否	接口调用失败的回调函数
complete	Function	否	接口调用结束的回调函数（调用成功、失败都会执行）

在pages/my/index.vue文件中新增的代码如下：

```
<template>
    <view>
        ...
        <button @click="setTabBarBadge()">右上角添加文本</button>
        <button @click="removeTabBarBadge()">删除右上角文本</button>
    </view>
</template>

<script>
    export default {
        name: "my",
        methods:{
            ...
            //右上角添加文本
            setTabBarBadge(){
                uni.setTabBarBadge({
                    index: 1,
                    text: '10'
                })
            },
            //删除右上角文本
            removeTabBarBadge(){
                uni.removeTabBarBadge({
                    index:1
                })
            }
        }
    }
</script>
```

点击“右上角添加文本”按钮，在tabBar的第二项右上角添加文本。在微信开发者工具中预览的效果如图10-15所示。

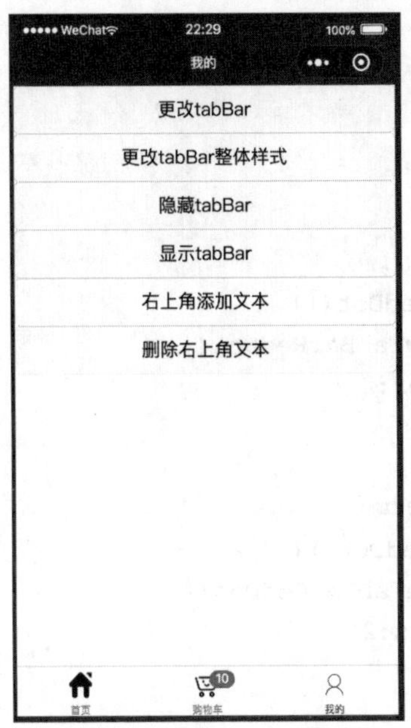

图 10-15　在 tabBar 第二项右上角添加文本预览效果

10.8.5　显示 / 隐藏 tabBar 某一项右上角的红点

使用uni.showTabBarRedDot可以显示tabBar某一项右上角的红点，使用uni.hideTabBarRedDot可以隐藏tabBar某一项的右上角的红点。其支持的平台有App、H5、微信小程序、支付宝小程序、百度小程序、QQ小程序，不支持的平台有字节跳动小程序。

showTabBarRedDot和hideTabBarRedDot的参数说明见表10-41。

表10-41　showTabBarRedDot和hideTabBarRedDot的参数说明

参数名	类　型	是否必填	说　明
index	Number	是	tabBar 的哪一项，从左边算起
success	Function	否	接口调用成功的回调函数
fail	Function	否	接口调用失败的回调函数
complete	Function	否	接口调用结束的回调函数（调用成功、失败都会执行）

在pages/my/index.vue文件中新增的代码如下：

```
<template>
    <view>
        ...
        <button @click="showTabBarRedDot()">显示右上角的红点</button>
        <button @click="hideTabBarRedDot()">隐藏右上角的红点</button>
    </view>
</template>
```

```
<script>
    export default {
        name: "my",
        methods:{
            ...
            //显示右上角的红点
            showTabBarRedDot(){
                uni.showTabBarRedDot({
                    index:2
                })
            },
            //隐藏右上角的红点
            hideTabBarRedDot(){
                uni.hideTabBarRedDot({
                    index:2
                })
            }
        }
    }
</script>
```

点击“显示右上角的红点”按钮，在tabBar第三项显示右上角红点。在微信开发者工具中预览的效果如图10-16所示。

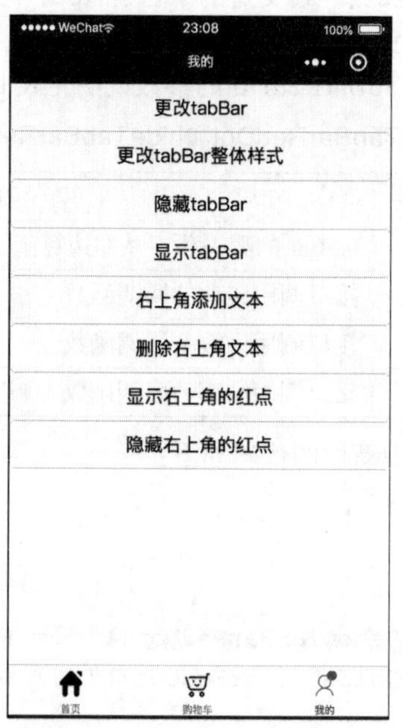

图 10-16　在 tabBar 第三项显示右上角红点预览效果

10.9 录音管理

扫一扫，看视频

录音通常用于开发聊天系统时的发送语音功能，也可以利用录音配合人工智能API实现语音识别功能。

录音管理官方文档地址:https://uniapp.dcloud.io/api/media/record-manager。

uni.getRecorderManager可以获取全局唯一的录音管理器，支持的平台有App、微信小程序、百度小程序、字节跳动小程序、QQ小程序，不支持的平台有H5、支付宝小程序。

getRecorderManager的方法说明见表10-42。

表10-42 getRecorderManager的方法说明

方法名	参数名	说 明	平台差异说明
start	options	开始录音	
pause		暂停录音	App 暂不支持
resume		继续录音	App 暂不支持
stop		停止录音	
onStart	callback	录音开始事件	
onPause	callback	录音暂停事件	
onStop	callback	录音停止事件，会回调文件地址	
onFrameRecorded	callback	已录制完指定帧大小的文件，会回调录音分片结果数据。如果设置了 frameSize，则会回调此事件	App 暂不支持
onError	callback	录音错误事件，会回调错误信息	

start(options)属性说明见表10-43。

表10-43 start(options)属性说明

属性名	类 型	是否必填	说 明
duration	Number	否	指定录音的时长，单位为ms。如果传入了合法的duration，在到达指定的duration后会自动停止录音，最大值为600000（10min），默认值为60000（1min）
sampleRate	Number	否	采样率，有效值为8000、16000、44100
numberOfChannels	Number	否	录音通道数，有效值为1、2
encodeBitRate	Number	否	编码码率，有效值见表10-44
format	String	否	音频格式，有效值为aac、mp3、wav、PCM
frameSize	String	否	指定帧大小，单位为KB。传入frameSize后，每录制指定帧大小的内容后，会回调录制的文件内容，不指定则不会回调。暂仅支持mp3格式

encodeBitRate有效值说明(采样率和编码码率有一定要求)见表10-44。

表10-44 encodeBitRate有效值说明

采样率	编码码率
8000	16000~48000
11025	16000~48000
12000	24000~64000
16000	24000~96000
22050	32000~128000
24000	32000~128000
32000	48000~192000
44100	64000~320000
48000	64000~320000

onStop(callback) 回调结果说明见表10-45。

表10-45 onStop(callback) 回调结果说明

属性名	类 型	说 明
tempFilePath	String	录音文件的临时路径

onFrameRecorded(callback) 回调结果说明见表10-46。

表10-46 onFrameRecorded(callback) 回调结果说明

属性名	类 型	说 明
frameBuffer	ArrayBuffer	录音分片结果数据
isLastFrame	Boolean	当前帧是否为正常录音结束前的最后一帧

onError(callback) 回调结果说明见表10-47。

表10-47 onError(callback) 回调结果说明

属性名	类 型	说 明
errMsg	String	错误信息

在pages文件夹下创建record/index.vue文件，该文件中的代码如下：

```
<template>
    <view>
        <button @click="startRecord">开始录音</button>
        <button @click="endRecord">停止录音</button>
        <button @click="playVoice">播放录音</button>
```

```
    </view>
</template>

<script>
    export default {
        name: "record",
        data() {
            return {
                voicePath: ''           //录音文件的临时路径
            }
        },
        onLoad() {
            //创建音频控制组件
            this.innerAudioContext = uni.createInnerAudioContext();
            //录音管理实例
            this.recorderManager = uni.getRecorderManager();
            //监听录音停止事件
            this.recorderManager.onStop( (res)=> {
                console.log('recorder stop' + JSON.stringify(res));
                //录音文件的临时路径
                this.voicePath = res.tempFilePath;
            });
        },
        methods: {
            startRecord() {
                console.log('开始录音');
                this.recorderManager.start();
            },
            endRecord() {
                console.log('录音结束');
                this.recorderManager.stop();
            },
            playVoice() {
                console.log('播放录音');
                if (this.voicePath) {
                    this.innerAudioContext.src = this.voicePath;
                    this.innerAudioContext.play();
                }
            }
        }
    }
</script>
```

最终录音的音频类型为aac格式，如果将录音文件上传到服务器，可以使用res.tempFilePath字段。

10.10 视频组件控制

扫一扫，看视频

视频组件控制可以控制<video>组件使用JS实现播放、暂停、全屏、弹幕等功能。

10.10.1 自由地控制视频

使用uni.createVideoContext创建并返回video上下文 videoContext 对象，可以操作组件内的<video> 组件。

uni.createVideoContext对象的方法说明见表10-48。

表10-48 createVideoContext的方法说明

方法名	参数名	说 明	平台差异说明
play		播放	
pause		暂停	
seek	position	跳转到指定位置，单位为 s	
stop		停止视频	微信小程序
sendDanmu	danmu	发送弹幕，danmu 包含两个属性：text、color	
playbackRate	rate	设置倍速播放，支持的倍速有 0.5、0.8、1.0、1.25、1.5。自微信基础库 2.6.3 起支持 2.0 倍速	
requestFullScreen		进入全屏，可传入 {direction} 参数，有效值为 0（正常竖向）、90（屏幕逆时针 90°）、-90（屏幕顺时针 90°）	
exitFullScreen		退出全屏	
showStatusBar		显示状态栏，仅在 iOS 全屏下有效	微信小程序、百度小程序、QQ 小程序
hideStatusBar		隐藏状态栏，仅在 iOS 全屏下有效	微信小程序、百度小程序、QQ 小程序

在pages/video/index.vue文件中的代码如下：

```
<template>
    <view>
        <video id="my-video" ref="my-video" :src="videoUrl" :controls =
         "false" @timeupdate="getTime"></video>
        当前播放时间:{{currentTime|formatSeconds}}/{{duration|formatSeconds}}
        <button @click="play()">播放</button>
        <button @click="pause()">暂停</button>
        <button @click="goTime()">前进15s</button>
        <button @click="backTime()">后退15s</button>
    </view>
```

```
</template>

<script>
    export default {
        name: "video-component",
        data(){
            return {
                videoUrl:"http://www.lucklnk.com/video/4.mp4",
                currentTime:"00:00",          //当前音频的播放时间
                duration:"00:00"              //音频总时长
            }
        },
        onReady: function (res) {
            this.videoContext = uni.createVideoContext('my-video')
        },
        methods:{
            //播放视频
            play(){
                this.videoContext.play();
            },
            //暂停视频
            pause(){
                this.videoContext.pause();
            },
            //前进15s
            goTime(){
                this.videoContext.seek(this.currentTime+15);
            },
            //后退15s
            backTime(){
                this.videoContext.seek(this.currentTime-15);
            },
            //通过视图中video组件@timeupdate事件获取视频播放时长与总时长
            getTime(e){
                this.currentTime=e.target.currentTime;     //视频当前播放时长
                this.duration=e.target.duration;           //视频总时长
            }
        },
        //局部过滤器
        filters:{
            //将秒转成03:30格式
            formatSeconds(value){
                let minute=parseInt(value/60)?parseInt(value/60):'0';//分钟
                let second=parseInt(value%60)?parseInt(value%60):'0';//秒
                if(minute<10){          //如果小于10，补0
                    minute="0"+minute;
                }
```

```
                    if(second<10){
                        second="0"+second;
                    }
                    return minute+":"+second;
                }
            }
        }
    </script>

    <style scoped>
        video{width:100%;height:500rpx;}
    </style>
```

以上代码注释很清晰，请读者仔细阅读，不再赘述。注意，为了平台兼容性，推荐同时设置video组件的id属性和ref属性。

10.10.2 发送弹幕

uni-app的video组件配合uni.createVideoContext API的sendDanmu()方法可以轻松实现弹幕功能。在pages/video/index.vue文件中新增的代码如下：

```
<template>
    <view>
        <video id="my-video" ref="my-video" :src="videoUrl" :controls="true"
         :enable-danmu="true" :danmu-btn="true" :danmu-list="danmuList"></video>
        <view>
            <input type="text" placeholder="弹幕内容" v-model="danmuValue" />
             <button type="primary" class="danmu-btn" @click="sendDanmu()">
             发送弹幕</button>
        </view>
    </view>
</template>

<script>
    export default {
        name: "video-component",
        data(){
            return {
                videoUrl:"http://www.lucklnk.com/video/4.mp4",
                danmuValue:"",
                //弹幕数据
                danmuList: [
                    {
                        text: '第 1s 出现的弹幕',
                        color: '#ff0000',
                        time: 1
                    },
```

```
                {
                    text: '第 3s 出现的弹幕',
                    color: '#ff00ff',
                    time: 3
                }
            ]
        }
    },
    onReady: function (res) {
        this.videoContext = uni.createVideoContext('my-video')
    },
    methods:{
        //发送弹幕
        sendDanmu(){
            this.videoContext.sendDanmu({
                text: this.danmuValue,
                color: this.getRandomColor()
            })
        },
        //弹幕文字随机颜色
        getRandomColor: function () {
            const rgb = []
            for (let i = 0; i < 3; ++i) {
                let color = Math.floor(Math.random() * 256).toString(16)
                color = color.length == 1 ? '0' + color : color
                rgb.push(color)
            }
            return '#' + rgb.join('')
        }
    }
}
</script>

<style scoped>
    video{width:100%;height:500rpx;}
    .danmu-btn{width:180rpx;height:60rpx;font-size:28rpx;}
    .danmu-btn:after{border:0 none;}
</style>
```

要发送弹幕，需要在video组件上设置enable-danmu属性值为true。在实际开发中，弹幕内容应让所有观看视频的用户看到，需要通过WebSocket实现。首先搭建服务端的socket环境，然后通过客户端的WebSocket连接通信。由于本课程没有服务端的socket环境，因此不再演示。企业中一般会搭建socket服务器供用户使用。

WebSocket官方文档地址:https://uniapp.dcloud.io/api/request/websocket。

10.11 网络状态

扫一扫，看视频

在实际开发中，通过获取网络状态可以实现观看视频时检测当前网络环境的功能。如果用户使用的是非Wi-Fi网络，则可以提示用户，提升用户的观看体验。

10.11.1 获取网络类型

使用uni.getNetworkType可以获取网络类型，其参数说明见表10-49。

表10-49 getNetworkType的参数说明

参数名	类 型	必 填	说 明
success	Function	是	接口调用成功，返回网络类型 networkType
fail	Function	否	接口调用失败的回调函数
complete	Function	否	接口调用结束的回调函数（调用成功、失败都会执行）

success返回参数说明见表10-50。

表10-50 success返回参数说明

参数名	说 明
networkType	网络类型

networkType有效值说明见表10-51。

表10-51 networkType有效值说明

值	说 明	平台差异说明
wifi	Wi-Fi 网络	
2g	2G 网络	
3g	3G 网络	
4g	4G 网络	
ethernet	有线网络	App
unknown	Android 下不常见的网络类型	
none	无网络	

代码示例如下：

```
uni.getNetworkType({
    success: function (res) {
        console.log(res.networkType);
    }
});
```

上述代码很简单，接下来开发一个当用户在非Wi-Fi网络下观看视频时提示网络环境并暂停播放的功能。

在pages文件夹下创建network/video.vue文件，该文件中的代码如下：

```
<template>
    <view>
        <view class="view-wrap">
            <video id="my-video" ref="my-video" :src="videoUrl" :controls =
             "true" :autoplay="true"></video>
            <view class="tip" v-show="isTip">当前网络环境为"<text> {{networkType}}
             </text>"</view>
            <view class="play-btn" @click="play()" v-show="isTip">继续播放</view>
        </view>
    </view>
</template>

<script>
    export default {
        name: "network-video",
        data(){
            return {
                videoUrl:"http://www.lucklnk.com/video/4.mp4",
                networkType:"",                //网络类型
                isTip:false                    //是否显示网络类型提示
            }
        },
        onReady(){
            this.videoContext = uni.createVideoContext('my-video');
            uni.getNetworkType({
                success: (res)=> {
                    this.networkType=res.networkType;
                    //如果为非Wi-Fi网络环境
                    if(this.networkType!='wifi'){
                        //暂停播放
                        this.videoContext.pause();
                        this.isTip=true;      //显示网络类型提示
                    }
                }
            });
        },
```

```
        methods:{
            //播放视频
            play(){
                this.videoContext.play();
                this.isTip=false;      //隐藏网络类型提示
            }
        }
    }
</script>

<style scoped>
    .view-wrap{width:100%;height:500rpx;position: relative;}
    .view-wrap video{width:100%;height:100%;}
    .view-wrap .tip{width:auto;height:auto;position: absolute;z-index:2;left:5
     0%;top:50%;transform: translate(-50%,-50%);padding:10rpx 30rpx;background-
     color:rgba(0,0,0,0.6);font-size:28rpx;color:#FFFFFF;border-radius: 4px;}
    .view-wrap  .play-btn{position: absolute;z-index:2;left:50%;top:60
     %;transform: translateX(-50%);padding:10rpx 30rpx;border-radius:
     4px;background-color:rgba(255,50,50,0.6);font-size:28rpx;color:#FFFFFF}
</style>
```

上述代码注释很清晰，不再赘述。可以在微信开发者工具中切换网络环境进行测试，如图10-17所示。

图 10-17　在微信开发者工具中切换网络环境

在微信开发者工具中预览的效果如图10-18所示。

图 10-18　提示网络环境的预览效果

10.11.2　监听网络状态变化

使用uni.onNetworkStatusChange可以监听网络状态变化，其参数说明见表10-52。

表10-52　onNetworkStatusChange参数说明

参数名	类　型	说　明	平台差异说明
isConnected	Boolean	当前是否有网络连接	字节跳动小程序不支持
networkType	String	网络类型	

代码示例如下：

```
onReady(){
    uni.onNetworkStatusChange(function (res) {
        console.log(res.isConnected);
        console.log(res.networkType);
    });
}
```

注意：

必须在网络环境发生变化时才能触发，可以在微信开发者工具中切换网络环境进行测试，如图10-17所示。

10.12　剪贴板

我们可以自己设置剪贴板内容和获取剪贴板内容，同时可以实现复制口令功能。

扫一扫，看视频

10.12.1 设置系统剪贴板内容

使用uni.setClipboardData可以设置系统剪贴板内容。除了H5平台不支持外，其他平台都支持此功能。setClipboardData的参数说明见表10-53。

表10-53 setClipboardData的参数说明

参数名	类　型	是否必填	说　明
data	String	是	需要设置的内容
success	Function	否	接口调用成功的回调函数
fail	Function	否	接口调用失败的回调函数
complete	Function	否	接口调用结束的回调函数（调用成功、失败都会执行）

在pages文件夹下创建clipboard/index.vue文件，该文件中的代码如下：

```
<template>
    <view>
        <button @click="setClipBoard()">复制口令</button>
    </view>
</template>

<script>
    export default {
        name: "clipboard",
        methods:{
            setClipBoard(){
                uni.setClipboardData({
                    data:"复制口令",
                    success:()=>{
                        console.log("内容已复制");
                    }
                })
            }
        }
    }
</script>
```

在微信开发者工具中预览的效果如图10-19所示。

图 10-19 设置系统剪贴板内容

10.12.2 获取系统剪贴板内容

使用uni.getClipboardData可以获取系统剪贴板内容，其参数说明见表10-54。

表10-54 getClipboardData的参数说明

参数名	类 型	是否必填	说 明
success	Function	否	接口调用成功的回调函数
fail	Function	否	接口调用失败的回调函数
complete	Function	否	接口调用结束的回调函数（调用成功、失败都会执行）

success返回参数说明见表10-55。

表10-55 success返回参数说明

参数名	类 型	说 明
data	String	剪贴板的内容

在pages/clipboard/index.vue文件中新增的代码如下：

```
<template>
    <view>
        ...
        <button @click="getClipBoard()">获取口令</button>
        <view>口令:{{content}}</view>
    </view>
</template>
```

```
<script>
    export default {
        name: "clipboard",
        data(){
            return {
                content:""
            }
        },
        methods:{
            ...
            getClipBoard(){
                uni.getClipboardData({
                    success:(res)=>{
                        console.log(res.data);
                        this.content=res.data;
                    }
                })
            }
        }
    }
</script>
```

10.13 拨打电话

拨打电话功能兼容所有平台，应用比较广泛，支持手机和座机。

使用uni.makePhoneCall可以实现拨打电话的功能，其参数说明见表10-56。

表10-56 makePhoneCall的参数说明

参数名	类 型	是否必填	说 明
phoneNumber	String	是	需要拨打的电话号码
success	Function	否	接口调用成功的回调函数
fail	Function	否	接口调用失败的回调函数
complete	Function	否	接口调用结束的回调函数（调用成功、失败都会执行）

代码示例如下：

```
uni.makePhoneCall({
    phoneNumber: '114'          //仅为示例
});
```

10.14 扫码

扫码功能支持使用相机直接扫码和从相册中选择图片进行扫码，也支持条码扫描。

使用uni.scanCode可以调出客户端扫码界面，扫码成功后返回对应的结果。该功能不支持H5平台，其他平台都支持。

scanCode的参数说明见表10-57。

表10-57 scanCode的参数说明

参数名	类 型	是否必填	说 明	平台差异说明
onlyFromCamera	Boolean	否	是否只能使用相机扫码，不允许从相册选择图片	字节跳动小程序不支持
scanType	Array	否	扫码类型，参数类型是数组，二维码是qrCode，一维码（条形码）是barCode，DataMatrix是datamatrix，PDF417是pdf417	
success	Function	否	接口调用成功的回调函数，返回内容详见返回参数说明（表10-58）	
fail	Function	否	接口调用失败的回调函数（识别失败、用户取消等情况下触发）	
complete	Function	否	接口调用结束的回调函数（调用成功、失败都会执行）	

success返回参数说明见表10-58。

表10-58 success返回参数说明

参数名	说 明	平台差异说明
result	扫码的内容	
scanType	扫码的类型	App、微信小程序、百度小程序、QQ小程序
charSet	扫码的字符集	App、微信小程序、百度小程序、QQ小程序
path	当所扫的码为当前应用的合法二维码时，会返回此字段，内容为二维码携带的path	App、微信小程序、百度小程序、QQ小程序

在pages文件夹下创建scancode/index.vue文件，该文件中的代码如下：

```
<template>
    <view>
        <button @click="scanCode1()">从相机和相册扫码</button>
        <button @click="scanCode2()">只允许通过相机扫码</button>
        <button @click="scanCode3()">只允许通过相册扫码</button>
        <view>条码类型:{{scanType}}</view>
        <view>条码内容:{{result}}</view>
    </view>
```

```
</template>

<script>
    export default {
        name: "scancode",
        data(){
            return {
                scanType:"",                          //条码类型
                result:""                             //条码内容
            }
        },
        methods:{
            //允许从相机和相册扫码
            scanCode1(){
                uni.scanCode({
                    success:  (res)=> {
                        console.log('条码类型:' + res.scanType);
                        console.log('条码内容:' + res.result);
                        this.scanType=res.scanType;
                        this.result=res.result;
                    }
                });
            },
            //只能从相机扫码
            scanCode2(){
                uni.scanCode({
                    onlyFromCamera: true,
                    success:  (res)=> {
                        console.log('条码类型:' + res.scanType);
                        console.log('条码内容:' + res.result);
                        this.scanType=res.scanType;
                        this.result=res.result;
                    }
                });
            },
            //调出条形码扫描
            scanCode3(){
                uni.scanCode({
                    scanType: ['barCode'],
                    success:  (res)=> {
                        console.log('条码类型:' + res.scanType);
                        console.log('条码内容:' + res.result);
                        this.scanType=res.scanType;
                        this.result=res.result;
```

```
                    }
                });
            },
        }
    }
</script>
```

该功能应在真机下测试。如果在真机下显示console.log()的内容，可在微信开发者工具中单击“真机调试”按钮，生成二维码，使用微信扫一扫测试，如图10-20所示。

图10-20 微信开发者工具真机调试

10.15 动画

uni-app的动画API可以实现复杂的动画效果，性能高于CSS动画，其动画队列可以轻松完成一组动画效果，方便而灵活。

扫一扫，看视频

使用uni.createAnimation创建一个动画实例 animation，然后通过调用实例的方法来描述动画，最后通过动画实例的export()方法导出动画数据，传递给组件的animation属性。

createAnimation的参数说明见表10-59。

表10-59 createAnimation的参数说明

参数名	类　型	是否必填	默认值	说　明
duration	Integer	否	400	动画持续时间，单位为 ms
timingFunction	String	否	"linear"	定义动画的效果
delay	Integer	否	0	动画延迟时间，单位为 ms
transformOrigin	String	否	"50% 50% 0"	设置 transform-origin

timingFunction有效值说明见表10-60。

表10-60 timingFunction有效值说明

值	说 明
linear	动画从头到尾的速度是相同的
ease	动画以低速开始，然后加快，在结束前变慢
ease-in	动画以低速开始
ease-in-out	动画以低速开始和结束
ease-out	动画以低速结束
step-start	动画第一帧就跳至结束状态直到结束
step-end	动画一直保持开始状态，最后一帧跳到结束状态

创建动画实例animation的代码示例如下：

```
let animation = uni.createAnimation({
    transformOrigin: "50% 50%",
    duration: 1000,
    timingFunction: "ease",
    delay: 0
})
```

animation动画实例可以调用表10-61~表10-66的方法来描述动画，调用结束后会返回自身，支持链式调用的写法。

表10-61 animation对象的方法列表的样式

方法名	参数名	说 明
opacity	value	透明度，参数范围为 0~1
backgroundColor	color	颜色值
width	length	长度值，如果传入 Number，则默认使用 px。可传入其他自定义单位的长度值
height	length	长度值，如果传入 Number，则默认使用 px。可传入其他自定义单位的长度值
top	length	长度值，如果传入 Number，则默认使用 px。可传入其他自定义单位的长度值
left	length	长度值，如果传入 Number，则默认使用 px。可传入其他自定义单位的长度值
bottom	length	长度值，如果传入 Number，则默认使用 px。可传入其他自定义单位的长度值
right	length	长度值，如果传入 Number，则默认使用 px。可传入其他自定义单位的长度值

表 10-62 旋转样式方法列表

方法名	参数名	说 明
rotate	deg	deg 的范围为 -180~180，从原点顺时针旋转一个 deg 角度
rotateX	deg	deg 的范围为 -180~180，在 X 轴旋转一个 deg 角度
rotateY	deg	deg 的范围为 -180~180，在 Y 轴旋转一个 deg 角度
rotateZ	deg	deg 的范围为 -180~180，在 Z 轴旋转一个 deg 角度
rotate3d	(x,y,z,deg)	deg 的范围为 -180~180，沿自定义轴旋转一个 deg 角度

表 10-63 缩放样式方法列表

方法名	参数名	说 明
scale	sx,[sy]	当为一个参数时，表示在 X 轴、Y 轴同时缩放 sx 倍数；当为两个参数时，表示在 X 轴缩放 sx 倍数，在 Y 轴缩放 sy 倍数
scaleX	sx	在 X 轴缩放 sx 倍数
scaleY	sy	在 Y 轴缩放 sy 倍数
scaleZ	sz	在 Z 轴缩放 sy 倍数
scale3d	(sx,sy,sz)	在 X 轴缩放 sx 倍数，在 Y 轴缩放 sy 倍数，在 Z 轴缩放 sz 倍数

表 10-64 偏移样式方法列表

方法名	参数名	说 明
translate	tx,[ty]	当为一个参数时，表示在 X 轴偏移 tx，单位为 px；当为两个参数时，表示在 X 轴偏移 tx，在 Y 轴偏移 ty，单位为 px
translateX	tx	在 X 轴偏移 tx，单位为 px
translateY	ty	在 Y 轴偏移 ty，单位为 px
translateZ	tz	在 Z 轴偏移 tz，单位为 px
translate3d	(tx,ty,tz)	在 X 轴偏移 tx，在 Y 轴偏移 ty，在 Z 轴偏移 tz，单位为 px

表 10-65 倾斜样式方法列表

方法名	参数名	说 明
skew	ax,[ay]	参数范围为 -180~180。当为一个参数时，Y 轴坐标不变，X 轴坐标顺时针倾斜 ax 度；当为两个参数时，分别在 X 轴倾斜 ax 度，在 Y 轴倾斜 ay 沿度
skewX	ax	参数范围为 -180~180，Y 轴坐标不变，X 轴坐标沿顺时针倾斜 ax 度
skewY	ay	参数范围为 -180~180，X 轴坐标不变，Y 轴坐标沿顺时针倾斜 ay 度

表 10-66 矩阵变形样式方法列表

方 法	参数名	说 明
matrix	(a,b,c,d,tx,ty)	同 CSS 样式的 transform-function matrix
matrix3d		同 CSS 样式的 transform-function matrix3d

在pages文件夹下创建animation/index.vue文件，该文件中的代码如下：

```
<template>
    <view>
        <view class="box" :animation="animationData"></view>
        <button @click="rotateAndScale">旋转同时放大</button>
        <button @click="rotateThenScale">先旋转后放大</button>
        <button @click="rotateAndScaleThenTranslate">先旋转同时放大，然后平移</button>
    </view>
</template>

<script>
    export default {
        name: "animation",
        data(){
            return {
                animationData:{}
            }
        },
        onLoad(){
            //创建animation实例
            this.animation = uni.createAnimation({
                duration: 2000,                //动画持续时间
                timingFunction: 'ease',        //动画效果
            })
            //链式调用实现动画队列，调用step()表示一组动画完成
            this.animation.scale(2,2).rotate(45).step();
            this.animation.scale(1,1).rotate(0).step({duration:1000});
            this.animationData=this.animation.export();
        },
        methods:{
            //旋转同时放大
            rotateAndScale(){
                this.animation.rotate(45).scale(2, 2).step();
                this.animationData = this.animation.export();
            },
            // 先旋转后放大
            rotateThenScale() {
                this.animation.rotate(45).step();
                this.animation.scale(2, 2).step();
                this.animationData = this.animation.export();
            },
```

```
            // 先旋转同时放大，然后平移
            rotateAndScaleThenTranslate() {
                this.animation.rotate(45).scale(2, 2).step();
                this.animation.translate(100, 100).step({ duration: 1000 });
                this.animationData = this.animation.export();
            }
        }
    }
</script>

<style scoped>
 .box{width:200rpx;height:200rpx;background-color:#FF0000}
</style>
```

注意上述加粗代码，在<view>组件添加animation属性，值为动画实例的export方法导出的动画数据，即this.animation.export()。调用动画操作方法后要调用step()来表示一组动画完成。可以在一组动画中调用任意多个动画方法，一组动画中的所有动画会同时开始，一组动画完成后才会进行下一组动画。step()可以传入一个与uni.createAnimation() 一样的配置参数，用于指定当前组动画的配置。

10.16 下拉刷新

下拉刷新功能在实际开发中经常用到，通常应用于首页、栏目页、订单页的数据刷新。

扫一扫，看视频

下拉刷新官方文档地址:https://uniapp.dcloud.io/api/ui/pulldown?id= onpulldownrefresh。

在pages.json文件中定义 onPullDownRefresh()处理函数（和onLoad()等生命周期函数同级），监听该页面用户下拉刷新事件。

使用onPullDownRefresh()函数前，需要在pages.json里找到当前页面的pages节点，并在style选项中开启enablePullDownRefresh，才能支持下拉刷新。

当处理完数据刷新后，使用uni.stopPullDownRefresh 可以停止当前页面的下拉刷新。

首先在pages.json文件中开启支持下拉刷新，新增代码如下：

```
{
    "pages": [
        ...
        {
            "path": "pages/downrefresh/index",
            "style": {
                "navigationBarTitleText": "下拉刷新",
                "enablePullDownRefresh": true        //开启下拉刷新
            }
```

```
        }
    ],
    ...
}
```

接下来，在pages文件夹下创建downrefresh/index.vue文件，该文件中的代码如下：

```
<template>
    <view>
        <view class="shop-wrap">
            <view class="shop-list" v-for="item in shops" :key="item.branch_
             shop_id">{{item.branch_shop_name}}</view>
        </view>
    </view>
</template>

<script>
    export default {
        name: "downrefresh",
        data(){
            return {
                shops:[]
            }
        },
        onLoad(){
                //开始下拉刷新
            uni.startPullDownRefresh({
              success:()=>{
                      this.getShopData();
              }
            });
        },
        //监听下拉刷新事件
        onPullDownRefresh() {
           //下拉刷新后，重新获取商铺数据列表
            this.getShopData(()=>{
                //数据加载完成后，停止下拉刷新
                uni.stopPullDownRefresh();
            });
        },
        methods:{
            //获取商铺数据列表
            getShopData(complete){
```

```
                uni.request({
                    url:"https://diancan.glbuys.com/api/v1/business/shop?pag
                        e=1&lng=113.123762&lat=39.836312"
                }).then(data=>{
                    let [err,res]=data;
                    if(res.data.code==200){
                        this.shops=res.data.data;
                    }else{
                        this.shops=[];
                    }
                    //回调函数
                    if(complete){
                        complete();
                    }
                });
            }
        }
    }
</script>

<style scoped>
    .shop-wrap{width:100%;}
    .shop-wrap .shop-list{width:100%;height:80rpx;border:1px solid #EFEFEF;}
</style>
```

在onLoad()钩子函数中使用uni.startPullDownRefresh可以实现在第一次访问页面时就进行下拉刷新，效果与用户手动下拉刷新一样。在onPullDownRefresh()函数中调用this.getShopData()方法重新获取服务端数据，数据获取完成后，在该方法的回调函数中调用uni.stopPullDownRefresh()停止当前页面下拉刷新，最终实现下拉刷新功能。

10.17 授权登录

授权登录可以实现多平台一站式登录。例如，获取微信小程序用户信息及openid，将openid存储到数据库中即可实现一站式登录。

授权登录官方文档地址:https://uniapp.dcloud.io/api/plugins/login。

扫一扫，看视频

10.17.1 登录

使用uni.login可以实现授权登录，如QQ登录、微博登录、微信登录等第三方平台登录。除了H5平台不支持该功能外，其他平台都支持。login的参数说明见表10-67。

表10-67 login的参数说明

参数名	类　型	是否必填	说　明	平台差异说明
provider	String	否	登录服务提供商	
scopes	String/Array	否	授权类型，默认为auth_base。其支持auth_base（静默授权）、auth_user（主动授权）、auth_zhima（芝麻信用）	支付宝小程序
timeout	Number	否	超时时间，单位为ms	微信小程序、百度小程序
success	Function	否	接口调用成功的回调函数	
fail	Function	否	接口调用失败的回调函数	
complete	Function	否	接口调用结束的回调函数（调用成功、失败都会执行）	

provider有效值说明见表10-68。

表10-68 provider有效值说明

值	说　明	备　注
weixin	微信登录	
qq	QQ登录	
sinaweibo	新浪微博登录	
xiaomi	小米登录	
apple	Apple登录	仅iOS 13、HBuilder X 2.4.7+支持

success返回参数说明见表10-69。

表10-69 success返回参数说明

参数名	说　明
authResult	登录服务商提供的登录信息，服务商不同，返回的结果不完全相同
code	小程序专有，用户登录凭证。开发者需要在开发者服务器后台使用code获取openid和session_key等信息
errMsg	描述信息

在pages文件夹下创建login/index.vue文件，该文件中的代码如下：

```
<template>
    <view>
        <button type="default" @click="doLogin()">微信登录</button>
    </view>
</template>

<script>
    export default {
        name: "login",
        methods:{
```

```
            //授权登录
            doLogin(){
                uni.login({
                    provider: 'weixin',
                    success:  (loginRes)=> {
                        let code=loginRes.code;      //用户登录凭证
                        //使用code获取openId，openId是微信用户唯一标识
                        let appId="wx1a2ccf9b9d1bb416"; //微信小程序的appId
                        let secret="8c1d726ecd003fa55208809a67021b0f";
                        //微信小程序的appSecret
                        uni.request({
                            url:"https://api.weixin.qq.com/sns/jscode2session?
                            appid="+appId+"&secret="+secret+"&js_code= "+code+"
                            &grant_ type=authorization_code"
                        }).then(data=>{
                            let [err,res]=data;
                            let openId=res.data.openid;
                            let sessionKey=res.data.session_key;
                            console.log(openId,sessionKey);
                        });
                    }
                });
            }
        }
    }
</script>
```

单击“微信登录”按钮，调用自定义方法doLogin()，在该方法内部调用uni.login API，provider参数的值为weixin。从success回调函数的参数中获取code，将code传入微信服务端API登录接口，该接口的地址为https://api.weixin.qq.com/sns/jscode2session?appid=APPID&secret=SECRET&js_code=JSCODE&grant_type=authorization_code。

使用uni.request API参数时，URL的值为服务端API登录接口地址，最终获取到的是openid和session_key。其实uni.login的最终目的就是获取openid和session_key。openid是用户唯一标识，将openid保存到数据库可以实现一站式登录，在项目实战开发中会详细讲解如何将其保存到数据库中以实现一站式登录。session_key是会话密钥，用于获取手机号，解密时需要使用session_key，如果忘记如何获取手机号可以复习9.6.2节。

服务端API登录接口的请求参数的属性说明见表10-70。

表10-70 请求参数的属性说明

属性名	类 型	说 明
appid	String	小程序 appId
secret	String	小程序 appSecret
js_code	String	uni.login 获取的 code
grant_type	String	授权类型，此处只需填写 authorization_code

服务端API登录接口的返回值是一个JSON数据包，表10-71为属性说明。

表10-71 JSON数据包的属性说明

属性名	类 型	说 明
openid	String	用户唯一标识
session_key	String	会话密钥
unionid	String	用户在开放平台的唯一标识符，在满足 UnionID 下发条件时会返回
errcode	Number	错误码
errmsg	String	错误信息

errcode的合法值说明见表10-72。

表10-72 errcode的合法值

值	说 明
-1	系统繁忙，此时请开发者稍后再试
0	请求成功
40029	code 无效
45011	频率限制，每个用户 100 次 /min

需要注意的是，使用“微信服务端API登录接口”时必须在微信开发者工具中勾选“不校验合法域名、web-view（全业务域名）、TLS版本以及HTTPS证书”复选框，否则无法使用此接口。在真机测试中无法直接使用此接口，必须使用服务端程序（如PHP、Java、Nodejs等语言）调用此接口。如果读者对服务端语言不太熟悉，可以使用本书的接口文档进行真机测试。关注“html5程序思维”公众号，发送“uniapp”即可获取接口文档下载地址。

获取微信小程序的AppId和AppSecret的流程：首先登录微信小程序管理平台，登录地址为https://mp.weixin.qq.com。登录后单击左侧导航栏中的“开发”按钮，进入开发页面，单击“开发设置”按钮，找到“开发者ID”选项，即可获取微信小程序的AppId和AppSecret，如图10-21所示。

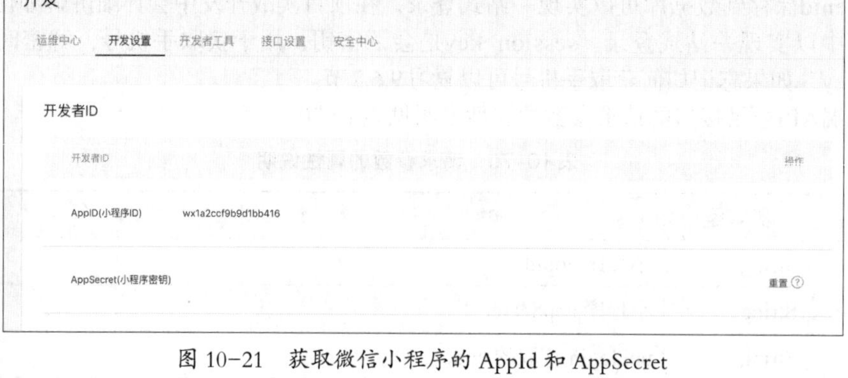

图 10-21 获取微信小程序的 AppId 和 AppSecret

10.17.2 获取用户信息

使用uni.getUserInfo可以获取用户信息。在微信小程序端，当用户未授权时，调用此接口不会出现授权弹窗，而会直接进入fail回调；当用户已授权时，调用此接口可成功获取用户信息。getUserInfo的参数说明见表10-73。

表10–73 getUserInfo的参数说明

参数名	类型	是否必填	说明	平台差异说明
provider	String	否	登录服务提供商，见10.17.1节provider有效值说明（表10-68）	
withCredentials	Boolean	否	是否带上登录态信息	微信小程序、字节跳动小程序
lang	String	否	指定返回用户信息的语言，有效值为zh_CN、zh_TW、en，默认为en	微信小程序
timeout	Number	否	超时时间，单位为ms	微信小程序
success	Function	否	接口调用成功的回调函数	
fail	Function	否	接口调用失败的回调函数	
complete	Function	否	接口调用结束的回调函数（调用成功、失败都会执行）	

注意：

在小程序withCredentials为true时或是在App调用uni.getUserInfo时，要求此前调用过uni.login且登录状态尚未过期。

success返回参数说明见表10-74。

表10–74 success返回参数说明

参数名	类型	说明	平台差异说明
userInfo	Object	用户信息对象	
rawData	String	不包括敏感信息的原始数据字符串，用于计算签名	
signature	String	使用sha1(rawData + sessionkey)得到字符串，用于校验用户信息	微信小程序、字节跳动小程序
encryptedData	String	包括敏感数据在内的完整用户信息的加密数据	微信小程序、字节跳动小程序
iv	String	加密算法的初始向量	微信小程序、字节跳动小程序
errMsg	String	描述信息	

userInfo参数说明见表10-75。

表10-75 userInfo参数说明

参数名	类 型	说 明	平台差异说明
nickName	String	用户昵称	
openId	String	该服务商唯一用户标识	App
avatarUrl	String	用户头像	

注意：

除了以上3个必有的信息外，不同服务供应商返回的其他信息会存在差异。

接下来实现一个微信授权登录并获取用户信息的功能。在pages/login/index.vue文件中的代码如下：

```
<template>
    <view>
        <button type="default" @click="doLogin()">微信登录</button>
    </view>
</template>

<script>
    export default {
        name: "login",
        methods:{
            //授权登录
            doLogin(){
                uni.login({
                    provider: 'weixin',
                    success:  (loginRes)=> {
                        let code=loginRes.code;            //用户登录凭证
                        //使用code获取openId，openId是微信用户唯一标识
                        let appId="wx1a2ccf9b9d1bb416";      //微信小程序appId
                        let secret="8c1d726ecd003fa55208809a67021b0f";
                        //微信小程序的appSecret
                        uni.request({
                            url:"https://api.weixin.qq.com/sns/jscode2session?
                            appid="+appId+"&secret="+secret+"&js_code="+code+
                            "&grant_type=authorization_code"
                        }).then(data=>{
                            let [err,res]=data;
                            let openId=res.data.openid;
                            let sessionKey=res.data.session_key;
```

```
                    });

                    //获取用户信息
                    uni.getUserInfo({
                        provider: 'weixin',
                        success: (infoRes)=> {
                            console.log(infoRes.userInfo.nickName) //用户昵称
                            console.log(infoRes.userInfo.avatarUrl)//头像地址
                            console.log(infoRes.userInfo.country)  //国家
                            console.log(infoRes.userInfo.province) //省份
                            console.log(infoRes.userInfo.city)     //城市
                            console.log(infoRes.userInfo.gender)
                            //性别。值为1代表男，值为2代表女
                        }
                    });

                }
            });
        }
    }
}
</script>
```

以上加粗代码为新增代码，获取的用户信息通常写在uni.login的回调函数success()内部。

10.18 微信支付

uni-app的在线支付功能非常强大，支持各平台的客户端支付API，如支付宝、微信、百度收银台、苹果等支付平台。

扫一扫，看视频

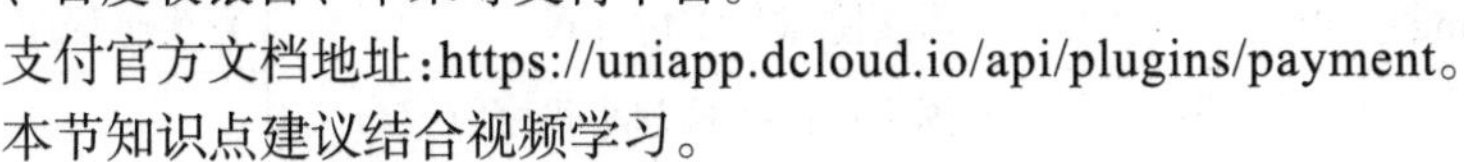

支付官方文档地址:https://uniapp.dcloud.io/api/plugins/payment。

本节知识点建议结合视频学习。

10.18.1 支付流程及思路

在开发微信支付功能之前，首先介绍支付流程。微信支付需要服务端和前端配合完成，在微信支付官网（https://pay.weixin.qq.com）申请并开通微信支付。只有企业可以申请微信支付，个人无法申请。完成微信支付申请流程后，根据提示操作即可。

本书已经申请好了微信支付，并且提供了服务端微信支付API接口。关注“html5程序思维”公众号，发送“uniapp”，即可获取接口文档下载地址。

微信支付开发流程如下：

(1)在微信小程序内调用登录接口，获取用户的openid。

（2）在微信小程序内使用uni.request，调用服务端“统一下单”API接口。

（3）在微信小程序内使用uni.requestPayment，完成微信支付。

服务端开发者需要了解“统一下单”API接口，其官方文档地址为https://pay.weixin.qq.com/wiki/doc/api/wxa/wxa_api.php?chapter=9_1&index=1。

微信支付官方代码示例下载地址：https://pay.weixin.qq.com/wiki/doc/api/jsapi.php?chapter=11_1。

目前微信官网提供的代码示例支持的服务端开发语言为Java、C#、PHP。

10.18.2 完成微信支付

uni.requestPayment是一个统一各平台的客户端支付API，支持各平台小程序和App，不支持H5平台。

requestPayment的参数说明见表10-76。

表10-76 requestPayment的参数说明

参数名	类 型	是否必填	说 明	平台差异说明
provider	String	是	服务提供商	
orderInfo	String/Object	微信小程序不用填	订单数据	App、支付宝小程序、百度小程序、字节跳动小程序
timeStamp	String	微信小程序必填	时间戳从1970年1月1日至今的秒数，即当前的时间	微信小程序
nonceStr	String	微信小程序必填	随机字符串，长度为32个字符以下	微信小程序
package	String	微信小程序必填	统一下单接口返回的prepay_id参数值，提交格式如prepay_id=xx	微信小程序
signType	String	微信小程序必填	签名算法，暂支持MD5	微信小程序
paySign	String	微信小程序必填	签名	
bannedChannels	Array<String>	否	需要隐藏的支付方式	百度小程序
service	Number	字节跳动小程序必填	固定值：1（拉起小程序收银台）。开发者如果不希望使用字节跳动小程序收银台，可将service设置为3或4，可以直接拉起微信或支付宝进行支付。service=3：微信API支付，不拉起小程序收银台；service=4：支付宝API支付，不拉起小程序收银台。其中service=3或4，仅在1.35.0.1+基础库（头条743+）中支持	字节跳动小程序

续表

参数名	类　型	是否必填	说　明	平台差异说明
_debug	Number	否	仅限调试用，上线前删除该参数。当 _debug=1 时，微信支付期间可以看到中间报错信息，方便调试	字节跳动小程序
getOrderStatus	Function	字节跳动小程序必填	商户前端实现的查询支付订单状态方法（该方法需要返回 Promise 对象）。service=3 或 4 时不需要传递	字节跳动小程序
success	Function	否	接口调用成功的回调函数	
fail	Function	否	接口调用失败的回调函数	
complete	Function	否	接口调用结束的回调函数（调用成功、失败都会执行）	

provider有效值说明见表10-77。

表10–77　provider有效值说明

值	说　明	备　注
alipay	支付宝支付	
wxpay	微信支付	
baidu	百度收银台	
appleiap	苹果应用内支付	iOS 应用打包后可获取

orderInfo注意事项如下：

(1) 百度小程序的 orderInfo 为 Object 类型，详细的数据结构可参考百度收银台支付，官方地址为https://smartprogram.baidu.com/docs/develop/api/open/payment_swan-requestPolymerPayment。

(2) 支付宝小程序的 orderInfo（支付宝的规范为 tradeNO）为 String 类型，表示支付宝交易号。

(3) 字节跳动小程序的orderInfo为Object类型，该支付接口已更新，大家可以使用tt.pay判断是否支持新版本接口。

(4) App端，支付宝支付 orderInfo 为 String 类型。

(5) App端，微信支付 orderInfo 为 Object 类型。

(6) App端，苹果应用内支付 orderInfo 为Object 类型，{productid: 'productid'}。

在开发微信支付功能之前，必须先在微信支付管理后台授权微信小程序的AppId。这里给读者提供一个已授权的微信支付AppId：wx1a2ccf9b9d1bb416，打开mainifest.json文件，修改AppId为此AppId即可。

在pages文件夹下创建pay/wx.vue文件，该文件中的代码如下：

```
<template>
    <view>
        金额:<input type="text" placeholder="请输入金额" class="price"
             v-model="price" /> 元
        <button type="primary" class="pay-btn" @click="goPay()">支付</button>
    </view>
</template>

<script>
    export default {
        name: "wxpay",
        data(){
            return {
                price:""
            }
        },
        methods:{
            //去支付
            goPay(){
                if(this.price.trim()==''){
                    uni.showToast({
                        title:"请输入支付金额",
                        duration: 2000,
                        icon:"none"
                    })
                    return;
                }
              //1.授权登录获取openId
                uni.login({
                    provider: 'weixin',
                    success:  (loginRes)=> {
                        let code=loginRes.code;      //用户登录凭证
                        //使用code获取openId，openId是微信用户唯一标识
                        let url="https://diancan.glbuys.com/api/v1/wechat_openid";
                        uni.request({
                            url:url,
                            method:"POST",
                            header:{
                                "content-type":"application/x-www-form-urlencoded"
                            },
                            data:{code:code}
```

```
}).then(data=>{
    let [err,res]=data;
    let openId=res.data.data.openid;   //获取openId

    //微信支付
    //2.调用服务端"统一下单"接口
    let orderUrl="https://diancan.glbuys.com/api/v1/
        wxpay/wechat_unifiedorder";
    uni.request({
        url:orderUrl,
        method: "POST",
        header:{
            "content-type":"application/x-www
             -form-urlencoded"
        },
        data:{
            open_id:openId,           //用户登录的openId
            ordernum:this.getOrderNum(),  //订单编号
            price:this.price,         //金额
            notify_url:"https://diancan.glbuys.com",
            //此回调地址为测试地址，不具有任何实际意义
            body:"测试支付"              //商品描述
        }
    }).then(data=>{
        let [err,res]=data;
        //通过“统一下单”接口获取以下值
        let orderRes=res.data.data;
        let timeStamp=orderRes.timeStamp; //时间戳
        let nonceStr=orderRes.nonceStr;    //随机字符串
        let packages=orderRes.package;
        //统一下单接口返回的prepay_id参数值
        let sign=orderRes.paySign;          //签名
        //3.调起客户端微信支付
        uni.requestPayment({
            provider: 'wxpay',
            timeStamp: timeStamp.toString(),
            //将时间戳转成字符串类型
            nonceStr: nonceStr,
            package: packages,
            signType: 'MD5',
            paySign: sign,
```

```
                                        success:(res)=>{
                                            console.log(res,"支付成功");
                                        }
                                    })
                                });

                            });


                        }
                    });
                },
                //生成订单编号
                getOrderNum(){
                    return new Date().getTime();
                }
            }
        }
    </script>

    <style scoped>
        .price{width:250rpx;height:50rpx;}
        .pay-btn{width:150rpx;height:70rpx;border-radius: 5px;font-size:28rpx;}
        .pay-btn:after{border:0 none;}
    </style>
```

以上代码为完整支付代码，注释很清晰，请读者仔细阅读。在开发时AppId必须为wx1a2ccf9b9d1bb416，因为此AppId是经过微信支付授权过的。特别提醒：**此支付为真实支付，请谨慎填写支付金额，建议填写0.01元。**

10.19 节点信息

扫一扫，看视频

在实际开发中，需要获取节点信息来得到元素的宽、高及布局的位置。在传统Vue中可以使用Refs获取DOM得到节点信息；在uni-app中为了兼容各种平台，采用小程序获取节点方式来获取节点信息。

节点信息官方文档地址:https://uniapp.dcloud.io/api/ui/nodes-info。

10.19.1 获取单个节点

使用uni.createSelectorQuery返回一个 SelectorQuery 对象实例。可以在该实例上使用select()方法获取单个节点信息。

在pages文件夹下创建selectquery/index.vue文件，该文件中的代码如下：

```
<template>
    <view>
        <view id="box" class="box">box1</view>
    </view>
</template>

<script>
    export default {
        name: "selectquery",
        onReady(){
            let selectQuery = uni.createSelectorQuery();
            selectQuery.select("#box").boundingClientRect((data)=>{
                console.log("得到布局位置信息:" + JSON.stringify(data));
                console.log("节点的上边界坐标:" + data.top);
                console.log("节点的下边界坐标:" + data.bottom);
                console.log("节点的左边界坐标:" + data.left);
                console.log("节点的右边界坐标:" + data.right);
                console.log("节点的宽:" + data.width);
                console.log("节点的高:" + data.height);
            }).exec();
        }
    }
</script>

<style scoped>
 .box{width:200rpx;height:200rpx;background-color:#007aff;margin-top:20rpx;text-
  align: center; line-height:200rpx;f ont-size:32rpx; color:#FFFFFF;}
</style>
```

注意加粗代码，select()方法表示在当前页面下选择第一个匹配选择器selector的节点，返回一个NodesRef对象实例，可以用于获取节点信息。selector类似于CSS的选择器，但仅支持下列语法：

（1）ID选择器:#the-id。

（2）class选择器（可以连续指定多个）:.a-class.another-class。

（3）子元素选择器:.the-parent > .the-child。

（4）后代选择器:.the-ancestor .the-descendant。

（5）跨自定义组件的后代选择器:.the-ancestor >>> .the-descendant。

（6）多选择器的并集:#a-node, .some-other-nodes。

boundingClientRect(callback)方法表示节点布局位置的查询请求，相对于显示区域，以px为单位，其功能类似于DOM的getBoundingClientRect。callback返回参数的属性说明见表10-78。

表10-78 callback返回参数的属性说明

属性名	类 型	说 明
id	String	节点的 ID
dataset	Object	节点的 dataset
left	Number	节点的左边界坐标
right	Number	节点的右边界坐标
top	Number	节点的上边界坐标
bottom	Number	节点的下边界坐标
width	Number	节点的宽度
height	Number	节点的高度

10.19.2 获取多个节点

在实际开发中经常会遇到一次获取多个节点信息的情况，可以使用selectAll()方法获取匹配选择器selector的所有节点，返回一个NodesRef对象实例，用于获取节点信息。

在pages/selectquery文件夹下创建selectall.vue文件，该文件中的代码如下：

```
<template>
    <view>
        <view id="box1" class="box">box1</view>
        <view id="box2" class="box">box2</view>
        <view id="box3" class="box">box3</view>
    </view>
</template>

<script>
    export default {
        name: "selectquery",
        onReady(){
            let selectQuery = uni.createSelectorQuery();
            selectQuery.selectAll(".box").boundingClientRect((data)=>{
                console.log(data);          //返回的是一个数组
                //循环输出
                for(let i=0;i<data.length;i++){
                    console.log("id:" + data[i].id);
                    console.log("节点的上边界坐标:" + data[i].top);
                    console.log("节点的下边界坐标:" + data[i].bottom);
                    console.log("节点的左边界坐标:" + data[i].left);
                    console.log("节点的右边界坐标:" + data[i].right);
```

```
                    console.log("节点的宽:" + data[i].width);
                    console.log("节点的高:" + data[i].height);
                }
            }).exec();
        }
    }
</script>

<style scoped>
  .box{width:200rpx;height:200rpx;background-color: #007aff; margin-top:20rpx;text-
    align: center; line-height: 200rpx; font-size:32rpx;color:#FFFFFF;}
</style>
```

selectAll和select选择器使用方式基本相同，唯一不同的是boundingClientRect的回调方法返回的参数是一个数组。上述代码中，exec(1)方法表示执行所有的请求。

10.19.3 在组件内获取节点信息

使用selectorQuery.in(component)将选择器的选取范围更改为在自定义组件 component 内，返回一个 SelectorQuery 对象实例。

在components文件夹下创建selectquery/index.vue文件，该文件中的代码如下：

```
<template>
    <view>
        <view class="box"></view>
    </view>
</template>

<script>
    export default {
        name: "selectquery",
        onReady(){
            let selectQuery = uni.createSelectorQuery().in(this);
            selectQuery.select(".box").boundingClientRect((data)=>{
                console.log("组件内的布局位置信息:" + JSON.stringify(data));
                console.log("组件内节点的上边界坐标:" + data.top);
                console.log("组件内节点的下边界坐标:" + data.bottom);
                console.log("组件内节点的左边界坐标:" + data.left);
                console.log("组件内节点的右边界坐标:" + data.right);
                console.log("组件内节点的宽:" + data.width);
                console.log("组件内节点的高:" + data.height);
            }).exec();
        }
```

```
    }
</script>

<style scoped>
    .box{width:100rpx;height:100rpx;background-color:#007aff;}
</style>
```

注意上述加粗代码，使用了in(this)获取组件内节点信息。

10.20 调试

扫一扫，看视频

在实际开发中，调试非常重要。调试的环境分为两种：开发（测试）环境和生产（正式）环境，开发环境可以在微信开发者工具中调试，也可以使用真机调试，都可以使用console.log()在控制台输出内容。

调试官方文档地址：https://uniapp.dcloud.io/api/other/set-enable-debug。

使用uni.setEnableDebug设置是否打开调试开关，此开关支持正式版和测试版。其支持的平台有微信小程序、百度小程序、QQ小程序，不支持的平台有App、H5、支付宝小程序、字节跳动小程序。

setEnableDebug的参数说明见表10-79。

表10-79 setEnableDebug的参数说明

属性名	类 型	是否必填	说 明	平台差异说明
enableDebug	Boolean	是	是否打开调试	
success	Function	否	接口调用成功的回调函数	微信小程序
fail	Function	否	接口调用失败的回调函数	微信小程序
complete	Function	否	接口调用结束的回调函数（调用成功、失败都会执行）	微信小程序

此代码需要写在App.vue文件中的onLaunch()生命周期钩子函数中，代码如下：

```
onLaunch: function() {
    //设置调试
    uni.setEnableDebug({
        enableDebug: true       //值为true表示打开调试，值为false表示关闭调试
    })
},
```

这样无论是在开发环境还是在生成环境中，用手机打开小程序都会显示console控制台，以便于调试。

10.21 小结

本章主要讲解了常用API，包括request发起请求、uploadFile文件上传、数据缓存、获取位置、获取系统信息、交互反馈、动态设置导航条、动态设置tabBar、录音管理、视频组件控制、网络状态、剪贴板、拨打电话、扫码、动画、下拉刷新、授权登录、微信支付、节点信息、调试等。其中，最常用且必会的API有request、获取位置、获取系统信息、交互反馈、动态设置tabBar、动态设置导航条、下拉刷新、授权登录、节点信息、微信支付等。本章内容比较多，建议看视频，多练习，记住它们的用法。

第11章

仿美团点餐小程序客户端开发

本章将进入项目实战阶段，会用到之前学过的知识点。项目实战会提高我们的各种能力，包括解决问题能力、逻辑思维能力和学习能力，同时也会帮助我们深度理解组件封装、Vuex异步数据流、在线支付、授权登录、获取手机号、获取位置查找附近商铺、前后端数据交互、获取节点信息等实用功能。

由于本项目难度比较高，因此强烈建议各位读者结合视频进行学习。

11.1　扫二维码欣赏完整项目

扫描图11-1所示的二维码，欣赏实战项目演示效果。

图 11-1　点餐小程序案例欣赏

此项目为真实运营项目，数据信息均真实有效，支持堂食、自提、配送点餐功能。小程序为客户端，H5公众号版为商家端，还有一个后台管理为平台运营商，本章实战项目为点餐小程序客户端。为了让读者体验开发项目的真实性，我们会搭建一个测试的服务器及数据库供读者使用，但其安全性可能会有所下降，请读者见谅。接口文档、项目所需素材（如图片和项目源码）可以关注 **“html5程序思维”公众号，然后发送“uniapp”，即可获取下载地址。**

注意：

此项目的UI参考地址为https://diancan.lucklnk.com。该地址是H5公众号版本的点餐系统，与点餐小程序功能一样，需要用微信开发者工具的“公众号网页调试”命令进行访问，如图11-2所示。

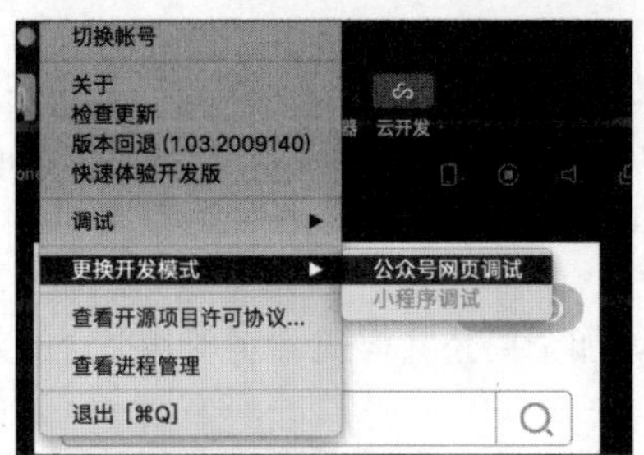

图 11-2　选择“公众号网页调试”命令

此项目的后台管理地址为https://diancan.glbuys.com/hadmin.php，用户名为demo，密码为123456。在开发中为了方便数据测试，可以自己进入后台修改或添加数据。

注意：

不要添加不良信息，不要随意删除数据或进行非法手段攻击。

11.2　首页

首页是小程序启动访问的第一个页面，其主要功能是获取位置并显示附近的商家。

首页的接口在“接口文档”文件夹下的“商家.docx”文件中。

扫一扫，看视频

11.2.1 首页布局

首先，使用HBuilder X创建一个新的项目，项目名称为ordering_app，在pages文件下创建main/main.vue文件，创建完成后配置路由。pages.json的代码如下：

```
{
    "pages": [
        {
            "path": "pages/main/main",
            "style": {
                "navigationBarTitleText": "点餐系统",
                "navigationStyle": "custom",
                "enablePullDownRefresh": true
            }
        }
    ],
    "globalStyle": {
        "navigationBarTextStyle": "white",
        "navigationBarBackgroundColor": "#E30019",
        "backgroundColor": "#FFFFFF",
        "onReachBottomDistance": 50
    }
}
```

navigationStyle属性值为custom，设置成自定义导航；enablePullDownRefresh属性值为true，表示支持下拉刷新；onReachBottomDistance属性值为50，表示上拉触底事件触发时距页面底部距离，单位为px。

其次，在App.vue文件中增加全局公共样式，实现各平台的兼容性，代码如下：

```
<script>
    export default {
        onLaunch: function() {

        },
        onShow: function() {

        },
        onHide: function() {

        }
    }
```

```
</script>

<style>
    /*每个页面公共CSS */
    page {
        width:100%;
        min-height: 100%;
        font-size: 16px;
        display: flex;
    }

    /*#ifdef MP-BAIDU*/
    page {
        width: 100%;
        height: 100%;
        display: block;
    }

    swan-template {
        width: 100%;
        min-height: 100%;
        display: flex;
    }

    /*原生组件模式下需要注意组件外部样式*/
    custom-component {
        width: 100%;
        min-height: 100%;
        display: flex;
    }

    /*#endif*/

    /*#ifdef MP-ALIPAY*/
    page {
        min-height: 100vh;
    }

    /*#endif*/

    /*原生组件模式下需要注意组件外部样式*/
    m-input {
```

```
        width: 100%;
        /*min-height: 100%;*/
        display: flex;
        flex: 1;
    }

    input,button{border:0 none;outline: none;border-radius: 0px;}
    button:after{border:0 none;outline: none;border-radius: 0px;}
    .status_bar {
        height: var(--status-bar-height);
        width: 100%;
    }
    .no-data{width:100%;text-align:center;font-size:32rpx;position:
     absolute;left:50%;top:30%;z-index:3;transform: translate(-50%,-80%)}
</style>
```

最后，开发首页布局，pages/main/main.vue文件中的代码如下：

```
<template>
    <view class="page">
        <view class="status_bar bg-color"></view>
        <view class="header">
            <view :class="{'search-header':true,'ipx':false}" >
                <view class="search-wrap">
                    <view class="icon"></view>
                    <view class="search">输入商家名或菜品</view>
                </view>
            </view>
        </view>
        <view :class="{'shop-main':true,ipx:false}">
            <view class="shop-list">
                <view class="shop-wrap">
                    <view class="image"><image src="https://diancan.lucklnk.
                    com/businessfiles/logo/1600743948.jpg" :lazy-load="true">
                     </image></view>
                    <view class="shop-info">
                        <view class="shop-name">精武鸭脖</view>
                        <view class="distance">0.5km</view>
                        <view class="address">山西怀仁商业步行街北口对面</view>
                        <view class="pack-btn">自提</view>
                    </view>
                </view>
            </view>
```

```
            <view class="shop-list">
                <view class="shop-wrap">
                    <view class="image"><image src="https://diancan.lucklnk.
                    com/businessfiles/logo/1600743948.jpg" :lazy-load="true">
                     </image></view>
                    <view class="shop-info">
                        <view class="shop-name">精武鸭脖</view>
                        <view class="distance">0.5km</view>
                        <view class="address">山西怀仁商业步行街北口对面</view>
                        <view class="pack-btn">自提</view>
                    </view>
                </view>
            </view>
            <view class="shop-list">
                <view class="shop-wrap">
                    <view class="image"><image src="https://diancan.lucklnk.
                    com/businessfiles/logo/1600743948.jpg" :lazy-load="true">
                     </image></view>
                    <view class="shop-info">
                        <view class="shop-name">精武鸭脖</view>
                        <view class="distance">0.5km</view>
                        <view class="address">山西怀仁商业步行街北口对面</view>
                        <view class="pack-btn">自提</view>
                    </view>
                </view>
            </view>
        </view>
    </view>
</template>

<script>
    export default {
        name: "main-component"
    }
</script>

<style scoped>
    .page{width:100%;min-height:100vh;overflow:hidden;}
    .status_bar.bg-color{background-color:#E30019;}
    .header{width:100%;background-color:#eb1625;overflow:hidden;position: fixed;
     left:0;top:0;z-index:90;}
```

```
    .search-header{width:100%;height:170rpx;display: flex;justify-content:
    center;align-items: flex-end;margin-top:40rpx;box-sizing:
    border-box;padding-bottom:20rpx;}
    .search-header.ipx{height:210rpx;}
    .header .search-wrap{width:80%;height:52rpx;background-color:rgba(255,255
    ,255,0.9);border-radius: 5px;display:flex;justify-content: start;align-
    items: center;}
    .header .search-wrap .icon{width:44rpx;height:44rpx;background-image:url("~@/
    static/images/main/search_icon.png");background-size:100%;background-
    repeat: no-repeat;background-position: center;margin:0 20rpx;}
    .header .search-wrap .search{font-size:28rpx;color:#999999;}
    .shop-main{width:100%;margin-top:220rpx;}
    .shop-main.ipx{margin-top:260rpx;}
    .shop-main .shop-list{width:100%;border-bottom: 1px solid #EFEFEF;box-
    sizing: border-box;padding:20rpx 0;}
    .shop-main .shop-list .shop-wrap{width:92%;height:auto;margin:0
    auto;display:flex;justify-content: start;}
    .shop-main .shop-list .shop-wrap .image{width:160rpx;height:160rpx;margi
    n-right:20rpx;}
    .shop-main .shop-list .shop-wrap .image image{width:100%;height:100%;bord
    er-radius: 5px;}
    .shop-main .shop-list .shop-info{width:72%;clear: both;}
    .shop-main .shop-list .shop-info .shop-name{font-size:32rpx;font-weight:
    bold;width:100%;height:45rpx;white-space: nowrap;overflow:hidden;text-
    overflow: ellipsis;}
    .shop-main .shop-list .shop-info .distance{font-size: 28rpx;color:
    #666666;margin-top:10rpx;}
    .shop-main .shop-list .shop-info .address{font-size: 28rpx;color:
    #666666;margin-top:10rpx;width:100%;height:45rpx;white-space:
    nowrap;overflow:hidden;text-overflow: ellipsis;}
    .shop-main .shop-list .shop-info .pack-btn{padding:10rpx 20rpx;font-size
    :28rpx;color:#FFFFFF;background-color:#eb1625;width:auto;height:auto;display:tab
    le;margin-top:10rpx;border-radius: 5px;float:right;}
    .shop-main .shop-list .shop-info .pack-btn.disable {color:
    #666666;background-color:#EFEFEF}
</style>
```

以上代码很简单，没有业务逻辑，只是一个首页布局。此项目为微信小程序，在微信开发者工具中预览的效果如图 11-3 所示。

图 11-3　点餐系统首页布局

11.2.2　异步数据流对接配合地图定位显示附近的商铺

扫一扫，看视频

首页布局完成，接下来对接服务端数据。在开发Vue的实战项目时都会将服务端请求的数据存放到Vuex中，然后从Vuex中渲染数据到视图。uni-app开发也是如此，uni-app已内置Vuex，无须安装即可使用。首先配置Vuex，在项目根目录创建store文件夹，在该文件夹内创建index.js文件，该文件中的代码如下：

```
import Vue from 'vue'
import Vuex from 'vuex';

Vue.use(Vuex);

const store = new Vuex.Store({
    modules:{}
});

export default store
```

接下来，将Vuex注册到Vue中，main.js文件中的代码如下：

```
import Vue from 'vue'
import App from './App'
```

```
import store from "./store";

Vue.config.productionTip = false

App.mpType = 'app'

const app = new Vue({
    store,
    ...App
})
app.$mount()
```

以上加粗代码为新增代码，这样Vuex就配置完成了。

下面封装request()方法，目的是更简便地请求服务端数据。在static文件夹下创建js/utils/request.js文件，该文件中的代码如下：

```
function request(url,method="GET",data={}){
    return new Promise((resolve, reject)=>{
        uni.request({
            url: url,
            method:method.toLocaleUpperCase(),
            data: data,
            header:{"Content-Type": "application/x-www-form-urlencoded"},
            success: (res) => {
                resolve(res.data);
            },
            fail:(res)=>{
                reject(res);
            }
        });
    })
}
export {
    request
}
```

上述代码使用Promise()方法封装了uni-app的uni.request API，自己封装的request使用起来会更方便、更灵活。上述封装代码比较简单，不再赘述。

接下来配置公共的参数，其中存放接口地址。在js文件中创建conf/config.js文件，该文件中的代码如下：

```
let baseApi="https://diancan.glbuys.com/api";
export default {
    baseApi:baseApi
}
```

变量baseApi的值为服务端接口请求地址。请求服务端数据，在项目根目录创建api文件夹，在该文件夹下创建business/index.js文件，该文件中的代码如下：

```
import config from "../../static/js/conf/config";
import {request} from "../../static/js/utils/request";

//显示首页商铺
export function getShopData(data){
    return request(config.baseApi+"/v1/business/shop","get",data);
}
```

getShopData()函数可以从服务端接口获取商铺信息。接下来与Vuex对接，在store文件夹下创建business/index.js文件，该文件中的代码如下：

```
//导入服务端接口的请求方法
import {getShopData} from "../../api/business";
export default {
    namespaced:true,
    state:{
        shops:[]
    },
    mutations: {
        //设置商铺数据
        ["SET_SHOPS"](state,payload){
            state.shops=payload.shops;
        },
        //设置商铺分页数据
        ["SET_SHOPS_PAGE"](state,payload){
            state.shops.push(...payload.shops);
        }
    },
    actions:{
        //获取首页商铺数据
        getShop(conText,payload){
            getShopData(payload).then(res=>{
                if(res.code==200){
                    conText.commit("SET_SHOPS",{shops:res.data});
                    //请求成功回调
                    if(payload.success){
                        //res.pageinfo.pagenum总页数
                        payload.success(res.pageinfo.pagenum);
                    }
                }else{
```

```
                    conText.commit("SET_SHOPS",{shops:[]});
                }
                //请求完成回调
                if(payload.complete){
                    payload.complete();
                }
            });
        },
        //获取首页商铺分页数据
        getShopPage(conText,payload){
            getShopData(payload).then(res=>{
                if(res.code==200){
                    conText.commit("SET_SHOPS_PAGE",{shops:res.data});
                }
            });
        }
    }
}
```

由于我们的首页是自定义导航，需要先解决iPhone X“刘海”的兼容性问题，才能将Vuex的数据渲染到视图中，在10.5.2节中已经学习了如何解决这一问题，这里不详细讲解。由于这个项目会有多个页面需要自定义导航，因此我们需要一个全局变量去识别是否在iPhone X中，全局首选Vuex，接下来在store文件夹下创建system/index.js文件，在该文件中的代码如下：

```
export default {
    namespaced: true,
    state: {
        isIpx:false,            //是否为iPhone X
    },
    mutations:{
        //设置isIpx
        ["SET_IPX"](state,payload){
            state.isIpx=payload.isIpx;
        }
    }
}
```

注册到Vuex的模块中，在store/index.js文件中新增代码如下：

```
import Vue from 'vue'
import Vuex from 'vuex';
import business from "../business";
import system from "../system";
```

```
Vue.use(Vuex);

const store = new Vuex.Store({
    modules:{
        business,
        system
    }
});

export default store
```

上述加粗代码为新增代码。接下来判断是否在iPhone X中进行，在App.vue文件中新增代码如下：

```
onLaunch: function() {
    //#ifndef H5
    uni.getSystemInfo({
        success: (res) =>{
            var name = 'iPhone X';
            //判断是否为iPhone X
            if(res.model.indexOf(name) > -1){          //如果在iPhone X中
                //调用store/system/index.js文件中mutations内部的
                //SET_IPX()方法，将isIpx设置为true
                this.$store.commit("system/SET_IPX",{isIpx:true});
            }
        }
    });
    //#endif
},
```

将Vuex的数据对接到pages/main/main.vue文件中，该文件中的代码如下：

```
<template>
    <view class="page">
        <view class="status_bar bg-color"></view>
        <view class="header">
            <view :class="{'search-header':true,'ipx':isIpx}">
                <view class="search-wrap">
                    <view class="icon"></view>
                    <view class="search">输入商家名或菜品</view>
                </view>
            </view>
        </view>
        <view :class="{'shop-main':true,ipx:isIpx}">
```

```
            <view class="shop-list" v-for="(item,index) in shops" :key
            ="index" @click="goPage('/pages/goods/index?branch_shop_
            id='+item.branch_shop_id+'')">
                <view class="shop-wrap">
                    <view class="image"><image :src="item.logo"
                    :lazy-load="true"></image></view>
                    <view class="shop-info">
                        <view class="shop-name">{{item.branch_shop_name}}</view>
                        <view class="distance">{{item.distance}}</view>
                        <view class="address">{{item.address}}</view>
                        <view class="pack-btn">自提</view>
                    </view>
                </view>
            </view>
        </view>
    </view>
</template>

<script>
    import {mapState,mapActions} from "vuex";
    export default {
        name: "main-component",
        onLoad() {
            this.maxPage=0;     //总页数
            this.curPage=1;     //当前页码
            this.lng=0;         //经度
            this.lat=0;         //纬度
        },
        onShow(){
            // #ifdef MP-WEIXIN
            //如果用户关闭了地理位置功能
            //getSetting()获取用户的当前设置
            ①uni.getSetting({
                success:(res)=> {
                    //如果没有开启地理位置
                    if(!res.authSetting['scope.userLocation']){
                        uni.showModal({
                            title: '开启获取地理位置',
                            content: '请打开"位置信息"权限，找到附近的店铺',
                            success: (res)=> {
                                if (res.confirm) {
                                    //openSetting()调起客户端小程序设置界面，返回
```

```
                    //用户设置的操作结果
                    uni.openSetting({
                        success:(res2)=> {
                            //如果打开了地理位置
                            if(res2.authSetting['scope.
                            userLocation']){
                                //获取地理位置
                                uni.getLocation({
                                    type: 'gcj02',
                                    complete: (res)=> {
                                        this.lng=res.longitude;
                                        //经度
                                        this.lat=res.latitude;
                                        //纬度
                                        //获取商铺信息
                                        this.getShop({page:1,
                                        lng:this.lng?this.lng:0,
                                        lat:this.lat?this.lat:0,
                                        success:(pageNum)=>{
                                        this.maxPage= pageNum;
                                        //总页码数
                                        }});
                                    }
                                });
                            }
                        }
                    });
                } else if (res.cancel) {
                    //console.log('用户点击取消');
                }
            }
        });
    }
}
})
//#endif

//如果用户没有关闭地理位置功能，直接获取位置并获取商铺
②uni.getLocation({
    type: 'gcj02',
    complete: (res)=> {
        this.lng=res.longitude;          //经度
```

```
                        this.lat=res.latitude;          //纬度
                        this.getShop({page:1,lng:this.lng?this.lng:0,lat:this.
                            lat?this.lat:0,success:(pageNum)=>{
                              this.maxPage=pageNum;    //总页码数
                          }});
                    }
                });

            },
            computed:{
                ...mapState({
                    isIpx:state=>state.system.isIpx,     //获取Vuex是否在iPhone X中
                    shops:state=>state.business.shops   //获取Vuex的商铺数据
                })
            },
            methods:{
                ...mapActions({
                    getShop:"business/getShop",//store/business/index.js文件中actions
                                                    //内部的getShop()方法
                    getShopPage:"business/getShopPage"//store/business/index.js文件中
                                                    //actions内部的getShopPage()方法
                }),
                //跳转页面
                goPage(url){
                  uni.redirectTo({
                      url: url
                  });
                }
            },
        }
    </script>

    <style scoped>
       //CSS样式省略
    </style>
```

以上加粗代码为新增或修改的代码，代码注释很清晰，请读者仔细阅读。上述代码中的注意事项如下：

（1）使用getSetting()方法获取用户的当前设置，利用返回值res.authSetting['scope.userLocation']判断用户是否开启地理位置。如果没有开启，使用openSetting()方法调起客户端小程序设置界面。用户开启地理位置后，使用getLocation()方法获取地理位置。注意，为了定位准确，务必使用GCJ-02获取坐标，最后调用getShop()方法获取商品列表数据。

(2)如果用户没有关闭地理位置功能，则可以直接获取地理位置并获取商铺列表数据。在微信开发者工具中预览的效果如图11-4所示。

图 11-4　点餐系统首页

11.2.3　实现下拉刷新

扫一扫，看视频

一般首页会有下拉刷新功能。10.16.1节已经介绍了下拉刷新。现在在实战项目中使用。在pages/main/main.vue文件中新增代码如下：

```
<template>
    ...
</template>

<script>
    import {mapState,mapActions} from "vuex";
    export default {
        name: "main-component",
        ...
        //下拉刷新
        onPullDownRefresh(){
            this.curPage=1;            //将当前页面设置为1
            //重新获取商铺数据
```

```
                this.getShop({page:1,lng:this.lng?this.lng:0,lat:this.lat?this.
                    lat:0,success:(pageNum)=>{
                        this.maxPage=pageNum;
                    },
                    complete:()=>{
                        //服务端请求完成后，停止下拉刷新
                        uni.stopPullDownRefresh();
                    }
                });
            },
        }
</script>
...
```

使用onPullDownRefresh()钩子函数监听下拉刷新，在该函数内部将当前页面设置为1；调用getShop()方法，重新获取商铺数据；在complete()回调函数内使用stopPullDownRefresh()方法停止下拉刷新。

11.2.4 实现上拉加载和分享小程序

扫一扫，看视频

由于首页的商铺会越来越多，因此采取上拉加载分页显示数据，可以大大提高性能及用户体验。我们也可以将小程序分享出去获取更多的流量。在pages/main/main.vue文件中新增代码如下：

```
<template>
    ...
</template>

<script>
    import {mapState,mapActions} from "vuex";
    export default {
        name: "main-component",
        ...
        //上拉加载数据
        onReachBottom(){
            //如果当前页码小于总页码
            if(this.curPage<this.maxPage){
                this.curPage++;          //当前页码+1
                //调用Vuex商铺分页显示数据的方法
                this.getShopPage({page:this.curPage,lng:this.lng?this.
                    lng:0,lat:this.lat?this.lat:0});
            }
        },
```

```
            //支持分享小程序
            onShareAppMessage(res){
                if (res.from === 'menu') {
                    return {
                        title: '点餐系统',
                        path: '/pages/main/main'
                    }
                }
            },
        }
</script>
...
```

使用onReachBottom()函数可以监听页面是否滚动到底部，如果滚动到底部，并且当前页码小于总页码，则当前页码加1，调用Vuex里商铺分页显示数据的getShopPage()方法，并将页码传递进去实现上拉加载数据。使用onShareAppMessage()函数可以实现分享小程序的功能。

11.3 菜品搜索组件

当点击首页的搜索框时，会弹出菜品搜索窗口。为了后期维护方便，可以将搜索功能封装成一个组件，该搜索组件的功能包括搜索菜品、最近搜索和热门搜索。搜索组件演示效果如图11-5所示。

图 11-5 搜索组件演示效果

菜品搜索组件的接口在“接口文档”文件夹下的“搜索.docx”文件中。

11.3.1 最近搜索

扫一扫，看视频

搜索组件分为最近搜索和热门搜索，最近搜索使用本地缓存实现，热门搜索的数据需要调用服务端接口获取。本小节实现最近搜索功能，在components文件夹下创建search/index.vue文件，此处省略页面布局代码，只讲解在Vuex中实现数据的逻辑，在store文件夹中创建search/index.js文件，该文件中的代码如下：

```
export default {
    namespaced:true,
    state:{
        //历史搜索关键词
        historyKwords:uni.getStorageSync('historyKwords')?JSON.parse(uni.ge
        tStorageSync('historyKwords')):[],
        kwords:"",          //搜索的关键词
    },
    mutations:{
        //设置历史关键词
        ["SET_HISTORY_KWRODS"](state,payload){
            if(payload.kwords){
                if(state.historyKwords.length>0){
                    for(let i=0;i<state.historyKwords.length;i++){
                        //如果历史搜索关键词有相同的关键词
                        if(state.historyKwords[i]==payload.kwords){
                            state.historyKwords.splice(i--,1);  //删除相同的关键词
                            break;
                        }
                    }
                }
                //从数组开头添加搜索关键词
                state.historyKwords.unshift(payload.kwords);
                //将关键词保存在缓存中，实现持久保存
                uni.setStorageSync("historyKwords",JSON.stringify(state.historyKwords));
            }
        },
        //设置关键词
        ["SET_KWORDS"](state,payload){
            state.kwords=payload.kwords;
        },
        //清除历史记录
        ["CLEAR_KWORDS"](state,payload){
```

```
            state.historyKwords=[];                    //将历史关键词设置为空
            uni.removeStorageSync("historyKwords"); //清除历史关键词缓存
        }
    }
}
```

state内部的historyKwords属性值可以获取本地缓存的历史搜索关键词，kwords属性值为保存的用户输入的搜索关键词。mutations内部的SET_HISTORY_KWRODS方法可以将搜索的关键词保存到本地缓存并且将搜索的关键词添加到state内部的historyKwords属性中，SET_KWORDS方法可以设置state内部的kwords属性值，CLEAR_KWORDS方法可以清除历史搜索关键词。

接下来注册到Vuex的模块中，在store/index.js文件中新增代码如下：

```
...
import search from './search';

Vue.use(Vuex);

const store = new Vuex.Store({
    modules:{
        ...
        search
    }
});

export default store
```

加粗代码为新增代码。接下来将Vuex对接到components/search/index.vue文件中，代码如下：

```
<template>
    <view class="page" v-show="show">
        <view class="status_bar"></view>
        <view :class="{'search-title':true,wechat:true,ipx:isIpx}">
            <view class="search-close" @click="close()"></view>
            <view class="title">搜索</view>
        </view>
        <view class="search-header">
            <view class="search-wrap">
                <view class="search-input"><input type="text" confirm-type="search"
                placeholder="请输入商家或菜品名称" @confirm="goSearch($event)"
                :focus="isFocus" :value="kwords" @input="setKwords($event)" /></view>
                <view class="search-btn" @click="goSearch($event)"></view>
            </view>
```

```
        </view>
        <view class="search-main">
            <view class="search-area" v-if="historyKwords.length>0">
                <view class="search-name-wrap">
                    <view class="search-name">最近搜索</view>
                    <view class="bin" @click="clearHistoryKeywords()"></view>
                </view>
                <view class="search-kwords">
                    <view class="kwords" v-for="(item,index) in historyKwords":
                    key="index"
                    @click="goSearch($event,item)">{{item}}</view>
                </view>
            </view>
            <view class="search-area">
                <view class="search-name-wrap">
                    <view class="search-name">热门搜索</view>
                </view>
                <view class="search-kwords">
                    <view class="kwords">可乐</view>
                    <view class="kwords">汉堡</view>
                </view>
            </view>
        </view>
    </view>
</template>

<script>
    import {mapState,mapMutations,mapActions} from "vuex";
    export default {
        name: "search",
        data(){
            return {
                isFocus:false           //input是否自动获取焦点
            }
        },
        props:{
            //是否显示搜索组件
            show:{
                type:Boolean,
                default:false
            },
            //是否在菜品搜索页面进行搜索
```

```
        isLocal:{
            type:Boolean,
            default:false
        }
    },
    methods:{
        ...mapMutations({
        //store/search/index.js文件中mutations内部的SET_KWORDS方法
            SET_KWORDS:"search/SET_KWORDS",
        //store/search/index.js文件中mutations内部的SET_HISTORY_KWRODS方法
            SET_HISTORY_KWRODS:"search/SET_HISTORY_KWRODS",
        //store/search/index.js文件中mutations内部的CLEAR_KWORDS方法
            CLEAR_KWORDS:"search/CLEAR_KWORDS"
        }),
        //跳转到菜品搜索页面
        goSearch(event,kwords){
            let keyWords=kwords || this.kwords;
            //如果有热门或历史搜索词，则获取此搜索词，否则为input组件输入的搜索词
            if(keyWords){
                this.close();              //隐藏搜索组件
                this.SET_HISTORY_KWRODS({kwords:keyWords});
                //添加搜索的关键词到Vuex中
                if(this.isLocal){          //如果是在菜品搜索页面进行搜索
                    //重定向跳转
                    uni.redirectTo({
                        url:'/pages/search/index?kwords='+keyWords+''
                    });
                }else {                    //如果在其他页面进行搜索，如首页
                    //打开新页面跳转
                    uni.navigateTo({
                        url: '/pages/search/index?kwords=' + keyWords + ''
                    });
                }
            }else{
                uni.showToast({
                    title:"请输入搜索词",
                    icon:"none",
                    duration:2000
                })
            }

        },
```

```
            //清除关键词
            clearHistoryKeywords(){
                uni.showModal({
                    title: '',
                    content: '确认要清除历史记录吗？ ',
                    success: (res)=> {
                        if (res.confirm) {
                            this.CLEAR_KWORDS();
                        } else if (res.cancel) {

                        }
                    }
                });
            },
            //隐藏搜索组件
            close(){
                this.$emit("close");
            },
            //设置关键词
            setKwords(e){
                this.SET_KWORDS({kwords:e.target.value});
            }
        },
        computed:{
            ...mapState({
                isIpx:state=>state.system.isIpx,            //是否在iPhone X中
                kwords:state=>state.search.kwords,          //搜索的关键词
                historyKwords:state=>state.search.historyKwords, //历史搜索词
            })
        },
        watch:{
            show(val){
                //延迟300ms，可以实现input组件获取焦点
                setTimeout(()=>{
                    this.isFocus=val;
                },300);
            }
        }
    }
</script>

<style scoped>
    .page{width:100%;height:100%;position: fixed; left:0; top:0;
     z-index:999;background-color:#FFFFFF;}
```

```
    .search-header{width:100%;height:100rpx;border-bottom: 1px solid
     #EFEFEF;display: flex;box-sizing: border-box;padding:0px 20rpx;}
    .search-title .search-close{width:50rpx;height:50rpx;background-image:url("~@/
     static/images/common/back2.png");background-size:100%;background-repeat: no-
     repeat;background-position: center;margin-right:20rpx;}
    .search-header .search-wrap{width:95%;height:68rpx;border:1px solid
     #b2b2b2;border-radius: 5px;display: flex;justify-content: space-
     between;align-items: center;margin:0 auto;}
    .search-header .search-wrap .search-input{width:85%;height:100%;border-
     right:1px solid #b2b2b2;}
    .search-header .search-wrap .search-input input{width:90%;
     height:100%;font-size:28rpx;padding-left:20rpx;}
    .search-header .search-wrap .search-btn{width:50rpx;height:50rpx;backg
     round-image:url("~@/static/images/main/search_icon.png");background-
     size:100%;background-repeat: no-repeat;background-position:
     center;margin-right:20rpx;}
    .search-title{width:100%;height:100rpx;display:flex;position:
     relative;z-index:1;align-items: center;margin-top:0rpx;box-sizing:
     border-box;padding:0 20rpx;}
    /*#ifdef MP-WEIXIN*/
    .search-title.wechat{margin-top:40rpx;}
    .search-title.wechat.ipx{margin-top:80rpx;}
    /*#endif*/
    .search-title .title{width:auto;position: absolute;z-index:1;left:50%;top:
     50%;transform: translate(-50%,-50%);text-align: center;font-size:36rpx;}

    .search-main{width:100%;}
    .search-main .search-area{width:93%;margin:0 auto;}
    .search-main .search-area .search-name-wrap{width:100%;display:flex;
     justify-content: space-between;margin-top:20rpx;}
    .search-main .search-area .bin{width:40rpx;height:40rpx;
     background-image:url("~@/static/images/main/bin.png");background-
     size:100%;background-repeat: no-repeat;background-position: center;}
    .search-main .search-area .search-name{font-size:28rpx;}
    .search-main .search-area .search-kwords{width:100%;display: flex;flex-
     wrap: wrap;margin-top:20rpx;}
    .search-main .search-area .search-kwords .kwords{width:170rpx;
     height:64rpx;background-color:#FFFFFF;border:1px solid #EFEFEF;border-
     radius: 20px;font-size:28rpx;color:#717376;text-align:center;line-
     height:64rpx;white-space: nowrap;overflow:hidden;text-overflow:
     ellipsis;padding:0 20rpx;margin:0 8rpx;margin-bottom:20rpx;}
</style>
```

以上代码注释很清晰，请读者仔细阅读。接下来在首页中使用该组件，在pages/main/main.vue文件中新增代码如下：

```
<template>
    <view class="page">
        <view class="status_bar bg-color"></view>
        <view class="header">
            <view :class="{'search-header':true,'ipx':isIpx}"
             @click="isShow=true">
                <view class="search-wrap">
                    <view class="icon"></view>
                    <view class="search">输入商家名或菜品</view>
                </view>
            </view>
        </view>
        ...
        <Search :show="isShow" @close="isShow=false"></Search>
    </view>
</template>

<script>
    import {mapState,mapActions} from "vuex";
    import Search from "../../components/search";
    export default {
        name: "main-component",
        data(){
            return {
                isShow:false            //是否显示搜索组件
            }
        },
        ...
        components:{
            Search
        }
    }
</script>
...
```

上述加粗代码为新增代码，导入Search组件，在components里注册组件，并在视图中调用Search组件，添加自定义属性show和自定义事件close，这样即可实现显示与隐藏Search组件。

11.3.2 热门搜索

热门搜索需要请求服务端数据，在api文件夹下创建search/index.js文件，该文件中的代码如下：

扫一扫，看视频

```
import config from "../../static/js/conf/config";
import {request} from "../../static/js/utils/request";

//获取热门搜索关键词
export function getHotKeywordsData(data){
    return request(config.baseApi+"/v1/search/hotkeywords","get",data);
}
```

getHotKeywordsData()函数可以从服务端接口获取热门搜索关键词数据。接下来对接到Vuex中，在store/search/index.js文件中新增代码如下：

```
import {getHotKeywordsData} from "../../api/search";
export default {
    namespaced:true,
    state:{
        ...
        hotKeywords:[]          //热门搜索关键词
    },
    mutations:{
        ...
        //设置热门搜索关键词
        ["SET_HOTKEYWORDS"](state,payload){
            state.hotKeywords=payload.hotKeywords;
        }
    },
    actions:{
        //显示热门搜索关键词
        getHotKeywords(conText,payload){
            getHotKeywordsData(payload).then(res=>{
                if(res.code==200){
                    conText.commit("SET_HOTKEYWORDS",{hotKeywords:res.data});
                }else{
                    conText.commit("SET_HOTKEYWORDS",{hotKeywords:[]});
                }
            })
        }
    }
}
```

以上加粗代码为新增代码，state内部的hotKeywords属性值可以存储热门搜索的关键词;mutations内部的SET_HOTKEYWORDS方法可以设置state内部hotKeywords的属性值;actions内部的getHotKeywords()方法内部的getHotKeywordsData()函数可以从服务端接口获取热门搜索关键词的数据，并使用commit提交SET_HOTKEYWORDS，将数据保存到state.hotKeywords中。

接下来对接到components/search/index.vue文件中，在该文件中新增代码如下：

```
<template>
    <view class="page" v-show="show">
        ...
        <view class="search-main">
            ...
            <view class="search-area" v-if="hotKeywords.length>0">
                <view class="search-name-wrap">
                    <view class="search-name">热门搜索</view>
                </view>
                <view class="search-kwords">
                    <view class="kwords" v-for="(item,index) in hotKeywords"
                     :key= "index" @click="goSearch($event,item.
                     title)">{{item.title}}</view>
                </view>
            </view>
        </view>
    </view>
</template>

<script>
    import {mapState,mapMutations,mapActions} from "vuex";
    export default {
        name: "search",
        ...
        mounted(){
            //热门搜索关键词
            this.getHotKeywords();
        },
        methods:{
            ...
            ...mapActions({
                getHotKeywords:"search/getHotKeywords"
            }),
            ...
        },
```

```
        computed:{
            ...mapState({
                ...
                hotKeywords:state=>state.search.hotKeywords//热门搜索词
            })
        },
        ...
    }
</script>
...
```

以上加粗代码为新增或修改的代码。注意，在uni-app中，为了兼容各种平台，初始化数据建议放在mounted()钩子函数中。

11.4　菜品搜索

11.3节开发了菜品搜索组件，在菜品搜索组件输入搜索的关键词，跳转到菜品搜索页面，并显示相关菜品，默认显示距离用户最近的商铺里的菜品。单击菜品，进入该商铺的菜品展示页面，下单即可。菜品搜索页面演示效果如图11-6所示。

扫一扫，看视频

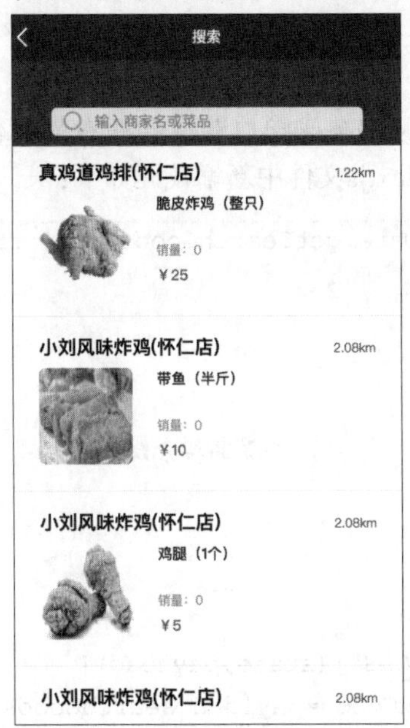

图 11-6　菜品搜索页面演示效果

菜品搜索的接口在“接口文档”文件夹下的“搜索.docx”文件中。

下面讲解对接数据显示附近商铺的菜品的开发流程。

在pages文件夹下创建search/index.vue文件，然后配置路由。在pages.json文件中新增代码如下：

```
{
    "pages": [
        ...
        {
            "path": "pages/search/index",
            "style": {
                "navigationBarTitleText": "搜索",
                "navigationStyle": "custom",
                "disableScroll": false
            }
        }
    ],
    ...
}
```

获取服务端请求，在api/search/index.js文件中新增代码如下：

```
...
//搜索菜品
export function getSearchGoodsData(data) {
    return request(config.baseApi+"/v1/search/goods","get",data);
}
```

getSearchGoodsData()可以从服务端接口获取到搜索菜品名称返回的菜品数据。接下来对接到Vuex中，在store/search/index.js文件中新增代码如下：

```
import {getHotKeywordsData,getSearchGoodsData} from "../../api/search";
export default {
    namespaced:true,
    state:{
        ...
        searchGoods:[],         //菜品搜索数据列表
    },
    mutations:{
        ...
        //设置菜品搜索数据
        ["SET_SEARCH_GOODS"](state,payload){
            state.searchGoods=payload.searchGoods;
        },
        //设置菜品搜索分页数据
        ["SET_SEARCH_GOODS_PAGE"](state,payload){
            state.searchGoods.push(...payload.searchGoods);
        }
    },
```

```
    actions:{
        ...
        //搜索菜品
        getSearchGoods(conText,payload){
            getSearchGoodsData(payload).then(res=>{
                if(res.code==200){
                    conText.commit("SET_SEARCH_GOODS",{searchGoods:res.data});
                    if(payload.success){
                        payload.success(res.pageinfo.pagenum);
                    }
                }else{
                    conText.commit("SET_SEARCH_GOODS",{searchGoods:[]});
                }
            })
        },
        //搜索菜品分页数据
        getSearchGoodsPage(conText,payload){
            getSearchGoodsData(payload).then(res=>{
                if(res.code==200){
                    conText.commit("SET_SEARCH_GOODS_PAGE",
                                    {searchGoods:res.data});
                }
            })
        },
    }
}
```

以上加粗代码为新增代码，state内部的searchGoods属性值可以存储搜索菜品返回的数据。mutations内部的SET_SEARCH_GOODS方法可以设置state内部的searchGoods的属性值;SET_SEARCH_GOODS_PAGE方法内部使用的push()方法可以对state.searchGoods数组添加新的值，用于上拉加载数据。actions内部的getSearchGoods()方法可以获取服务端返回的数据，并使用commit提交到SET_SEARCH_GOODS中，将数据存储到state.searchGoods中；getSearchGoodsPage()方法可以获取服务端的数据，使用commit提交到SET_SEARCH_GOODS_PAGE中，在state.searchGoods数组中添加新的值。

接下来将Vuex对接到pages/search/index.vue文件中，该文件中的代码如下：

```
<template>
    <view class="page">
        <view class="status_bar bg-color"></view>
        <view :class="{nav:true,wechat:true,ipx:isIpx}">
            <view class="back" @click="back()"></view>
            <view class="title">搜索</view>
```

```
            </view>
            <view :class="{header:true,ipx:isIpx}">
                <view :class="{'search-header':true}">
                    <view class="search-wrap" @click="isShow=true">
                        <view class="icon"></view>
                        <view class="search">输入商家名或菜品</view>
                    </view>
                </view>
            </view>
            <view :class="{main:true,ipx:isIpx}" v-if="searchGoods.length>0">
                <view class="goods-wrap" v-for="(item,index) in searchGoods"
                 :key="index" @click="replacePage('/pages/goods/index?branch_
                 shop_id='+item.branch_shop_id+'&gid='+item.gid+'')">
                    <view class="shop">
                        <view class="shop-name">{{item.branch_shop_name}}</view>
                        <view class="distance">{{item.distance}}</view>
                    </view>
                    <view class="goods-list">
                        <view class="image"><image :src="item.image"></image></view>
                        <view class="goods-info">
                            <view class="goods-title">{{item.title}}</view>
                            <view class="sales">销量:{{item.sales}}</view>
                            <view class="price-wrap">
                                <view class="price">￥{{item.price}}</view>
                            </view>
                        </view>
                    </view>
                </view>
            </view>
            <view class="no-data" v-if="searchGoods.length<=0">没有相关菜品! </view>
            <Search :show="isShow" @close="isShow=false" :isLocal="true"></Search>
        </view>
    </template>

    <script>
        import {mapState,mapActions} from 'vuex';
        import Search from "../../components/search";
        export default {
            name: "search",
            data(){
                return {
                    isShow:false                              //是否显示搜索组件
```

```
        }
    },
    onLoad(opts){
        this.kwords=opts.kwords?opts.kwords:""; //获取搜索的关键词
        this.maxPage=0;                          //总分页数
        this.curPage=1;                          //当前页码
        this.lng=0;                              //经度
        this.lat=0;                              //纬度
    },
    onShow(){
        //获取地理位置的坐标
        uni.getLocation({
            type: 'gcj02',
            complete: (res)=> {
                this.lng=res.longitude;          //经度
                this.lat=res.latitude;           //纬度
                //获取菜品搜索的数据
                this.getSearchGoods({page:1,kwords:this.kwords,lng:this.lng?this.
                    lng:0,lat:this.lat?this.lat:0,success:(pageNum)=>{
                        this.maxPage=pageNum;    //获取总分页数
                    }});
            }
        });
    },
    //上拉加载数据
    onReachBottom(){
        //如果当前页面小于总页数
        if(this.curPage<=this.maxPage){
            this.curPage++;                      //当前页码+1
            //调用Vuex菜品搜索分页显示数据的方法
            this.getSearchGoodsPage({page:this.curPage,kwords:this.
                kwords,lng:this.lng?this.lng:0,lat:this.lat?this.lat:0});
        }
    },
    components:{
        Search                                   //搜索组件
    },
    computed:{
        ...mapState({
            isIpx:state=>state.system.isIpx,             //是否在iPhone X中
            searchGoods:state=>state.search.searchGoods //菜品搜索的数据
        })
```

```
        },
        methods:{
            ...mapActions({
                getSearchGoods:"search/getSearchGoods",
                //store/search/index.js文件中actions内部的getSearchGoods方法
                getSearchGoodsPage:"search/getSearchGoodsPage"
                //store/search/index.js文件中actions内部的getSearchGoodsPage方法
            }),
            //返回上一页
            back(){
                uni.navigateBack({
                    detail:1
                })
            },
            //跳转页面
            replacePage(url){
                uni.redirectTo({
                    url:url
                })
            }
        }
    }
</script>

<style scoped>
    .page{width:100%;background-color:#FFFFFF;overflow:hidden;}
    .status_bar.bg-color{background-color:#E30019;}
    .nav{width:100%;height:88rpx;background-color:rgb(227, 0, 25);position:
    fixed;z-index:91;left:0;top:0;display: flex;justify-content: space-
    between;align-items: center;}
    .nav .back{width:40rpx;height:40rpx;background-image:url("~@/static/
    images/common/back.png");background-size:100%;background-repeat: no-
    repeat;background-position: center;}
    .nav .title{position: absolute;z-index:1;left:50%;top:50%;transform:
    translate(-50%,-50%);color:#FFFFFF;font-size:28rpx;}
    .header{width:100%;background-color:#eb1625;overflow:hidden;position:
    fixed;left:0;top:88rpx;z-index:90;}
    .search-header{width:100%;height:auto;display: flex;justify-
    content: center;align-items: flex-end;margin-top:88rpx;box-sizing:
    border-box;padding-bottom:20rpx;}
    .header .search-wrap{width:80%;height:52rpx;background-color:rgba(255,255,
    255,0.9);border-radius: 5px;display:flex;justify-content: start;align-
    items: center;}
```

```
    .header .search-wrap .icon{width:44rpx;height:44rpx;background-image:url("~@/static/images/main/search_icon.png");background-size:100%;background-repeat: no-repeat;background-position: center;margin:0 20rpx;}
    .header .search-wrap .search{font-size:28rpx;color:#999999;}

    .main{width:100%;margin:0 auto;margin-top:280rpx;overflow:hidden;}
    .goods-wrap{width:100%;overflow:hidden;margin-bottom:40rpx;border-bottom:1px solid #EFEFEF;box-sizing: border-box; padding-left: 50rpx;padding-right:50rpx;padding-bottom:50rpx;}
    .shop{width:100%;height:36rpx;display: flex;justify-content: space-between;align-items: center;overflow:hidden;}
    .shop .shop-name{font-size:36rpx;font-weight: bold;width:80%;height:54rpx;overflow:hidden;white-space: nowrap;text-overflow: ellipsis;}
    .shop .distance{font-size:24rpx;color:#666666;}
    .goods-list{width:100%;height:auto;display:flex;justify-content:space-between;margin-top:20rpx;}
    .goods-list .image{width:180rpx;height:180rpx;border-radius: 6px;}
    .goods-list .image image{width:100%;height:100%;border-radius: 6px;}
    .goods-list .goods-info{width:62%;height:auto;margin-right:20rpx;}
    .goods-list .goods-title{width:100%;height:80rpx;overflow:hidden;font-size:28rpx;color:#333333;font-weight: bold;}
    .goods-list .sales{font-size:24rpx;color:#999999;margin-top:10rpx;}
    .goods-list .price{font-size:28rpx;color:#fb4e44;font-weight: bold;}
    .goods-list .price-wrap{width:100%;margin-top:10rpx;display:flex;justify-content: space-between;}

    /*#ifdef MP-WEIXIN*/
    .nav.wechat{padding-top:70rpx;}
    .nav.wechat.ipx{padding-top:90rpx;}
    .nav.wechat .title{top:65%}
    .nav.wechat.ipx .title{top:75%}
    .header.ipx{top:170rpx;}
    .main.ipx{margin-top:330rpx;}
    /*#endif*/
</style>
```

在onShow()函数内使用getLocation()函数获取地理位置的坐标，并在complete回调函数内部调用getSearchGoods()方法，将搜索的关键词、地理位置的经度和纬度以对象形式的参数传递进去，获取附近商铺的菜品。在onReachBottom()函数内部可以实现上拉加载数据。

11.5 点餐

点餐页面是整个项目的核心部分，此页面的功能有选择单品或套餐、加入购物车计算总价、商家信息展示，并在地图上显示商家所在位置。如果是堂食点餐，用微信扫二维码直接进入此页面即可；如果是自提点餐，可以从首页点击商家进入此页面或者从菜品搜索页面点击菜品进入此页面。商家菜品展示页面演示效果如图11-7所示。

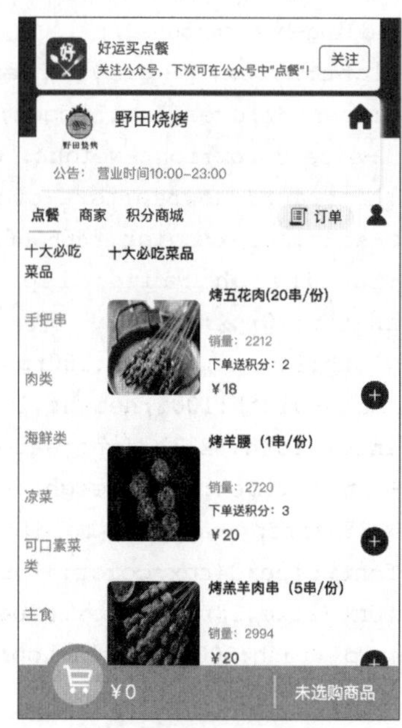

图 11-7 商家菜品展示页面演示效果

图11-7演示的是正式运营的项目，功能一直在迭代开发，所以有积分商城功能。由于本书实战开发项目是该程序的初版，因此没有积分商城功能。

商家菜品展示页面的接口在“接口文档”文件夹下的“菜品展示.docx”文件中。

11.5.1 开发点餐页面

扫一扫，看视频

首先开发静态页面和所需要的js效果，完成静态页面；然后对接服务端数据。在pages文件夹下创建goods/index.vue文件，创建完成后配置路由，在pages.json文件中新增代码如下：

```
{
    "pages": [
        ...
        {
            "path": "pages/goods/index",
```

```
            "style": {
                "navigationBarTitleText": "点餐",
                "navigationStyle": "custom",
                "disableScroll": true,
                "navigationBarTextStyle": "black"
            }
        }
    ],
    ...
}
```

路由配置完成，接下来开发点餐页面，在pages/good/index.vue文件中的代码如下：

```
<template>
    <view class="page">
        <view class="status_bar">
            <view class="bar-bg"></view>
        </view>
        <view :class="{header:true,ipx:false,official:true}">
            <view :class="{'mp-wrap':true,ipx:false}">
                <view class="mp-logo">
                    <image src="../../static/images/common/logo.jpg"></image>
                </view>
                <view class="info">
                    <view class="name">好运买点餐</view>
                    <view class="desc">关注公众号，下次可在公众号中"点餐"！ </view>
                    <view  class="focus"  @click="pushPage('/pages/main/
                     focus')">关注</view>
                </view>
            </view>
        </view>
        <view class="business-wrap">
            <view class="business-info">
                <view class="business-logo">
                    <image  src="https://diancan.lucklnk.com/businessfiles/
                     logo/1593255093.png"></image>
                </view>
                <view class="business-name">野田烧烤</view>
                <view class="notice">公告：营业时间10:00-23:00</view>
                <view class="home" @click="replacePage('/pages/main/main')"></view>
            </view>
        </view>
        <view class="tags">
```

```
            <view class="item-wrap">
                <view :class="{item:true,active:goodsAction}" @
                click="changeTab(true,false)">点餐</view>
                <view :class="{item:true, active:businessAction}" @
                click="changeTab(false,true)">商家</view>
            </view>
            <view class="tags-handle">
                <view class="myorder">
                    <view class="myorder-icon"></view>
                    <view class="text">订单</view>
                </view>
                <view class="line"></view>
                <view class="my" @click="pushPage('/pages/my/index?branch_shop_id=
                '+branch_shop_id+'&table_code='+table_code+'')"></view>
            </view>
        </view>
        <!--菜品展示组件-->
        <goods :branchShopId="branch_shop_id" :isShow="goodsAction"
        :gid="gid"></goods>
        <!--商家信息组件-->
        <business-info :isShow="businessAction" :branchShopId=
        "branch_shop_id"></business-info>
        <!--只有在菜品展示组件显示的情况下，才显示购物车导航-->
        <view class="cart-main" v-show="goodsAction">
            <view class="handle-cart">
                <view class="cart-cle">
                    <view class="cart-cle-2">
                        <view class="cart-icon" id="cart-icon">
                            <view class="badge">5</view>
                        </view>
                    </view>
                </view>
                <view class="total">￥100</view>
            </view>
            <view class="line"></view>
            <view class="submit">
                <template v-if="false">提交订单 &gt;</template>
                <template v-else>未选购商品</template>
            </view>
        </view>
    </view>
</template>
```

```
<script>
    import Goods from '../../components/goods';
    import BusinessInfo from '../../components/business/info';
    export default {
        name: "goods",
        data(){
            return {
                branch_shop_id:"",   //分店的id
                table_code:"",       //桌号
                goodsAction:true,    //是否显示菜品展示组件
                businessAction:false, //是否显示商家信息组件
                gid:"",              //菜品的id
            }
        },
        onLoad(opts){
            this.branch_shop_id=opts.branch_shop_id?opts.branch_shop_id:"";
            //分店的id
            this.table_code=opts.table_code?opts.table_code:"";  //桌号
            this.gid=opts.gid?opts.gid:"";//菜品的id，从搜索菜品页面进入此页面会有gid
        },
        components:{
            Goods,
            BusinessInfo
        },
        //分享此页面
        onShareAppMessage(res){
            if (res.from === 'menu') {
                return {
                    title: '好运买点餐',
                    path: '/pages/goods/index?branch_shop_id='+this.branch_
                          shop_id+'&table_code='+this.table_code+''
                }
            }
        },
        methods:{
            //打开新页面跳转
            pushPage(url){
                uni.navigateTo({
                    url:url
                })
            },
```

```
                //重定向跳转
                replacePage(url){
                    uni.redirectTo({
                        url:url
                    })
                },
                //菜品组件和商家信息组件切换显示
                changeTab(goodsAction,businessAction){
                    this.goodsAction=goodsAction;
                    this.businessAction=businessAction;
                }
            }
        }
</script>

<style scoped>
    .page{width:100%;height:100vh;overflow:hidden;position: relative;}
    .header{width:100%;height:230rpx;background-color:#E30019;position:relative
    ;z-index:1;}
    /* #ifndef H5 */
    .header.official{height:280rpx;}
    /* #endif */
    .business-wrap{width:100%;height:100rpx;position:relative;z-index:1;}
    .header.ipx.official{height:320rpx;}
    .status_bar{background-color:#E30019}
    .status_bar .bar-bg{width:90%;height:100%;background-color:#FFFFFF;
    margin:0 auto;}
    .business-wrap .business-info{width:90%;height:180rpx;background-
    color:#FFFFFF;position: absolute;z-index:1;left:50%;top:
    -80rpx;transform: translateX(-50%);box-shadow: 3px 3px 5px
    #efefef;border-radius: 5px;}
    .business-wrap .home{width:50rpx;height:50rpx;position:absolute;
    right:15rpx;top:15rpx;z-index:1;background-image: url("~@/static/
    images/goods/home.png");background-size:100%;background-repeat:
    no-repeat;background-position:center;}
    .business-wrap .business-logo{width:100rpx;height:100rpx;position:absolute;
    z-index:1;left:35rpx;top:15rpx;}
    .business-wrap .business-logo image{width:100%;height:100%;border-
    radius: 8px;}
    .business-wrap .business-name{width:450rpx;height:100rpx;font-
    size:36rpx;color:#333333;position: absolute;z-index:1;top:20rpx;left:156rpx;
    overflow:hidden;white-space: nowrap;text-overflow: ellipsis;}
```

```
.business-wrap .promotion{width:75%;height:60rpx;position:absolute;z-index:1;
left:156rpx;top:80rpx;font-size:26rpx;color:#E30019;overflow:hidden;white-
space: nowrap;text-overflow: ellipsis;text-decoration: underline;}
.business-wrap .notice{width:630rpx;height:60rpx;position:absolute;z-index:1;
left:35rpx;top:130rpx;font-size:26rpx;color:#999999;overflow:hidden;white-
space: nowrap;text-overflow: ellipsis;}
.tags{width:100%;height:100rpx;display:flex;justify-content: space-
between;align-items: center;padding:0px 25rpx;box-sizing: border-box;}
.tags .item-wrap{width:auto;height:auto;display:flex;}
.tags .item-wrap .item{width:auto;height:auto;color:#333333;font-
size:28rpx;padding:10rpx 5rpx;margin-right:30rpx;}
.tags .item-wrap .item.focus{color:#E30019}
.tags .item-wrap .item.active{border-bottom: 2px solid #ffd161;
font-weight:bold;}
.tags .myorder{width:150rpx;height:100%;border-radius: 15px;background-
color:#f5f5f5;display: flex;justify-content: center;align-items:
center; margin-right:10rpx;}
.tags .myorder-icon{width:40rpx;height:40rpx;background-image:url("~@/
static/images/goods/myorder.png");background-size:100%;background-
repeat: no-repeat;background-position: center;margin-right:10rpx;}
.tags .myorder .text{color:#333333;font-size:28rpx;}
.tags .tags-handle{display:flex;height:50rpx;}
.tags .tags-handle .line{width:3px;height:100%;background-color:
rgba(245,245,245,0.6);}
.tags .tags-text{font-size:28rpx;color:#FFFFFF;margin-left:10rpx;border-
radius: 5px;background-color:#E30019;padding:5rpx 10rpx;}
.tags .my{width:40rpx;height:40rpx;background-image: url("~@/static/
images/main/my.png");background-size:100%;background-repeat: no-
repeat;background-position: center;margin-left:10rpx;}

.cart-main{width:100%;height:100rpx;background-color:#f6ab00;position:
fixed;z-index:90;left:0px;bottom:0px;}
.cart-main .cart-cle{width:130rpx;height:130rpx;border-radius:
100%;background-color:#f6ab00;position:absolute;left:56rpx;top:-40rpx;}
.cart-main .cart-cle-2{width:100rpx;height:100rpx;border-radius:
100%;position: absolute;left:50%;top:50%;transform: translate(-50%,
-50%);background-color:#F8CE86}
.cart-main .cart-icon{width:60rpx;height:60rpx;background-image:url("~@/
static/images/main/cart.png"); background-size: 100%;background-
position:center;background-repeat: no-repeat;position: absolute;z-index:1;
left:50%;top:50%;transform: translate(-50%,-50%); }
```

```
    .cart-main .badge{width:50rpx;height:50rpx;border-radius:
    100%;background-color:#ff2400;position:absolute;right:-35rpx;top:
    -35rpx;font-size:24rpx;color:#FFFFFF;text-align: center;line-
    height:50rpx;white-space: nowrap;overflow:hidden;}
    .cart-main .total{font-size:36rpx;color:#FFFFFF;width:auto;height:auto;
    position:absolute;left:180rpx;top:50%;transform: translateY(-50%);}
    .cart-main .line{width:2px;height:62rpx;background-color:rgba(255,255,
    255,0.6);position:absolute;z-index:1;left:507rpx;top:50%;transform:
    translateY(-50%);}
    .cart-main .submit{width:175rpx;height:60rpx;position:absolute;z-index
    :1;right:30rpx;top:50%;transform: translateY(-50%); font-size: 32rpx;
    color:#FFFFFF;line-height:60rpx;}
    .cart-main .handle-cart{width:65%;height:100%;}

    .mp-wrap{width:90%;margin:0 auto;height:180rpx;background-color:#FFF
    FFF;display:flex;justify-content: space-between;align-items: flex-
    end;box-sizing: border-box;padding:0px 20rpx;border-bottom-left-radius:
    5px;border-bottom-right-radius:5px;overflow:hidden;}
    /* #ifdef H5 */
    .mp-wrap{height:140rpx;}
    /* #endif */
    .mp-wrap.ipx{height:220rpx;}
    .mp-wrap .mp-logo{width:80rpx;height:80rpx;margin-bottom:20rpx;}
    .mp-wrap .mp-logo image{width:100%;height:100%;border-radius: 8px;}
    .mp-wrap .info{width:84%;height:80rpx;position:
    relative;z-index:1;margin-bottom:20rpx;}
    .mp-wrap .info .name{width:auto;height:auto;font-size:28rpx;color:#5A5A5A;
    position:absolute;z-index:1;left:0px;top:0px;}
    .mp-wrap .info .desc{width:auto;height:auto;font-size:24rpx;color:#5A5A5A;
    position:absolute;z-index:1;left:0px;top:45rpx;}
    .mp-wrap .info .focus{width:auto;height:auto;border:1px solid
    #28A622;border-radius: 4px;font-size:28rpx;color:#28A622;position:absolute;
    z-index:1;right:0px;top:14rpx;padding:5rpx 20rpx;}

    /* #ifdef MP-WEIXIN */
    .mp-wrap2{width:100%;margin:0 auto;height:230rpx;background-
    color:#FFFFFF;display:flex;justify-content: center;align-items:
    flex-end;box-sizing: border-box;border-bottom-left-radius:
    5px;border-bottom-right-radius:5px;overflow:hidden;margin-top:20rpx;}
    .mp-wrap2.ipx{height:250rpx;}
    /* #endif */
</style>
```

以上代码实现了静态布局，可以看到此代码中有两个组件，分别是Goods（菜品展示组件）和BusinessInfo（商家信息组件）。接下来在components文件夹下分别创建goods/index.vue和business/info.vue文件。

11.5.2　开发菜品展示组件

扫一扫，看视频

菜品展示组件预览效果如图11-8所示。

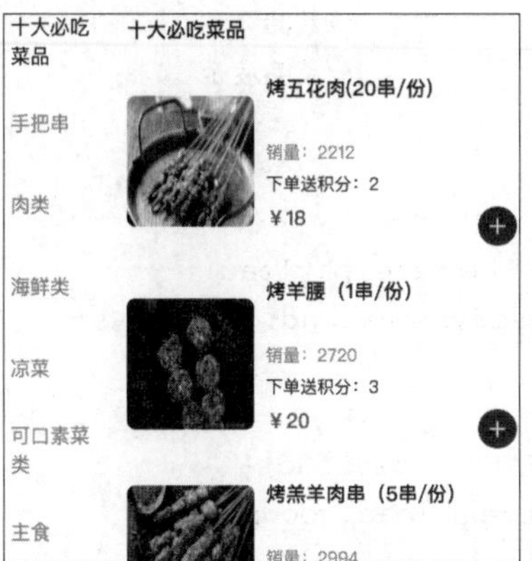

图 11-8　菜品展示组件预览效果

菜品展示组件左侧是菜品分类数据，右侧是菜品展示数据。这里为了开发方便，直接从服务端接口请求数据。为了组件后期的可维护性及扩展性，尽量不要在组件内部请求数据，可以在父组件请求数据，然后传值到子组件。这里在pages/goods/index.vue文件中请求菜品数据，首先从服务端接口请求数据，在api文件夹下创建goods/index.js文件，该文件中的代码如下：

```
import config from "../../static/js/conf/config";
import {request} from "../../static/js/utils/request";

//菜品分类
export function getClassifyData(branch_shop_id){
    return request(config.baseApi+"/v1/goods/classify", "post",
    {branch_shop_id:branch_shop_id});
}

//菜品展示
export function getGoodsData(branch_shop_id){
    return request(config.baseApi+"/v1/goods/show", "post", {branch_shop_
    id: branch_shop_id});
}
```

getClassifyData()函数可以从服务端接口获取菜品分类的数据，getGoodsData()函数可以从服务端接口获取菜品数据。接下来对接到Vuex中，在store文件夹下创建goods/index.js文件，该文件中的代码如下：

```
import Vue from "vue";
import {getClassifyData,getGoodsData} from "../../api/goods";
export default {
    namespaced:true,
    state:{
        classifys:[],          //菜品分类
        goods:[]               //菜品展示
    },
    mutations:{
        //设置菜品分类
        ["SET_CLASSIFY"](state,payload){
            state.classifys=payload.classifys;
        },
        //设置菜品展示
        ["SET_GOODS"](state,payload){
            state.goods=payload.goods;
        },
        //设置菜品数量
        ["SET_AMOUNT"](state,payload){
            level1:
            for(let i=0;i<state.goods.length;i++){
                level2:
                for(let j=0;j<state.goods[i].goods.length;j++){
                    //源数据套餐属性
                    let newMealItems=(state.goods[i].goods[j].new_meal_
                        items && state.goods[i].goods[j].new_meal_items.
                        length)>0?JSON.stringify(state.goods[i].goods[j].
                        new_meal_items):"";
                    //传入的套餐属性
                    let curNewMealItems=(payload.new_meal_items && payload.
                        new_meal_items.length)>0?JSON.stringify(payload.
                        new_meal_items):"";
                    //如果有相同的单品或套餐
                    if(state.goods[i].goods[j].gid==payload.gid &&
                      newMealItems==curNewMealItems){
                        //更新数量
                        state.goods[i].goods[j].amount=payload.amount;
                        //验证数量格式必须为数字，这里需要使用setTimeout延迟，否则无法验证
                        setTimeout(()=>{
```

```
                    state.goods[i].goods[j].amount=state.goods[i].
                        goods[j].amount.toString().replace(/[^\d]/g,"");
                },30);
                //如果数量为空或为0，强制设置数量为1
                if(!state.goods[i].goods[j].amount || state.goods[i].
                  goods[j].amount=='0'){
                    state.goods[i].goods[j].amount=1;
                }
                break level1;           //结束最外层循环
            }
        }
    }
},
//显示数量input
["SHOW_AMOUNT_INPUT"](state,payload){
    state.goods[payload.index].goods[payload.index2].
        isAmountInput=true;           //显示input
    state.goods[payload.index].goods[payload.index2].
        isAmountInputFocus=true;      //input获取焦点
    Vue.set(state.goods[payload.index].goods,payload.index2,state.
        goods[payload.index].goods[payload.index2]);
},
//隐藏数量input
["HIDE_AMOUNT_INPUT"](state,payload){
    state.goods[payload.index].goods[payload.index2].isAmountInput=false;
    state.goods[payload.index].goods[payload.index2].isAmountInputFocus
        =false;
    Vue.set(state.goods[payload.index].goods,payload.index2,state.
        goods[payload.index].goods[payload.index2]);
},
//设置某一个菜品的位置
["SET_GOODS_POISTION"](state,payload){
    state.goods[payload.index].position=payload.position;
    state.goods[payload.index].top=payload.top;
    state.goods[payload.index].left=payload.left;
}
},
actions:{
    //显示左侧分类
    getCalssifys(conText,payload){
        getClassifyData(payload.branch_shop_id).then(res=>{
            if(res.code==200){
```

```
                for(let i=0;i<res.data.length;i++){
                    if(i==0){
                        res.data[i].active=true; //选中分类时的样式
                    }else {
                        res.data[i].active = false;
                    }
                    //goodsTop、goodsHeight、top可以实现点击分类定位到相关的菜品
                    res.data[i].goodsTop=0;      //该分类对应菜品距顶部的距离
                    res.data[i].goodsHeight=0;   //该分类对应菜品的高
                    res.data[i].top=0;           //该分类距顶部的距离
                }
                conText.commit("SET_CLASSIFY",{classifys:res.data});
            }else{
                conText.commit("SET_CLASSIFY",{classifys:[]});
            }
            if(payload.success){
                payload.success();
            }
        })
    },
    //显示新上架或下架的菜品
    getNewGoods(conText,payload){
        getGoodsData(payload.branch_shop_id).then(res=>{
            if(res.code==200){
                if(payload.success){
                    payload.success(res.data);
                }
            }
        })
    },
    //显示菜品
    getGoods(conText,payload){
        getGoodsData(payload.branch_shop_id).then(res=>{
            if(res.code==200){
                for(let i=0;i<res.data.length;i++){
                    res.data[i].top=0;    //CSS样式position的top
                    res.data[i].left=0;   //CSS样式position的left
                    res.data[i].position="static";//动态改变CSS样式position的值
                 for(let j=0;j<res.data[i].goods.length;j++){
                        res.data[i].goods[j].itemTop=0;  //每个菜品的top
                        res.data[i].goods[j].amount=0;    //菜品数量
                        res.data[i].goods[j].new_meal_items=[];
```

```
                        //套餐规格
                        res.data[i].goods[j].isAmountInput=false;
                        //是否显示数量输入框
                        res.data[i].goods[j].isAmountInputFocus=false;
                        //数量输入框焦点
                        //重组套餐菜品中的规格，用于判断购物车中的套餐是否相同
                        if(res.data[i].goods[j].meal_items){
                            for(let k=0;k<res.data[i].goods[j].meal_items.length;k++){
                                res.data[i].goods[j].new_meal_items.
                                push({"gid":res.data[i].goods[j].meal_
                                items[k].gid})
                            }
                        }
                    }
                }
                conText.commit("SET_GOODS",{goods:res.data});
                if(payload.success){
                    payload.success(res.data);
                }
            }else{
                conText.commit("SET_GOODS",{goods:[]});
            }
        })
    }
  }
}
```

state内部的属性classifys为菜品分类，goods为菜品展示，其值为数组类型;mutations内部的方法可以响应式地实现设置菜品分类、菜品展示、菜品数量、显示和隐藏数量input输入框以及某一个菜品的位置;actions内部的方法可以异步请求服务端数据。接下来注册到Vuex的模块中，在store/index.js文件中新增代码如下：

```
...
import goods from '../goods';

...

const store = new Vuex.Store({
    modules:{
        ...
        goods
    }
```

```
});

export default store
```

接下来，将Vuex对接到pages/goods/index.vue文件中，新增代码如下：

```
<template>
    <view class="page">
        ...
        <!--菜品展示组件-->
        <goods :branchShopId="branch_shop_id" :isShow="goodsAction"
         :gid="gid" :amount="amount" @setAmount="setAmount" @showAmountInp
         ut="showAmountInput" :classifysData="classifys" :goodsData="goods"
         :shelfGoods="shelfGoods"></goods>
        ...
    </view>
</template>

<script>
    import {mapState,mapActions,mapMutations} from 'vuex';
    ...
    export default {
        name: "goods",
        data(){
            return {
                ...
                amount:0                //菜品数量
                shelfGoods:false        //是否有上架或下架的菜品
            }
        },
        onLoad(opts){
            this.branch_shop_id=opts.branch_shop_id?opts.branch_shop_id:"";
            //分店的id
            this.table_code=opts.table_code?opts.table_code:"";  //桌号
            this.gid=opts.gid?opts.gid:"";
            //菜品的id，从搜索菜品页面进入此页面会有gid
            this.getCalssifys({branch_shop_id:this.branch_shop_id,success:()=>{
                    //要在这里调用显示菜品，解决异步获取dom的问题
                    this.getGoods({branch_shop_id:this.branch_shop_id});
                }});
        },
onShow(){
            if(this.goods && this.goods.length>0) {
```

```
            this.getNewGoods({
                branch_shop_id: this.branch_shop_id, success: (data) => {
                    for (let i = 0; i < this.goods.length; i++) {
                        //是否有新上架或下架的菜品
                        if (this.goods[i].goods.length != data[i].goods.length) {
                            this.shelfGoods=true;//有上架或下架的菜品
                            //先清空Vuex中的菜品数据，再重新获取菜品数据，解决
                            //scroll-view无法重新计算高度的问题
                            this.SET_GOODS({goods:[]});//清空Vuex中的菜品数据
                            //有新上架或下架的菜品，重新获取数据
                            this.getGoods({branch_shop_id:this.branch_shop_id});
                            break;
                        }
                    }
                }
            });
        }
    },
    computed:{
        ...mapState({
            "classifys":state=>state.goods.classifys,  //菜品分类
            "goods":state=>state.goods.goods            //菜品展示
        })
    },
    ...
    methods:{
        ...mapMutations({
            SET_GOODS:"goods/SET_GOODS",                         //设置菜品
            SET_GOODS_AMOUNT:"goods/SET_AMOUNT",                 //设置数量
            SHOW_AMOUNT_INPUT:"goods/SHOW_AMOUNT_INPUT",         //显示input数量输入框
            HIDE_AMOUNT_INPUT:"goods/HIDE_AMOUNT_INPUT"          //隐藏input数量输入框
        }),
        ...mapActions({
            "getCalssifys":"goods/getCalssifys",        //获取分类的方法
            "getGoods":"goods/getGoods",                //获取菜品的方法
            "getNewGoods":"goods/getNewGoods"           //获取上架或下架的菜品
        }),
        ...
        //显示数量输入框
        showAmountInput(index,index2){
            this.SHOW_AMOUNT_INPUT({index,index2});
        },
```

```
                //修改数量
                setAmount(e,gid,new_meal_items,meal_items,index,index2){
                    let amount=e.target.value;
                    this.amount=amount;
                    //设置数量
                    this.SET_GOODS_AMOUNT({amount:e.target.value, gid, new_meal_
                        items: new_meal_items});
                    if(index!=undefined && index2!=undefined){
                        this.HIDE_AMOUNT_INPUT({index,index2}) //隐藏输入框
                    }
                },
            }
        }
</script>

<style scoped>
    ...
</style>
```

以上加粗代码为新增代码。接下来开发菜品展示组件，在components/goods/index.vue文件中的代码如下：

```
<template>
    <view class="goods-area" v-show="isShow">
        <view class="goods-main">
            <!--菜品分类-->
            <scroll-view id="classify" scroll-with-animation="true" scroll-
             y="true" :scroll-top="classifyScrollTop" class="classify">
                <view :class="{item:true, active:item.active}" v-for="(item,index)
                 in classifys" :key="item.cid" @click="clickClassify(index)">
                    <view>{{item.title}}</view>
                </view>
                <view style="height:170rpx;"></view>
            </scroll-view>
            <!--菜品展示-->
            <scroll-view scroll-y="true" id="goods-wrap" class="goods-
             wrap" scroll-with-animation="true" @scroll="eventGoodsScroll"
             :scroll-top="goodsScrollTop">
                <view class="goods-classify">
                    <block v-for="(item,index) in goods" :key="item.cid">
                        <view class="goods-classify-wrap">
                        <view class="classify-name" :style="{position:item.
                         position,left:item.left+'px',top:item.
                         top+'px',zIndex:10}">{{item.title}}</view>
```

```
                <view class="goods-list-wrap">
                    <view class="goods-list" v-for="(item2,index2) in
                    item.goods" :key="item2.gid">
                        <view class="image"><image :lazy-load="true"
                        :src="item2.image"></image></view>
                        <view class="goods-info">
                            <view class="goods-title">{{item2.
                            title}}</view>
                            <view class="sales">销　量:{{item2.sales}}
                            </view>
                            <!--如果是套餐-->
                            <view class="meal" v-if="item2.is_meal=='1'">
                                <view class="price"> ¥{{item2.price}}
                                </view>
                                <view class="meal-buy">购买</view>
                            </view>
                            <!--如果非套餐-->
                            <view class="price-wrap" v-else>
                                <view class="price"> ¥{{item2.
                                price}}</view>
                                <view class="amount">
                                <!--如果没有库存-->
                                <template v-if="item2.stock<=0">
                                    <view class="handle-amount">
                                        <text>已售光</text>
                                    </view>
                                </template>
                                <!--
                                如果菜品售卖在全部时间段
                                -->
                                <template v-else-if="item2.
                                sell_time_type=='0'">
                                    <view class="handle-amount">
                                        <!--如果数量大于0,显示减号-->
                                            <view class="dec" v-
show="parseInt(item2.amount)>0?true:false"></view>
                                            <!--为了防止scroll-view组件
内使用input卡顿的问题,使用view组件显示数量。单击该组件,显示input输入框,并隐藏view组
件。输入完成后使用@blur事件,隐藏input组件,显示view组件-->
                                                <view class="text"
v-if="!item2.isAmountInput" @click.stop="showAmountInput(index,index2)">
{{item2.amount}}</view><input v-if="item2.isAmountInput" :value="item2.
amount" :focus="item2.isAmountInputFocus" type="number" @click.stop=""
@blur="setAmount($event,item2.gid,item2.new_meal_items,item2.meal_
items,index,index2)" />
```

```
                                                                    <view class="inc">
                                                                </view>
                                                                </view>
                                                            </template>
                                                            <!--
                                                            如果菜品售卖在自定义时间段
                                                            -->
                                                            <template v-else>
                                                                <!--在售卖期-->
                                                                <template v-if="item2.
                                                                 sell_status=='1'">
                                                                    <view class="handle-
                                                                     amount">
                                                                        <view class="dec"
v-show="parseInt(item2.amount)>0?true:false"></view>
                                                                        <view class="text"
v-if="!item2.isAmountInput">{{item2.amount}}</view><input v-if="item2.
isAmountInput" :focus="item2.isAmountInputFocus" :value="item2.amount"
type="number" @click.stop="" @blur="setAmount($event,item2.gid,item2.new_
meal_items,item2.meal_items,index,index2)" />
                                                                                    <view
class="inc"></view>

                                                                    </view>
                                                                </template>
                                                                <template v-else>
                                                                    <text>非可售时间</text>
                                                                </template>
                                                            </template>
                                                        </view>
                                                    </view>
                                                </view>
                                            </view>
                                        </view>
                                    </view>
                                </block>
                            </view>
                            <view style="width:100%;height:190rpx;"></view>
                        </scroll-view>
                    </view>
                        <goods-details :show="isShowGoodsDetails" :branchShopId=
"branchShopId"></goods-details>
        </view>
```

```
</template>

<script>
    import {mapMutations} from "vuex";
    import GoodsDetails from './details';
    export default {
        name: "goods",
        data(){
            return {
                goodsScrollTop:0,
                //视图中id值为goods-wrap的scroll-view组件属性scroll-top的值
                classifyScrollTop:0,
                //视图中id值为classify的scroll-view组件属性scroll-top的值
                isShowGoodsDetails:false,    //是否显示商品详情
                classifys:[],                //左侧分类数据
                goods:[]                     //菜品数据
            }
        },
        props:{
            //是否显示菜品组件
            isShow:{
                type:Boolean,
                default:true
            },
            //分店的id
            branchShopId:{
                type:String,
                default:""
            },
            //菜品分类数据
            classifysData:{
                type:Array,
                default: ()=>[]
            },
            //菜品展示
            goodsData:{
                type:Array,
                default:()=>[]
            },
            //菜品的gid
            gid:{
                type:String,
```

```
            default:""
        },
        //菜品数量
        amount:{
            type:Number,
            default:0
        },
         //是否有上架或下架的菜品
        shelfGoods:{
            type:Boolean,
            default:false
        }
    },
    components:{
        GoodsDetails
    },
    mounted(){
        this.goodsWrapTop=0;//视图中id值为goods-wrap的scroll-view组件的top值
        this.goodsWrapLeft=0;//视图中id值为goods-wrap的scroll-view组件的left值
        this.goodsWrapHeight=0;//视图中id值为goods-wrap的scroll-view组件的高
        this.tempGoodsScrollTop=0;
        //记录滚动的位置，解决点击分类之后向上滑动，再次点击不滚动的bug
        this.goodsObj={};                    //存放菜品的id属性和值
        this.firstWatchGoods=true;           //第一次监听watch的goodsData发生变化
        //如果父组件的goodsData有菜品，直接赋值给goods
           if(this.goodsData && this.goodsData.length>0){
                 this.goods=this.goodsData;
            }
          //如果父组件的classifysData有菜品分类，直接赋值给classifys
           if(this.classifysData && this.classifysData.length>0){
                  this.classifys=this.classifysData;
             }
    },
    watch:{
        //监听菜品分类数据变化
        classifysData(val){
            this.classifys=val;
            //将菜品分类数据赋值给data()方法内部的classifys属性
        },
        goodsData(val){
```

```
            this.goods=val;     //将菜品的数据赋值给data()方法内部的goods属性
        },
        //监听上架或下架的菜品
        shelfGoods(val){
            if(val==true){ //如果有上架或下架的菜品
                //将this.goodsScrollTop和this.classifyScrollTop的值设置
                //为0，可以解决更新上架或下架菜品数据后，从微信的发现栏进入小程序，
                //scroll-view无法滚动到顶部的问题
                this.goodsScrollTop=0;
                this.classifyScrollTop=0;
            }
        },
        goods(val){
            if(this.firstWatchGoods){
                this.firstWatchGoods=false;
                //只允许第一次监听goodsData数据发生变化定位菜品的位置，如后面加入购物
                //车时更改菜品的数量，不再执行菜品定位this.getGoodsPosition()
                //方法，避免造成定位错误
                this.getGoodsPosition();          //菜品的位置
            }
        },
        classifys(val){
            //获取菜品分类的位置
            this.getClassifyPosition();
        }
    },
    methods:{
        ...mapMutations({
            SET_GOODS_POISTION:"goods/SET_GOODS_POISTION"//设置菜品的位置
        }),
        //获取菜品分类的位置
        getClassifyPosition(){
            let classify=uni.createSelectorQuery().in(this).select("#classify");
            classify.boundingClientRect(data=>{
                let classifyTop=data.top;//获取视图中id为classify的组件的top值
                let classifys=uni.createSelectorQuery().in(this).
                    selectAll("#classify .item");
                classifys.boundingClientRect(data=>{
                    if(data.length>0){
                        for(let i=0;i<data.length;i++){
                            //计算视图中class为item的组件top
```

```
                    this.classifys[i].top=data[i].top-classifyTop;
                }
            }
        }).exec();
    }).exec();
},
//获取菜品的位置
getGoodsPosition(){
    //获取商品最外面的元素高
    let goodsWrap=uni.createSelectorQuery().in(this).
        select("#goods-wrap");
    goodsWrap.boundingClientRect(data=>{
        this.goodsWrapTop=data.top-1;
        //视图中id为goods-wrap组件的top值
        this.goodsWrapLeft=data.left;
        //视图中id为goods-wrap组件的left值
        this.goodsWrapHeight=data.height;
        //视图中id为goods-wrap组件的高
        let classifyName = uni.createSelectorQuery().in(this).
            selectAll(".goods-classify-wrap");
        classifyName.boundingClientRect(data=>{
            let goodsTop=0;
            if(data.length>0){
                for(let i=0;i<data.length;i++){
                    //计算class为goods-classify-wrap的组件距id为
                    //goods-wrap的组件的top值
                    goodsTop=data[i].top-this.goodsWrapTop;
                    this.goods[i].top=goodsTop;
                    this.classifys[i].goodsTop=goodsTop;
                    //视图中class为goods-classify-wrap组件的top值
                    this.classifys[i].goodsHeight=data[i].height;
                    //视图中class为goods-classify-wrap组件的高
                }
            }

            let goodsItems=uni.createSelectorQuery().in(this).
                selectAll(".goods-list");
            goodsItems.boundingClientRect(data=>{
                if(data.length>0){
                    let itemIndex=0;
                    for(let i=0;i<this.goods.length;i++){
                        for(let j=0;j<this.goods[i].goods.length;j++){
```

```
                                    //视图中class为goods-list组件的top值
                                    this.goods[i].goods[j].itemTop=
                                        data[itemIndex].top;
                                    itemIndex++;
                                    let gid=this.goods[i].goods[j].gid;
                                    //将菜品的gid和值存放到this.goodsObj中，目
                                    //的是从菜品搜索页面进入点餐页面后的菜品定位
                                    this.goodsObj[gid]=this.goods[i].goods[j];
                                }
                            }
                            //定位某个菜品的位置
                            this.setItemPosition();
                        }
                    }).exec();
                }).exec();
            }).exec();
        },
        //定位某个菜品的位置
        setItemPosition(){
            //如果路由参数中存在gid，如从菜品搜索页面进入
            if(this.gid && this.goods.length>0){
                let itemTop=this.goodsObj[this.gid].itemTop;
                //获取视图中class为goods-list组件的top值
                //延迟可以解决scroll-view有时出现的不滚动的bug
                setTimeout(()=>{
                    this.goodsScrollTop=itemTop-this.goodsWrapTop*1.2;
                },600)
            }
        },
        //监听视图中id值为goods-wrap的scroll-view组件的scroll事件
        eventGoodsScroll(e){
            if(this.classifys.length>0){
                //记录滚动的位置，解决点击分类之后向上滑动，再次点击不滚动的bug
                this.tempGoodsScrollTop=e.detail.scrollTop;
                for(let i=0;i<this.classifys.length;i++){
                    //判断是否滚动到相应的菜品分类
                    if(e.detail.scrollTop>=this.classifys[i].goodsTop-90
                      && e.detail.scrollTop<=this.classifys[i].
                      goodsTop+this.classifys[i].goodsHeight){
                        //视图中的class的值为item的组件，如果存在active类
                        //样式，删除active
                        for(let key in this.classifys){
```

```
            if(this.classifys[key].active){
                this.classifys[key].active=false;
                break;
            }
        }
        //如果滚动到底部
        if(parseInt(e.detail.scrollTop/e.detail.
          scrollHeight*100)>=88){
            //视图中的class的值为item的组件，添加active类样式
            this.classifys[this.classifys.length-1].active=true;
        }else {          //如果没有到底部
            //视图中的class的值为item的组件，添加active类样式
            this.classifys[i].active = true;
        }
        //菜品分类滚动到相应的菜品位置
        this.classifyScrollTop=this.classifys[i].top;

        if(this.goods && this.goods.length>0){
            for(let key in this.goods){
                //如果菜品的标题是固定定位，则更改成静态定位
                if(this.goods[key].position=='fixed'){
                    this.goods[key].position="static";
                    this.goods[key].top=0;
                    this.goods[key].left=0
                //将更改的属性值同步到Vuex中，解决加入购物车时定
                //位错误问题
                 this.SET_GOODS_POISTION({index:key,
                     position:"static",top:0,left:0});
                    break;
                }
            }
            //将菜品的标题设置为固定定位
            this.goods[i].position="fixed";
            this.goods[i].top=this.goodsWrapTop;
            this.goods[i].left=this.goodsWrapLeft;
            //将更改的属性值同步到Vuex中，解决加入购物车时定位
            //错误问题
            this.SET_GOODS_POISTION({index:i,position:
            "fixed",top:this.goodsWrapTop,left:this.
            goodsWrapLeft})
        }
    }
```

```
            }

            //如果到顶部，则清除所有固定定位
            if(e.detail.scrollTop<=100){
                if(this.goods && this.goods.length>0) {
                    for (let key in this.goods) {
                        //如果菜品的标题是固定定位，则更改成静态定位
                        if (this.goods[key].position == 'fixed') {
                            this.goods[key].position = "static";
                            this.goods[key].top = 0;
                            this.goods[key].left = 0;
                            //将更改的属性值同步到Vuex中，解决加入购物车时定
                            //位错误问题
                            this.SET_GOODS_POISTION({index:key,
                                 position:"static",top:0,left:0});
                            break;
                        }
                    }
                }
            }
        };
    },
    //点击左侧分类
    clickClassify(index){
        if(this.classifys.length>0){
            for(let i=0;i<this.classifys.length;i++){
                //视图中的class的值为item的组件，如果存在active类样式，删除active
                if(this.classifys[i].active){
                    this.classifys[i].active=false;
                    break;
                }
            }
            //视图中的class的值为item的组件，添加active类样式
            this.classifys[index].active=true;
            this.$set(this.classifys,index,this.classifys[index]);
            if(index==0){
                this.goodsScrollTop=0;
            }else {
                //解决点击分类之后向上滑动，再次点击不滚动的bug
                this.goodsScrollTop=this.tempGoodsScrollTop;
                //使用延迟解决视图中id为goods-wrap的有时不滚动的bug
                setTimeout(()=>{
```

```
                        this.goodsScrollTop=this.classifys[index].goodsTop-40;
                    },50);
                }
            }
        },
        //显示数量输入框
        showAmountInput(index,index2){
           //执行pages/goods/index.vue文件中的showAmountInput方法
            this.$emit("showAmountInput",index,index2);
        },
        //修改数量
        //gid:菜品的id
        //new_meal_items:套餐规格，用于判断购物车是否有相同的菜品
        //meal_items:完整套餐规格，用于显示套餐中的菜品明细
        setAmount(e,gid,new_meal_items,meal_items,index,index2){
        //执行pages/goods/index.vue文件中的setAmount方法
          this.$emit("setAmount",e,gid,new_meal_items,meal_items,
          index,index2);
        }
    }
  }
</script>

<style scoped>
    .goods-area{width:100%;}
    .goods-main{width:100%;height:70vh;overflow:hidden;display:flex;justify-
     content: space-between;}
    .goods-main.classify{height:90%;width:156rpx;background-color:#fafafa;}
    .goods-main.classify.item{width:100%;height:72rpx;overflow:hidden;margin-
     bottom:40rpx;position: relative;z-index:1;}
    .goods-main.classify.item view{width:80%;font-size:28rpx;color:
     #999999;position:absolute;z-index:1;left:50%;top:50%;transform:
     translate(-50%,-50%)}
    .goods-main.classify.item.active{background-color:#FFFFFF;}
    .goods-main.classify.item.active view{color:#333333}

    .goods-main.goods-wrap{width:76%;height:90%;}
    .goods-main.goods-classify{width:100%;}
    .goods-main.goods-classify.classify-name{width:100%;height:auto;font-
     size:28rpx;color:#333333;font-weight: bold;background-color:#FFFFFF;}
    .goods-main.goods-list-wrap{width:100%;height:auto;margin-
     bottom:20rpx;}
```

```
    .goods-main.goods-list-wrap.goods-list{width:100%;height:auto;display:flex;j
     ustify-content: space-between;margin-top:40rpx;align-items: center}
    .goods-main.goods-list-wrap.goods-list.image{width:180rpx;height:180rpx;
     border-radius: 6px;}
    .goods-main.goods-list-wrap.goods-list.image image{width:100%;height:
     100%;border-radius: 6px;}
    .goods-main.goods-list.goods-info{width:62%; height:auto; margin-right:
     20rpx;}
    .goods-main.goods-list.goods-title{width:100%;height:80rpx;overflow:hidden;
     font-size:28rpx;color:#333333;font-weight: bold;}
    .goods-main.goods-list.sales{font-size:24rpx;color:#999999;
     margin-top:10rpx;}
    .goods-main.goods-list.points{font-size:24rpx;color:#D6001C;
     margin-top:10rpx;}
    .goods-main.goods-list.price{font-size:28rpx;color:#fb4e44;font-weight:
     bold;}
    .goods-main.goods-list.price-wrap{width:100%;margin-top:10rpx;display:
     flex;justify-content: space-between;}
    .goods-main.goods-list.meal{width:100%;margin-top:10rpx;display:flex;
     justify-content:space-between;}
    .goods-main.goods-list.meal.meal-buy{width:110rpx;height:50rpx;backg
     round-color:#D6001C;color:#FFFFFF;text-align:center;line-height:50rpx;font-
     size:28rpx;border-radius: 4px;}
    .goods-main.goods-list.amount{width:auto;height:auto;display:flex;}
    .goods-main.goods-list.amount>text{font-size:24rpx;color:#999999;}
    .goods-main.goods-list.handle-amount{width:auto;height:60rpx;display:flex;
     align-items: center;justify-content:flex-end}
    .oods-main.goods-list.handle-amount.dec{width:55rpx;height:55rpx;
     background-image:url("~@/static/images/main/dec.png");background-
     size:100%;background-position: center;background-repeat: no-repeat;}
    .goods-main.goods-list.handle-amount.inc{width:55rpx;height:55rpx;
     background-image:url("~@/static/images/main/inc.png");background-
     size:100%;background-position: center;background-repeat: no-repeat;}
    .goods-main.goods-list.handle-amount.text{font-size:28rpx;color:
     #333333;margin-right:10rpx;margin-left:10rpx;width:50rpx;text-align:
     center;}
    .goods-main.goods-list.start-sell{width:100%;height:auto;
     font-size:24rpx;color:#999999;}
    .goods-main.goods-list .handle-amount input{width:50rpx;height:100%;text-
     align: center;margin-right:10rpx;margin-left:10rpx;}
</style>
```

此组件的逻辑比较复杂，但代码注释很清晰，请读者仔细阅读。强烈建议结合视频教程学习。需要注意，this.firstWatchGoods全局变量的作用是只允许第一次监听goodsData数据发生变化定位菜品的位置，如加入购物车时更改菜品的数量后，不再执行菜品定位this.getGoodsPosition()方法，避免造成定位错误。其代码在watch监听的goods数据内实现。

上述代码中有一个goods-details菜品详情组件，11.5.3节中即对此进行开发。

扫一扫，看视频

11.5.3 开发菜品详情组件

菜品详情组件演示效果如图11-9所示。

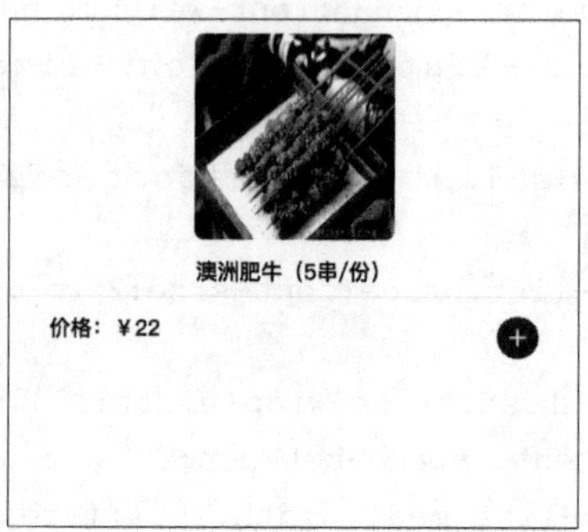

图 11-9 菜品详情组件演示效果

在components/goods文件夹下创建details.vue文件，由于该组件比较简单，重用性不是很强，因此可以在该组件中使用Vuex获取服务端数据。首先请求服务端数据，在api/goods/index.js文件中新增代码如下：

```
...
//菜品详情
export function getGoodsDetailsData(data){
    return request(config.baseApi+"/v1/goods/details","post",data);
}
```

getGoodsDetailsData()函数可以从服务端接口获取菜品详情数据，接下来对接到Vuex中，在store/goods/index.js文件中新增代码如下：

```
import Vue from "vue";
import {getClassifyData,getGoodsData,getGoodsDetailsData} from "../../api/goods";
export default {
    namespaced:true,
    state:{
        ...
        goodsDetails:{},            //菜品详情数据
    },
    mutations:{
```

```
        ...
        //设置菜品详情
        ["SET_GOODS_DETAILS"](state,payload){
            state.goodsDetails=payload.goodsDetails;
        }
    },
    actions:{
        ...
        //菜品详情
        getGoodsDetails(conText,payload){
            getGoodsDetailsData(payload).then(res=>{
                if(res.code==200){
                    conText.commit("SET_GOODS_DETAILS",{goodsDetails:res.data});
                }
            })
        }
    }
}
```

以上加粗代码为新增代码，state内部的goodsDetails属性值可以存储菜品详情数据；mutations内部的SET_GOODS_DETAILS方法可以设置state内部goodsDetails的属性值；actions内部的getGoodsDetails()方法内部使用getGoodsDetailsData()方法，可以从服务端获取菜品详情数据，并使用commit提交到SET_GOODS_DETAILS，将数据保存到state.goodsDetails中。

接下来将Vuex对接到components/goods/details.vue文件中，该文件中的代码如下：

```
<template>
    <view>
        <view class="mask" v-show="show" @click="close"></view>
        <view :class="{details:true, show:show}">
            <view class="goods">
                <view class="image">
                    <image :src="goodsDetails.image"></image>
                </view>
                <view class="text">{{goodsDetails.title}}</view>
            </view>
            <view class="goods-info">
                <view class="price-amount">
                    <view class="price">价格:¥{{goodsDetails.price}}</view>
                    <view class="handle-amount">
                        <view class="dec" v-show="amount>0?true:false"></view>
                        <view class="text" v-if="!isAmountInput" @click="show
                         AmountInput">{{amount}}</view>
                        <input v-if="isAmountInput" :focus=" isAmountInputFocus"
                         :value="amount" type="number" @blur="setAmount" />
```

```
                    <view class="inc"></view>
                </view>
            </view>
            <scroll-view scroll-y="true" class="content"><rich-text
             :nodes="goodsDetails.content"></rich-text></scroll-view>
        </view>
    </view>
  </view>
</template>

<script>
  import {mapActions,mapState} from "vuex";
  export default {
    name: "goods-details",
    data(){
        return {
            isAmountInput:false,          //是否显示input数量输入框
            isAmountInputFocus:false      //input数量输入框获取焦点
        }
    },
    props:{
        //是否显示组件
        show:{
            type:Boolean,
            default: false
        },
        //菜品的id
        gid:{
            type: String,
            default: ""
        },
        //分店的id
        branchShopId:{
            type:String,
            default:""
        },
        //菜品数量
        amount:{
            type:Number,
            default:0
        }
    },
    watch:{
        //监听菜品id的变化
        gid(val){
```

```
                //获取菜品详情数据
                this.getGoodsDetails({branch_shop_id:this.branchShopId,gid:val});
            }
        },
        computed:{
            ...mapState({
                goodsDetails:state=>state.goods.goodsDetails  //菜品详情数据
            })
        },
        methods:{
            ...mapActions({
                getGoodsDetails:"goods/getGoodsDetails"    //获取菜品详情数据
            }),
            //隐藏组件
            close(){
                this.$emit("close");
            },
            //修改数量
            setAmount(e){
                //执行components/goods/index.vue文件中的setAmount方法
                this.$emit("setAmount",e,this.gid,"","");
                this.hideAmountInput();
            },
            //显示数量输入框
            showAmountInput(){
                this.isAmountInput=true;
                this.isAmountInputFocus=true;
            },
            //隐藏数量输入框
            hideAmountInput(){
                this.isAmountInput=false;
                this.isAmountInputFocus=false;
            }
        }
    }
</script>

<style scoped>
    .mask{width:100%;height:100%;position: fixed;z-index:88;bottom:0px;left:
     0px;background-color:rgba(0,0,0,0.6)}
    .details{width:100%;height:795rpx;background-color:#FFFFFF;position:fixed;bottom
     :0px;left:0px;z-index:89;transform: translateY(100%);transition:all 0.3s;}
    .details.goods{width:100%;height:330rpx;border-bottom:1px solid #EFEFEF}
    .details.image{width:260rpx;height:260rpx;margin:0 auto;margin-top:20rpx;}
    .details.image image{width:100%;height:100%;border-radius: 6px;}
```

```
    .details.goods.text{width:auto;height:auto;margin:0 auto;margin-top:
     20rpx;white-space: nowrap; font-size: 28rpx;color: #333333; text-
     align:center;font-weight: bold;}

    .details.goods-info{width:85%;margin:0 auto;font-size:28rpx; color:
     #333333;font-weight: bold;}
    .details.goods-info.price-amount{width:100%;display:flex;justify-
     content: space-between;margin-top:20rpx;}
    .details.goods-info.handle-amount{width:auto;height:60rpx;display:flex;
     align-items: center;justify-content:flex-end}
    .details.goods-info.handle-amount .dec{width:55rpx;height:55rpx;
     background-image:url("~@/static/images/main/dec.png");background-
     size:100%;background-position: center;background-repeat: no-repeat;}
    .details.goods-info.handle-amount .inc{width:55rpx;height:55rpx;
     background-image:url("~@/static/images/main/inc.png");background-
     size:100%;background-position: center;background-repeat: no-repeat;}
    .details.goods-info.handle-amount input{width:50rpx;height:100%;text-
     align: center;margin-right:10rpx;margin-left:10rpx;}
    .details.goods-info.handle-amount .text{font-size:28rpx; color:
     #333333;margin-right:10rpx;margin-left:10rpx;width:50rpx;text-align:center;}
    .details.goods-info.content{width:100%;height:200rpx;margin-
     top:20rpx;overflow:hidden;}
    .details.show{transform: translateY(0%)}
</style>
```

该组件的代码比较简单，注释也很清晰，请读者仔细阅读。需要注意，props属性内部的gid选项为菜品编号，该编号是从服务端接口异步获取的，所以需要使用watch监听gid值的变化。如果有变化并获取到了gid的值，则调用this.getGoodsDetails()方法获取菜品详情数据并渲染到视图中。

接下来在components/goods/index.vue文件中添加操作菜品详情组件的代码，如下所示：

```
<template>
    <view class="goods-area" v-show="isShow">
        ...
        <goods-details :show="isShowGoodsDetails" :branchShopId ="branchShopId"
         :gid="gid" :amount="amount" @close="isShowGoodsDetails=false" @
         setAmount="setAmount"></goods-details>
    </view>
</template>

<script>
    import GoodsDetails from './details';
    export default {
        name: "goods",
        ...
```

```
        methods:{
            ...
            //显示商品详情
            //gid:菜品的id
            //isMeal:是否套餐
            //amount:菜品数量
            //sell_status:售卖状态,1为售卖期，0为非售卖时间
            //stock:库存
            showGoodsDetails(gid,isMeal,amount,sell_status,stock){
                //如果在售卖期并且有库存
                    if(sell_status=='1' && stock>0){
                    if(isMeal!='1'){                        //如果不是套餐
                        this.isShowGoodsDetails=true;   //显示菜品详情组件
                    }
                    //执行pages/goods/index.vue文件中的showGoodsDetails()方法
                    this.$emit("showGoodsDetails",gid,isMeal,amount,sell_status,stock);
                    }
            }
        }
    }
</script>
```

在显示菜品详情组件方法showGoodsDetails()内部使用this.$eimt()方法调用pages/goods/index.vue文件中的showGoodsDetails()方法，实现显示/隐藏菜品详情组件。

在pages/goods/index.vue文件中新增代码如下：

```
<template>
    <view class="page">
        ...
        <!--菜品展示组件-->
        <goods :branchShopId="branch_shop_id" :isShow="goodsAction"
         :gid="gid" :amount="amount" @setAmount="setAmount" @showAmountInput
         ="showAmountInput" :classifysData="classifys" :goodsData="goods" @s
         howGoodsDetails="showGoodsDetails"></goods>
        ...
</template>

<script>
    ...
    export default {
        name: "goods",
        ...
        methods:{
            ...
            //显示商品详情
```

```
                showGoodsDetails(gid,isMeal,amount,sell_status,stock){
                    if(isMeal=='1'){                          //如果是套餐
                        //跳转到套餐页面
                        this.pushPage('/pages/goods/meal?gid='+gid+"&branch_
                            shop_id="+this.branch_shop_id+"&table_code="+this.
                            table_code);
                    }else{
                        this.gid=gid;                         //菜品的id
                        this.amount=parseInt(amount);         //菜品数量
                    }
                },
            }
        }
    </script>
```

点餐页面中的showGoodsDetails()方法实现显示商品详情组件的逻辑，如果是套餐，则进入套餐页面；如果是单品，则将gid和amount传递给菜品展示组件，再从菜品展示组件传递给菜品详情组件。

扫一扫，看视频

11.5.4　开发套餐页面

如果菜品是套餐，则单击菜品可以进入套餐页面，在套餐页面可以选择自己想要的菜品，组成一个新的套餐。套餐页面的预览效果如图11-10所示。

图 11-10　套餐页面的预览效果

先在pages/goods文件夹下创建meal.vue文件，然后配置路由。在pages.json文件中新增代码如下：

```
{
    "pages": [
        ...
        {
            "path": "pages/goods/meal",
            "style": {
                "navigationBarTitleText": "套餐",
                "disableScroll": false
            }
        }
    ],
    ...
}
```

接下来请求服务端数据，在api/goods/index.js文件中新增代码如下：

```
...

//显示套餐里可替换菜品数据
export function getMealData(data){
    return request(config.baseApi+"/v1/meal/replace","post",data);
}

//显示套餐详情
export function getMealDetailsData(data){
    return request(config.baseApi+"/v1/meal/details","post",data);
}
```

getMealData()函数可以从服务端接口获取套餐里的数据，getMealDetailsData()函数可以从服务端接口获取套餐详情数据。接下来对接到Vuex中，在store文件夹下创建meal/index.js文件，该文件中的代码如下：

```
import {getMealData,getMealDetailsData} from "../../api/goods";
export default {
    namespaced:true,
    state:{
        title:"",              //套餐标题
        mealData:[]            //可替换套餐菜品列表
    },
    mutations:{
        //设置可替换套餐菜品列表
        ["SET_MEAL_DATA"](state,payload){
            state.mealData=payload.mealData;
        },
```

```
            //设置套餐标题
            ["SET_TITLE"](state,payload){
                state.title=payload.title;
            }
        },
        actions:{
            //获取套餐可替换数据
            getMeal(conText,payload){
                getMealData(payload).then(res=>{
                    if(res.code==200){
                        conText.commit("SET_MEAL_DATA",{mealData:res.data});
                        if(payload.success){
                            payload.success()
                        }
                    }
                })
            },
            //获取套餐详情
            getMealDetails(conText,payload){
                getMealDetailsData(payload).then(res=>{
                    if(res.code==200){
                        conText.commit("SET_TITLE",{title:res.data.title});
                        if(payload.success){
                            payload.success();
                        }
                    }
                })
            }
        }
    }
```

state内部的title属性值为套餐标题，mealData可以存储套餐菜品数据。mutations内部的SET_MEAL_DATA方法可以设置state内部mealData的属性值，SET_TITLE方法可以设置state内部title的属性值。actions内部的getMeal()方法内部使用getMealData()函数，可以从服务端获取套餐数据，并使用commit提交到SET_MEAL_DATA，将获取的数据保存到state.mealData中。getMealDetails()方法内部使用getMealDetailsData()函数,可以从服务端获取套餐详情数据,并使用commit提交到SET_TITLE，将获取的数据保存到state.title中。

接下来注册到Vuex的模块中，在store/index.js文件中新增代码如下：

```
...
import meal from './meal';
```

```
...

const store = new Vuex.Store({
    modules:{
        ...
        meal
    }
});

export default store
```

开发套餐页面，在pages/goods/meal.vue文件中的代码如下：

```
<template>
    <view class="page">
        <view class="main">
            <view class="goods-main" v-for="(item,index) in mealData" :key="index">
                <view class="nav">
                    <view class="icon"></view>
                    <view class="title">{{item.title}} (可选1份)</view>
                </view>
                <view class="goods-wrap">
                    <view class="goods-list" v-for="(item2,index2) in item.
                     goods" :key="index2" @click="selectGoods(index,index2)">
                        <view class="image">
                            <image :src="item2.image" alt=""></image>
                        </view>
                        <view class="text">{{item2.title}}</view>
                        <view class="price" v-if="item2.offset_price!=0">
                            {{item2.offset_price>0?'+'+item2.offset_
                             price:item2.offset_price}} 元
                        </view>
                        <view :class="{elc:true, active:item2.active}"></view>
                    </view>
                </view>
            </view>
        </view>
        <view style="width:100%;height:120rpx;"></view>
        <view class="cart-main">
            <view class="total">小计:¥{{total}}</view>
            <view class="line"></view>
            <view class="cart">
                <view class="cart-icon"></view>
```

```
                <view class="text">加入购物车 ></view>
            </view>
        </view>
    </view>
</template>

<script>
    import {mapActions,mapState} from 'vuex';
    export default {
        name: "meal",
        onLoad(opts){
            this.gid=opts.gid;                                          //套餐的id
            this.branch_shop_id=opts.branch_shop_id;        //分店的id
            this.table_code=opts.table_code;                    //桌号
            this.mealItems=[];                                          //套餐里的菜品明细
            this.simpleMealItems=[];
            //套餐里的菜品的gid，用于添加购物车时判断是否有相同的套餐
            //获取可替换菜品
            this.getMeal({gid:this.gid,branch_shop_id:this.branch_shop_id,
                 success:()=>{
                    //设置套餐里的菜品明细数据
                    this.setMealItems();
                }});
            //获取菜品详情
            this.getMealDetails({branch_shop_id:this.branch_shop_id,
                 gid:this.gid,success:()=>{
                    //动态设置导航栏标题
                    uni.setNavigationBarTitle({
                        title: this.title
                    });
                }});
        },
        computed:{
            ...mapState({
                mealData:state=>state.meal.mealData,        //可替换菜品的数据
                title:state=>state.meal.title
            }),
            //计算选择菜品的金额
            total(){
                let total=0;
                if(this.mealData.length>0){
                    for(let i=0;i<this.mealData.length;i++){
```

```
                    for(let j=0;j<this.mealData[i].goods.length;j++){
                        if(this.mealData[i].goods[j].active){
                            total+=parseFloat(this.mealData[i].
                                goods[j].meal_price);
                        }
                    }
                }
                total=parseFloat(total.toFixed(2));
            }
            return total;
        }
    },
    methods:{
        ...mapActions({
            "getMeal":"meal/getMeal",                    //获取可替换菜品的数据
            "getMealDetails":"meal/getMealDetails"       //获取套餐详情
        }),
        //重新组装mealItems套餐菜品明细字段数据
        setMealItems(){
            let mealItems=[];
            if(this.mealData.length>0){
                for(let i=0;i<this.mealData.length;i++){
                    for(let j=0;j<this.mealData[i].goods.length;j++){
                        //如果选择了菜品
                        if(this.mealData[i].goods[j].active){
                            //组装套餐菜品明细字段数据
                            /*
                            gid:菜品的gid
                            title:菜品标题
                            price:菜品价格
                            pack_price:包装费
                            dis_amount:初始化数量，购物车里不可更改的数量
                             */
                             mealItems.push({gid:this.mealData[i].goods[j].
gid,title:this.mealData[i].goods[j].title,price:this.mealData[i].goods[j].meal_
price,amount:1,pack_price:this.mealData[i].goods[j].pack_price,dis_amount:1});
                        }
                    }
                }
                //计算数量
                if(mealItems.length>0){
                    let miObj={};
```

```
                let newMealItems=[];            //套餐里菜品明细
                let simpleMealItems=[];         //套餐里菜品的gid
                for(let i=0;i<mealItems.length;i++){
                    if(!miObj[mealItems[i]["gid"]]){//如果没有重复的菜品
                        miObj[mealItems[i]["gid"]]=true;
                            newMealItems.push({gid:mealItems[i]["gid"],
title: mealItems[i]["title"],price:mealItems[i]["price"],amount:mealItems[i]
["amount"],pack_price:mealItems[i]['pack_price'],dis_amount:mealItems[i]
["dis_amount"]});
                        simpleMealItems.push({gid:mealItems[i]["gid"]})
                    }else{                          //如果有重复的菜品
                        if(newMealItems.length>0){
                            for(let j=0;j<newMealItems.length;j++){
                                //如果套餐里的默认菜品与选择的菜品相同
                                if(newMealItems[j]["gid"]==
                                mealItems[i]["gid"]){
                                    //数量加1
                                        newMealItems[j]["amount"]=
                                        ++newMealItems[j] ["amount"];
                                    //设置购物车里不可更改的数量
                                    newMealItems[j]["dis_amount"]
                                    =newMealItems[j]["amount"];
                                    break;
                                }
                            }
                        }
                    }
                }
                this.mealItems=newMealItems;
                //将组装好的套餐菜品明细赋值给全局变量this.mealItems
                this.simpleMealItems=simpleMealItems;
                //将组装好的菜品的gid赋值给全局变量this.simpleMealItems
            }
        }
    },
    //选择菜品
    selectGoods(index,index2){
        for(let i=0;i<this.mealData[index].goods.length;i++){
            if(this.mealData[index].goods[i].active){
                this.mealData[index].goods[i].active=false;
                break;
            }
```

```
                }
                this.mealData[index].goods[index2].active=true;
                this.$set(this.mealData[index].goods,index2,this.
                    mealData[index].goods[index2]);
                this.setMealItems();
            }
        }
    }
</script>

<style scoped>
    .page{width:100%;min-height:100%;margin:0 auto;background-color:
    #EFEFEF;}
    .page.main{width:95%;height:auto;background-color:#FFFFFF;margin:0
    auto;margin-top:30rpx;border-radius: 6px;box-shadow: 3px 3px 5px
    #e7e7e7;overflow:hidden;}
    .page.goods-main{width:90%;margin:0 auto; margin-top:20rpx;
    margin-bottom:20rpx;}
    .page.main.nav{width:100%;height:70rpx;display: flex;align-items: center;box-
    sizing: border-box;border-bottom:1px solid #EFEFEF;margin:0 auto;}
    .page.main.nav.icon{width:20rpx;height:45rpx;background-
    color:#f6ab00;border-radius: 20px;margin-right:20rpx;}
    .page.main.nav.title{font-size:28rpx;color:#333333;font-weight: bold;}
    .page.main.goods-wrap{width:100%;display:flex;flex-wrap:wrap;}
    .page.main.goods-list{width:200rpx;height:317rpx;box-shadow: 0px 3px
    5px #e7e7e7;border-radius: 6px;margin-top:30rpx;margin-left:1%;margin-
    right:1%;position: relative;}
    .page.main.goods-list.image{width:160rpx;height:160rpx;margin:0
    auto;margin-top:20rpx;}
    .page.main.goods-list.image image{width:100%;height:100%;border-radius: 6px;}
    .page.main.goods-list.text{width:160rpx;font-size:24rpx;text-align:
    center;white-space: nowrap;color:#333333;overflow:hidden;text-overflow:
    ellipsis;margin:0 auto;margin-top:20rpx;}
    .page.main.goods-list.price{text-align:center;font-size:24rpx; color:
    #ff0000;margin-top:5rpx;}
    .page.main.goods-list.elc{width:50rpx;height:50rpx;border-radius:
    100%;border:2px solid #EFEFEF;position: absolute;right:0px;top:0px;
    background-color:#FFFFFF;}
    .page.main.goods-list .elc.active{background-image:url("~@/static/
    images/goods/right.png");background-size:100%;background-repeat: no-
    repeat;background-position: center;}
```

```
    .cart-main{width:100%;height:100rpx;background-color:#f6ab00;position:
    fixed;z-index:90;left:0px;bottom:0px;display:flex;align-items:
    center;padding:0px 20rpx;box-sizing: border-box;}
    .cart-main.total{font-size:28rpx;color:#FFFFFF;width:408rpx;}
    .cart-main.line{width:2px;height:60rpx;background-color:rgba(255,255,255,
    0.4);margin-right:37rpx;}
    .cart-main.cart{width:auto;height:auto;display:flex;align-items: center;}
    .cart-main.cart.cart-icon{width:60rpx;height:60rpx;background-image:url("~@/
    static/images/goods/cart2.png");background-size:100%;background-
    position:center;background-repeat: no-repeat;margin-right:25rpx;}
    .cart-main.cart.text{font-size:28rpx;color:#FFFFFF;}
</style>
```

以上代码实现了显示套餐里的菜品，用户可以自由选择可替换的菜品，并且计算出价格的功能。加入购物车功能会在11.7节中讲解。

扫一扫，看视频

11.5.5 开发商家信息组件

商家信息组件预览效果如图11-11所示。

137XXXXXXXX

北京房山区XXXXXXXXXXXXXXX层08

营业时间10:00-23:00

图 11-11 商家信息组件预览效果

商家信息组件的数据是从pages/goods/index.vue文件中传递过去的，点击手机号可以拨打电话，点击地址可以查看商家所在位置。接下来先从服务端请求商家信息，在api/business/index.js文件中新增代码如下：

```
...
//显示商铺信息
export function getShopInfoData(branch_shop_id){
    return request(config.baseApi+"/v1/business/info", "get",{
    branch_shop_id:branch_shop_id});
}
```

getShopInfoData()函数可以从服务端获取商品信息，接下来对接到Vuex中，在store/business/index.js文件中新增代码如下：

```
import {getShopData,getShopInfoData} from "../../api/business";
export default {
    namespaced:true,
    state:{
       ...
        shopInfo:{},            //显示商铺信息
    },
    mutations: {
       ...
        //设置商铺信息
        ["SET_SHOP_INFO"](state,payload){
            state.shopInfo=payload.shopInfo;
        }
    },
    actions:{
        ...
        //显示商铺信息
        showShopInfo(conText,payload){
            getShopInfoData(payload.branch_shop_id).then(res=>{
                if(res.code==200){
                    conText.commit("SET_SHOP_INFO",{shopInfo:res.data});
                    if(payload.success){
                        payload.success(res.data);
                    }
                }
            })
        },
    }
}
```

以上加粗代码为新增代码，state内部的shopInfo属性值可以存储商铺信息;mutations内部的SET_SHOP_INFO方法可以设置state内部shopInfo的属性值;actions内部的showShopInfo()方法内部使用getShopInfoData()函数，可以从服务端获取商品信息数据，并使用commit提交到SET_SHOP_INFO，将获取的数据保存到state.shopInfo中。

将Vuex对接到pages/goods/index.vue文件中，该文件中的新增代码如下：

```
<template>
    <view class="page">
        ...
        <!--商家信息组件-->
        <business-info :isShow="businessAction" :branchShopId="branch_shop_
         id" :infoData="shopInfo"></business-info>
```

```
        ...
    </view>
</template>

<script>
    ...
    export default {
        name: "goods",
        ...
        onShow(){
            ...
            //商家信息
            this.showShopInfo({branch_shop_id:this.branch_shop_id});
        },
        computed:{
            ...mapState({
                ...
                "shopInfo":state=>state.business.shopInfo //商家信息
            })
        },
        ...
        methods:{
            ...mapActions({
                ...
                "showShopInfo":"business/showShopInfo"     //获取商家信息
            }),
            ...
        }
    }
</script>
```

以上加粗代码为新增代码。接下来开发商家信息组件，在components/business/info.vue文件中的代码如下：

```
<template>
    <scroll-view scroll-y="true" class="business-info-main" v-show="isShow">
        <view class="row-area" v-if="infoData.phone?true:false">
            <view class="col1 cellphone-icon"></view>
            <view class="col2"><text @click="makePhoneCall(''+infoData.
            phone+'')">{{infoData.phone}}</text></view>
        </view>
```

```
        <view class="row-area" v-if="infoData.address?true:false" @click=
         "businessMap()">
            <view class="col1 address-icon"></view>
            <view class="col2"><text selectable="true">{{infoData.
             address}}</text></view>
        </view>
        <view class="row-area" v-if="infoData.full_notice?true:false">
            <view class="col1 notice-icon"></view>
            <view class="col2">
                <rich-text :nodes="infoData.full_notice"></rich-text>
            </view>
        </view>
    </scroll-view>
</template>

<script>
    export default {
        name: "business-info",
        props:{
            //是否显示组件信息
            isShow:{
                type:Boolean,
                default:false
            },
            //商家信息数据
            infoData:{
                type:Object,
                default:()=>{}
            },
            //分店的id
            branchShopId:{
                type:String,
                default:""
            }
        },
        methods:{
            //拨打电话
            makePhoneCall(number){
                //在H5平台拨打电话
                // #ifdef H5
                window.location.href="tel:"+number+"";
                // #endif
                //在小程序中拨打电话
                // #ifndef H5
                uni.makePhoneCall({
```

```
                    phoneNumber: number
                });
                // #endif
            },
            //跳转到商家位置页面
            businessMap(){
                if(this.infoData.lng && this.infoData.lat){
                    this.pushPage('/pages/business_map/index?branch_shop_
                        id='+this.branchShopId+'&lng='+this.infoData.
                        lng+'&lat='+this.infoData.lat+'');
                }
            },
            //跳转页面
            pushPage(url){
                uni.navigateTo({
                    url:url
                })
            },
        }
    }
</script>

<style scoped>
    .business-info-main{width:100%;height:55vh;font-size:28rpx;}
    .business-info-main.row-area{width:100%;height:auto;display:flex;box-
     sizing: border-box;padding:10rpx 22rpx;}
    .business-info-main.col1{width:35rpx;height:35rpx;}
    .business-info-main.col2{width:700rpx;overflow:hidden;}
    .business-info-main.col1.cellphone-icon{background-image:url("~@/
     static/images/goods/cellphone.png");background-size:100%;background-
     repeat: no-repeat;background-position: center;margin-right:10rpx;}
    .business-info-main.col1.address-icon{background-image:url("~@/static/
     images/goods/address.png");background-size:100%;background-repeat: no-
     repeat;background-position: center;margin-right:10rpx;}
    .business-info-main.col1.notice-icon{background-image:url("~@/static/
     images/goods/notice.png");background-size:100%;background-repeat: no-
     repeat;background-position: center;margin-right:10rpx;}

    .points-tip{width:75%;height:25px;position:fixed;z-index:1;left:23%;bottom
     :95rpx;background-color:#FFFFFF;text-align: center;line-height:25px;font-
     size:28rpx;color:#E30019;border-radius: 5px;border:1px solid
     #f6ab00;overflow:hidden;white-space: nowrap;text-overflow: ellipsis;}
</style>
```

以上代码注释很清晰，功能也很简单。点击地址详情后调用businessMap()方法，进入商家位置页面，可以在地图上看到商家所在的地理位置。

11.5.6　查看商家位置

扫一扫，看视频

可以使用map组件查看商家位置，预览效果如图11-12所示。

图 11-12　商家位置预览效果

首先在pages文件夹下创建business_map/index.vue文件，然后配置路由。在pages.json文件中新增代码如下：

```
{
    "pages": [
        ...
        {
            "path": "pages/business_map/index",
            "style": {
                "navigationBarTitleText": "商家位置",
                "disableScroll": false
            }
        }
    ],
    ...
}
```

在pages/business_map/index.vue文件中的代码如下：

```
<template>
    <view class="page">
        <map class="map" :latitude="lat" :longitude="lng" :markers="markes"></map>
    </view>
</template>

<script>
    import {mapState} from 'vuex';
    export default {
        name: "business-map",
        data(){
            return {
                lng:"",                                     //商家的经度
                lat:"",                                     //商家的纬度
                markes:[
                    {
                        id:1,
                        latitude: 0,                        //图标的纬度
                        longitude: 0,                       //图标的经度
                        iconPath: '/static/images/business_map/map_pos.png',
                        //显示的图标
                        width:30,                           //图标的宽
                        height:30,                          //图标的高
                        callout:{                           //自定义标记点上方的气泡窗口
                            fontSize:14,                    //字体大小
                            content:"",                     //文本内容
                            color:"#000000",                //文本颜色
                            bgColor:"#FFFFFF",              //背景色
                            padding:8,                      //文本边缘留白
                            borderWidth:1,                  //边框
                            borderColor:"#CCCCCC",          //边框颜色
                            borderRadius:4,                 //callout边框圆角
                            display:"ALWAYS"                //常显
                        }
                    }
                ]
            }
        },
        onLoad(opts){
            this.branch_shop_id=opts.branch_shop_id?opts.branch_shop_id:"";
            //分店的id
            this.lng=opts.lng?opts.lng:"";              //商家的经度
            this.lat=opts.lat?opts.lat:"";              //商家的纬度
            this.markes[0].latitude=this.lat;           //将商家的纬度赋值给图标的纬度
            this.markes[0].longitude=this.lng;          //将商家的经度赋值给图标的经度
```

```
            //callout的文本内容
            this.markes[0].callout.content=this.shopInfo.branch_shop_
                name+"\r\n"+this.shopInfo.address;
        },
        computed:{
            ...mapState({
                "shopInfo":state=>state.business.shopInfo          //商家信息
            })
        }
    }
</script>

<style scoped>
    .page{width:100%;height:100vh;overflow:hidden;}
    .map{width:100%;height:100%;}
</style>
```

data()函数内部返回的markes属性值为标记点的配置选项；在onLoad()函数内部可以动态设置标记点内部参数的值，如经度、纬度和气泡窗口（callout）等内容。

11.5.7　在小程序中关注公众号

由于在微信小程序中无法自由地实现关注公众号功能，因此可以写一篇关注公众号的文章，使用web-view组件实现关注公众号的功能。在点餐页面点击“关注”按钮，跳转到关注公众号页面，预览效果如图11-13所示。

扫一扫，看视频

图 11-13　关注公众号页面预览效果

首先在pages/main文件夹下创建focus.vue文件，然后配置路由。在pages.json文件新增代码如下：

```
{
    "pages": [
        ...
        {
            "path": "pages/main/focus",
            "style": {
                "navigationBarTitleText": "关注公众号",
                "navigationStyle": "custom",
                "disableScroll": true
            }
        }
    ],
    ...
}
```

在pages/main/focus.vue文件中的代码如下：

```
<template>
    <view class="page">
        <web-view:webview-styles="webviewStyles" src="https://mp.weixin.
         qq.com/s?__biz=MzI0NDcxNzM4MQ==&mid=2247483659&idx=1&sn=33532f3d353f
         f99e41e938637a937735&chksm=e958c321de2f4a3750e2c4c0082bc23db97ef72a935
         fd8724d525882393b737db2b6cab751a8&mpshare=1&scene=23&srcid&sharer_
         sharetime=1579340139968&sharer_shareid=01f30d86ddfab21e70c219e68b0471
         55%23rd"></web-view>
    </view>
</template>

<script>
    export default {
        name: "focus",
        data(){
            return {
                webviewStyles: {
                    progress: {
                        color: '#43A243'
                    }
                }
            }
        }
    }
```

```
</script>

<style scoped>
    .page{width:100%;height:100vh;overflow:hidden;}
</style>
```

web-view组件的src属性值是微信公众号文章的地址，将公众号的文章分享出来即可获取该文章的URL地址。

11.6 会员登录

微信小程序最常用的功能就是微信授权会员登录，这样可以自动获取会员信息并存入数据库中。会员登录页面预览效果如图11-14所示。

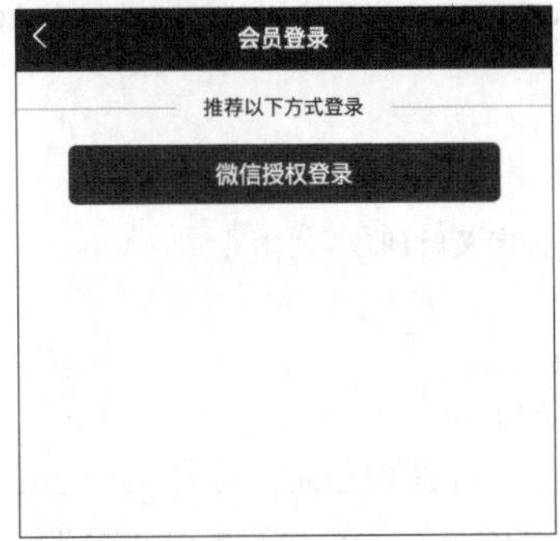

图 11-14 会员登录页面预览效果

会员登录页面的接口在“接口文档”文件夹下的“会员登录、注册、信息.docx”文件中。

11.6.1 微信授权实现一站式登录

扫一扫，看视频

在菜品展示组件中点击添加菜品按钮，需要验证会员是否登录，如果登录则将菜品加入购物车，如果未登录则跳转到登录页面。在components/goods/index.vue文件中新增代码如下：

```
<template>
        <view class="goods-area" v-show="isShow">
            ...

            <!--
            如果菜品售卖于全部时间段
            -->
            <template v-else-if="item2.sell_time_type=='0'">
                <view class="handle-amount">
                    <!--如果数量大于0，显示减号-->
```

```
                    <view class="dec" v-show="parseInt(item2.amount)>0?
                     true:false"></view>
                    <!--
                    为了防止scroll-view组件内使用input卡顿的问题，先使用view组件显
                    示数量。点击该组件，显示input输入框，并隐藏view组件，输入完成后使
                    用@blur事件隐藏input组件，显示view组件
                    -->
                      <view class="text" v-if="!item2.isAmountInput" @click.stop
="showAmountInput(index,index2)">{{item2.amount}}</view><input v-if="item2.
isAmountInput" :value="item2.amount" :focus="item2.isAmountInputFocus"
type="number" @click.stop="" @blur="setAmount($event,item2.gid,item2.new_
meal_items,item2.meal_items,index,index2)" />
                      <view class="inc" @click.stop="incAmount($event,item2.
gid,item2.new_meal_items,item2.meal_items)"></view>
                </view>
            </template>
            <!--
            如果菜品售卖于自定义时间段
            -->
            <template v-else>
                <!--在售卖期-->
                <template v-if="item2.sell_status=='1'">
                    <view class="handle-amount">
                        <view class="dec" v-show="parseInt(item2.amount)>0?
                         true:false"></view>
                           <view class="text" v-if="!item2.isAmountInput">{{item2.
amount}}</view><input v-if="item2.isAmountInput" :focus="item2.isAmountInputFocus"
:value="item2.amount" type="number" @click.stop="" @blur="setAmount($event,item2.
gid,item2.new_meal_items,item2.meal_items,index,index2)" />
                        <view class="inc" @click.stop="incAmount($event,item2.
gid,item2.new_meal_items,item2.meal_items)"></view>
                    </view>
                </template>
                <template v-else>
                    <text>非可售时间</text>
                </template>
            </template>
        ...
</template>

<script>
    import GoodsDetails from './details';
```

```
    export default {
        name: "goods",
       ...
        methods:{
            ...
            //添加菜品数量
            incAmount(e,gid,new_meal_items,meal_items){
                //触发pages/goods/index.vue文件中的incAmount()方法
                this.$emit("incAmount",e,gid,new_meal_items,meal_items)
            },
        }
    }
</script>
```

以上加粗代码为新增代码，在incAmount()方法内部使用this.$eimt()触发pages/goods/index.vue文件中的incAmount()方法，在pages/goods/index.vue文件中新增代码如下：

```
<!--菜品展示组件-->
<goods :branchShopId="branch_shop_id" :isShow="goodsAction" :gid="gid"
:amount="amount" @setAmount="setAmount" @showAmountInput="showAmountInput"
:classifysData="classifys" :goodsData="goods" @showGoodsDetails="showGoodsDe
tails" @incAmount="incAmount"></goods>
```

在视图中找到<goods>组件，添加incAmount事件，在js中添加以下代码：

```
computed:{
    ...mapState({
        'isLogin':state=>state.user.isLogin          //会员是否登录
    })
},
methods:{
    //添加菜品数量
    incAmount(e,gid,new_meal_items,meal_items){
        //如果会员已登录
        if(this.isLogin){
            //加入购物车
        }else{              //如果会员未登录
            //跳转到登录页面
            this.pushPage(`/pages/login/index?branch_shop_id=${this.branch_
                shop_id}&table_code=${this.table_code}`);
        }
    }
}
```

在methods内部添加incAmount()方法，使用this.isLogin判断如果会员没有登录，则跳转到登录页面。this.isLogin是Vuex中的属性，这时还没有开发Vuex，因此可能会报错。可以先注释Vuex相关的代码，随便写一个判断值，目的是跳转到登录页面，如可以将if(this.isLogin)改为if(false)进行测试。

接下来请求会员登录相关的服务端接口。在api文件夹下创建user/index.js文件，该文件中的代码如下：

```
import config from "../../static/js/conf/config";
import {request} from "../../static/js/utils/request";

//获取微信小程序登录的openid和unionid
export function getWeChatOpenIdData(data){
    return request(config.baseApi+"/v1/wechat_openid","post",data);
}

//微信小程序会员登录后保存到数据库
export function setWeChatUserData(data){
    return request(config.baseApi+"/v1/wechat_login_save","post",data);
}
```

getWeChatOpenIdData()函数可以获取微信小程序登录的openid和unionid，setWeChatUserData()函数可以将微信登录后的用户信息保存到数据库中。接下来对接到Vuex中，在store文件夹下创建user/index.js文件，该文件中的代码如下：

```
import {getWeChatOpenIdData,setWeChatUserData} from '../../api/user';
export default {
    namespaced:true,
    state:{
        //会员的唯一标识
        uid:uni.getStorageSync("uid")?uni.getStorageSync("uid"):"",
        //会员是否登录，true为登录，false为未登录
        isLogin:uni.getStorageSync("isLogin")?Boolean(uni.getStorageSync("isLogin")):false,
        //会员验证的token值
        token:uni.getStorageSync("token")?uni.getStorageSync("token"):"",
        //微信用户的唯一标识
        openId:uni.getStorageSync("openId")?uni.getStorageSync("openId"):"",
    },
    mutations: {
        //设置会员登录信息
        ["LOGIN"](state, payload) {
            state.uid=payload.uid;
            state.token=payload.token;
            state.openId=payload.openId;
```

```
            state.isLogin=true;
            uni.setStorageSync("uid",state.uid);
            uni.setStorageSync("token",state.token);
            uni.setStorageSync("isLogin",state.isLogin);
            uni.setStorageSync("openId",state.openId);
        }
    },
    actions:{
        //获取微信小程序登录的openid和unionid
        getWeChatOpenId(conText,payload){
            return getWeChatOpenIdData(payload).then(res=>{
                return res;
            })
        },
        //微信小程序会员登录后保存到数据库
        setWeChatUser(conText,payload){
            return setWeChatUserData(payload).then(res=>{
                return res;
            })
        }
    }
}
```

接下来注册到Vuex的模块中，在store/index.js文件中新增代码如下：

```
...
import user from '../user';

Vue.use(Vuex);

const store = new Vuex.Store({
    modules:{
        user
    }
});

export default store
```

开发会员登录页面，在pages文件夹下创建login/index.vue文件，该文件中的代码如下：

```
<template>
    <view class="page">
        <view class="other-login-text">
            <view class="line"></view>
```

```
                <view class="text">推荐以下方式登录</view>
                <view class="line"></view>
            </view>
            <view class="other-login">
                <!--#ifdef MP-WEIXIN-->
                <button class="wx-login" @click="wxLogin()">微信授权登录</button>
                <!--#endif-->
            </view>
        </view>
    </template>

    <script>
        import {mapActions,mapMutations} from "vuex";
        export default {
            name: "login",
            data(){
                return {
                    isBindMobile:false,            //是否显示绑定手机号组件
                    sessionKey:""                  //会话密钥
                }
            },
            onLoad(opts){
                //分店的id
                this.branch_shop_id=opts.branch_shop_id?opts.branch_shop_id:"";
                //桌号
                this.table_code=opts.table_code?opts.table_code:"";
            },
            methods:{
                ...mapMutations({
                    "login":"user/LOGIN"           //设置会员登录信息
                }),
                ...mapActions({
                    getWeChatOpenId:"user/getWeChatOpenId",
                    //获取微信小程序登录的openid
                    setWeChatUser:"user/setWeChatUser"
                    //将微信授权后的用户信息保存到数据库
                }),
                //微信授权登录
                wxLogin(){
                    // #ifdef MP-WEIXIN
                    uni.login({
                        provider: 'weixin',
                        success: async (loginRes)=> {
                            let code=loginRes.code;
                            //获取微信小程序登录的openid
```

```
                    let openData=await this.getWeChatOpenId({code:code});
                    if(openData.code==200){
                     //会话密钥
                       this.sessionKey=openData.data.session_key;
                        // 获取用户信息
                        uni.getUserInfo({
                            provider: 'weixin',
                            success:async (infoRes) =>{
                                 let userinfo={"nickname":infoRes.userInfo
['nickName'],"sex":infoRes.userInfo["gender"],"province":infoRes.userInfo
["province"],"city":infoRes.userInfo["city"],"headimgurl":infoRes.userInfo
["avatarUrl"],"openid":openData.data.openid};
                                //保存用户信息
                                 let userInfoRes=await this.setWeChatUser
({userinfo:JSON.stringify(userinfo),branch_shop_id:this.branch_shop_id,
table_code:this.table_code});
                                if(userInfoRes.code==200){
                                this.login({uid:userInfoRes.data.uid,
token:userInfoRes.data.token,openId:userInfoRes.data.open_id});
                                    //如果绑定手机
                                    if(userInfoRes.data.isbind==1){
                                        //返回上一页
                                        uni.navigateBack({
                                            delta: 1
                                        });
                                    }else{        //没有绑定手机
                                        //显示绑定手机号组件
                                        this.isBindMobile=true;
                                    }
                                }else{
                                    uni.showToast({
                                        title:""+userInfoRes.data+"",
                                        icon:"none",
                                        duration:2000
                                    })
                                }
                            }
                        });
                    }
                }
            });
            // #endif
        }
    },
}
```

```
</script>

<style scoped>
    .page{width:100%;min-height:100vh;overflow:hidden;backgroun
     d-color:#FFFFFF;}
    .other-login-text{width:100%;height:100rpx;display:flex;justify-
     content: space-between;align-items: center;}
    .other-login-text .line{width:30%;height:1px;background-color:#CCCCCC;}
    .other-login-text .text{font-size:28rpx;color:#333333}

    .other-login{width:100%;display:flex;justify-content: center;}
    .other-login  .wx-login{width:80%;height:80rpx;background-
     color:#21B319;margin:0 auto;border-radius: 5px;color:#FFFFFF;font-
     size:32rpx;text-align:center;line-height:80rpx;}
</style>
```

10.17.1节中已经介绍了授权登录，这里仅作简单介绍。在wxLogin()函数内部实现微信授权登录，首先调用uni.login()方法获取用户的code，然后调用getWeChatOpenId()函数获取用户的session_key，只有获取到session_key才能获取本机手机号。判断用户是否绑定手机号，如果绑定手机号，则直接登录并返回上一页；否则显示绑定手机号组件。

扫一扫，看视频

11.6.2 获取用户手机号进行绑定

9.6.2节介绍过如何获取手机号，现在在实战中进行实际应用。绑定手机号预览效果如图11-15所示。

图 11-15 绑定手机号预览效果

首先请求绑定手机号服务端接口，在api/user/index.js文件中新增代码如下：

```
...
//微信小程序解密获取手机号和用户信息
export function getDewxbizdataData(data){
    return request(config.baseApi+"/v1/dewxbizdata","post",data);
}

//微信小程序登录后绑定手机号
export function bindWechatLoginBindPhoneNumberData(data){
    return request(config.baseApi+"/v1/user/wechat_login_bindcellphone","post",data);
}
```

getDewxbizdataData()函数可以解密获取手机号和用户信息，bindWechatLoginBindPhoneNumberData()函数可以将获取的手机号保存到数据库中。接下来对接到Vuex中，在store/user/index.js文件中新增代码如下：

```
import {getWeChatOpenIdData,setWeChatUserData,getDewxbizdataData,bindWechatLog
inBindPhoneNumberData} from '../../api/user';
export default {
    ...
    actions:{
        ...
        //微信小程序解密获取手机号和用户信息
        getDewxbizdata(conText,payload){
            return getDewxbizdataData(payload).then(res=>{
                return res;
            })
        },
        //微信小程序登录后绑定手机号
        bindWechatLoginBindPhoneNumber(conText,payload){
            /*
            bindWechatLoginBindPhoneNumberData参数说明
            uid:会员的唯一标识
            token:会员的token
            platform:平台类型，1为微信小程序，2为微信公众号
            */
             return bindWechatLoginBindPhoneNumberData({uid:conText.rootState.
user.uid,token:conText.rootState.user.token,platform:conText.rootState.
system.platform,...payload}).then(res=>{
                return res;
            })
        },
```

```
    }
}
```

上述加粗代码为新增代码，bindWechatLoginBindPhoneNumberData()方法对象参数的platform属性的值在store/system/index.js文件中设置，该文件中的新增代码如下：

```
export default {
    namespaced: true,
    state: {
        ...
        platform:1              //平台类型，1为微信小程序，2为微信公众号
    },
    ...
}
```

开发绑定手机号的组件。在components文件夹下创建bind_mobile/index.vue文件，该文件中的代码如下：

```
<template>
    <view class="mobile-mask" v-show="show">
        <button class="bind-mobile-btn" open-type="getPhoneNumber" @
         getphonenumber="getPhoneNumber">绑定手机</button>
    </view>
</template>

<script>
    import {mapActions} from "vuex";
    export default {
        name: "bind-mobile",
        props:{
            //是否显示组件
            show:{
                type:Boolean,
                default:false
            },
            //会话密钥
            sessionKey:{
                type:String,
                required:true
            }
        },
        methods:{
            ...mapActions({
                getDewxbizdata:"user/getDewxbizdata",
```

```
            //微信小程序解密获取手机号和用户信息
            bindWechatLoginBindPhoneNumber:"user/bindWechatLoginBindPhoneNumber"

            }),
            //获取手机号
            async getPhoneNumber(e){
                let encrypteddata=e.detail.encryptedData;
                //包括敏感数据在内的完整用户信息的加密数据
                let iv=e.detail.iv;           //加密算法的初始向量
                //解密encryptedData
                let pnRes=await this.getDewxbizdata({encrypteddata,iv,session_
                    key:this.sessionKey});
                if(pnRes.code==200){
                    let phoneNumber=pnRes.data.phoneNumber;       //手机号
                    //保存手机号到数据库
                    let bpRes=await this.bindWechatLoginBindPhoneNumber({cellp
                        hone:phoneNumber});
                    if(bpRes.code==200){      //如果保存手机号
                        this.$emit("close"); //隐藏组件
                        //返回上一页
                        uni.navigateBack({
                            delta: 1
                        });
                    }else{                   //未保存成功
                        uni.showToast({
                            title:bpRes.data,
                            icon:"none",
                            duration:2000
                        })
                    }
                }else{                       //解密encryptedData失败
                    uni.showToast({
                        title:pnRes.data,
                        icon:"none",
                        duration:2000
                    })
                }
            }
        }
    }
</script>
```

```
<style scoped>
    .mobile-mask{width:100%;height:100%;position: fixed;z-index:99;
     left:0;top:0;background-color:rgba(0,0,0,0.8)}
    .bind-mobile-btn{width:70%;height:80rpx;position: absolute;left:50%;top:
     50%;transform: translate(-50%,-50%);background-color:#28A622;text-align:
     center;line-height:80rpx;font-size:32rpx;color:#FFFFFF;border-radius: 5px;}
</style>
```

9.6.2节已经介绍了获取手机号，这里不再详细讲解，其中bindWechatLoginBindPhoneNumber()函数可以将获取到的手机号保存到数据库中。接下来在pages/login/index.vue文件中使用该组件，新增代码如下：

```
<template>
    <view class="page">
        ...
        <bind-mobile :show="isBindMobile" @close="isBindMobile=false" :sessionKey
         ="sessionKey"></bind-mobile>
    </view>
</template>

<script>
    import {mapActions,mapMutations} from "vuex";
    import BindMobile from "../../components/bind_mobile";
    export default {
        ...
        components:{
            BindMobile
        },
        ...
    }
</script>
```

以上加粗代码为新增代码，bind-mobile组件的show属性可以显示组件；自定义close事件可以隐藏组件，sessionKey属性可以将获取的session_key传递给组件内部，必须有此值才能实现获取手机号的功能。

11.7 购物车

购物车是电商必备功能，使用Vuex配合服务端API接口完成。点餐系统的购物车支持单品和套餐，并且支持共享购物车功能。共享购物车就是多个用户可以看到同一桌号下购物车里的菜品。购物车预览效果如图11-16所示。

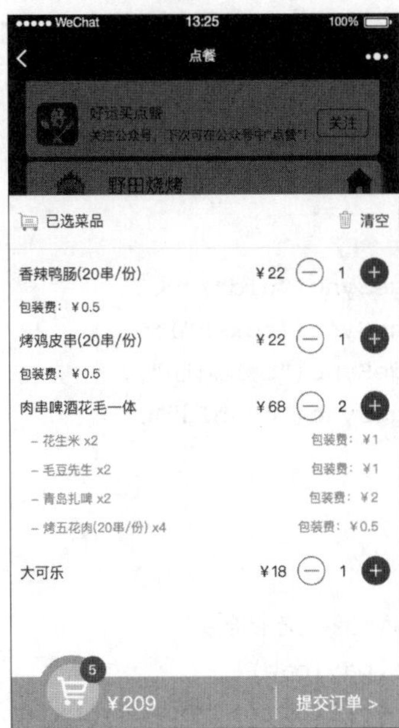

图 11-16　购物车预览效果

购物车接口在“接口文档”文件夹下的“购物车.docx”文件中。

11.7.1　会员 token 认证

将菜品添加到购物车之前，需要先验证会员是否登录，如果未登录则跳转到登录页面。在点餐页面添加会员token认证代码，首先请求服务端会员token认证接口，接口地址在“会员登录、注册、信息.docx”文件中。在api/user/index.js文件中新增代码如下：

```
...

//检测会员登录获取的token是否合法
export function safeUserData(data){
    return request(config.baseApi+"/v1/safe_user","post",data);
}
```

接下来对接到Vuex中，在store/user/index.js文件中新增代码如下：

```
import {getWeChatOpenIdData,setWeChatUserData,getDewxbizdataData,bindWechatLogin
BindPhoneNumberData,safeUserData} from '../../api/user';
export default {
    ...
    mutations: {
        ...
```

```
        //安全退出
        ["OUT_LOGIN"](state) {
            state.uid = "";
            state.isLogin = false;
            state.token="";
            state.openId="";
            uni.removeStorageSync("uid");
            uni.removeStorageSync("token");
            uni.removeStorageSync("isLogin");
            uni.removeStorageSync("openId");
        },
    },
    actions:{
        ...
        //检测会员登录获取的token是否合法
        safeUser(conText,payload){
            safeUserData({uid:conText.rootState.user.uid,token:conText.
            rootState.user.token,platform:conText.rootState.system.
            platform,...payload}).then(res=>{
                if(payload && payload.success){
                    payload.success(res);
                }
            })
        },
    }
}
```

safeUser()方法会在很多地方使用，为了使用方便，这里将该方法封装到公共文件中。在js/utils文件夹下创建index.js文件，该文件中的代码如下：

```
//检测会员登录获取的token是否合法
function safeUser(pThis,branch_shop_id="",table_code="",callback){
    //调用store/user/index.js文件中actions内部的safeUser()方法
    pThis.$store.dispatch("user/safeUser", {branch_shop_id: branch_shop_id,
        success:(res)=>{
            //如果token验证不成功
            if(res.code!==200){
                //退出登录
                pThis.$store.commit("user/OUT_LOGIN");
                //跳转到登录页面
                uni.navigateTo({
                    url:"/pages/login/index?branch_shop_id=" + branch_shop_
                        id + "&table_code=" + table_code
```

```
                    })
                }
                //如果token验证成功
                if(res.code==200){
                    if(callback){           //如果回调函数存在
                        //执行回调函数
                        callback()
                    }
                }
            }});
    }

    export default {
        safeUser
    }
```

将该文件导入main.js文件中，将其中的方法注册到Vue原型上，设置为全局属性。在main.js文件中新增代码如下：

```
...
//导入utils
import utils from "./static/js/utils";
...

//将utils注册到Vue原型上
Vue.prototype.$utils=utils;
...
```

在pages/goods/index.vue文件中实现菜品加入购物车时，验证会员是否登录功能。在pages/goods/index.vue文件中新增代码如下：

```
...
<script>
    ...
    export default {
        ...
        //添加菜品数量
        incAmount(e,gid,new_meal_items,meal_items){
            //如果会员已登录
            if(this.isLogin){
                //调用static/js/utils/index.js文件中的safeUser()方法，如果token验证成功
                this.$utils.safeUser(this,this.branch_shop_id,this.
                    table_code,()=>{
                    //加入购物车
```

```
                    console.log("加入购物车");
                })

            }else{      //如果会员未登录
                //跳转到登录页面
                this.pushPage('/pages/login/index?branch_shop_id=${this.
                    branch_shop_id}&table_code=${this.table_code}');
            }
        }
    }
    }
</script>
```

以上加粗代码为新增代码，这样就实现了点击加号按钮，自动检测会员token是否合法，如果不合法则跳转到登录页面的功能。

扫一扫，看视频

11.7.2 将菜品加入购物车

要将菜品加入购物车中，应先请求添加购物车服务端接口。在api文件夹下创建cart/index.js文件，该文件中的代码如下：

```
import config from "../../static/js/conf/config";
import {request} from "../../static/js/utils/request";

//添加购物车
export function addCartData(data){
    return request(config.baseApi+"/v1/user/cart/add","post",data);
}
```

接下来对接到Vuex中，在store文件夹下创建cart/index.js文件，该文件中的代码如下：

```
import {addCartData} from "../../api/cart";
export default {
    namespaced:true,
    state:{
        cartData:[],            //购物车里的菜品
        cartCount:0,            //购物车里菜品的总数量
        isPack:false            //是否自提
    },
    mutations:{
        //加入购物车
        ["ADD_CART_DATA"](state,payload){
            let isSame=false;            //是否有相同的菜品
```

```
        state.cartCount=0;          //购物车里菜品的数量
        if(state.cartData.length>0){
            for(let i=0;i<state.cartData.length;i++){
                //购物车里套餐菜品明细的gid
                let newMealItems=(state.cartData[i]["new_meal_items"] &&
                    state.cartData[i]["new_meal_items"].length)>0?JSON.
                    stringify(state.cartData[i]["new_meal_items"]):"";
                //从addCart传递过来的套餐菜品明细的gid
                let curNewMealItems=(payload.cartData.new_meal_items
                    && payload.cartData.new_meal_items.length)>0?JSON.
                    stringify(payload.cartData.new_meal_items):"";
                //如果菜品相同
                if(state.cartData[i]["gid"]==payload.cartData.gid &&
                  newMealItems==curNewMealItems){
                    isSame=true;
                    //增加购物车里菜品的数量
                    state.cartData[i].amount=parseInt(state.cartData[i].amount)+1;
                    //如果套餐里包含菜品
                    if((state.cartData[i].meal_items && state.cartData[i].
                      meal_items.length)>0){
                        for(let j=0;j<state.cartData[i].meal_items.length;j++){
                            //增加套餐里菜品的数量
                            state.cartData[i].meal_items[j].amount=state.
                            cartData[i].meal_items[j].dis_amount*state.
                            cartData[i].amount;
                        }
                    }
                }
                //计算购物车里菜品的总数量
                state.cartCount+=parseInt(state.cartData[i].amount);
            }
        }
        //如果购物车里没有相同的菜品
        if(!isSame){
            state.cartCount=parseInt(state.cartCount)+parseInt(payload.
                cartData.amount);
            //将菜品加入购物车
            state.cartData.unshift(payload.cartData);
        }
    }
},
actions:{
```

```
        //添加购物车
        addCart(conText,payload){
            let data={
                "uid":conText.rootState.user.uid,           //会员的唯一标识
                "token":conText.rootState.user.token,       //会员登录时的token
                "platform":conText.rootState.system.platform,     //平台类型
                "branch_shop_id":payload.cartData.branch_shop_id, //分店的id
                "table_code":payload.cartData.table_code,  //桌号
                "gid":payload.cartData.gid,                 //菜品的唯一标识
                "amount":payload.cartData.amount,           //菜品数量
                "is_meal":payload.cartData.is_meal,         //是否套餐
                "meal_type":payload.cartData.meal_type,
                //套餐类型，0为普通套餐，1为自定义套餐
                "meal_items":(payload.cartData.meal_items && payload.cartData.
meal_items.length>0)?JSON.stringify(payload.cartData.meal_items):"",
                //套餐里菜品的明细
                 "new_meal_items":(payload.cartData.new_meal_items && payload.cartData.
new_meal_items.length>0)?JSON.stringify(payload.cartData.new_meal_items):"",
                //套餐里菜品的gid
            };

            //添加购物车到数据库
            addCartData(data).then(res=>{
                if(res.code==200){
                    conText.commit("ADD_CART_DATA",{cartData:payload.cartData});
                    conText.commit("SET_GOODS_CART_AMOUNT",{goods:conText.
                    rootState.goods.goods});
                    if(payload.success){
                        payload.success()
                    }
                }
            });

        }
    },
    getters:{
        //计算总价
        total(state){
            let total=0;                  //菜品总价格
            let mealTotal=0;              //套餐总价
            let packTotal=0;              //包装费总价
            let footTotal=0;              //单品总价
```

```
            let packFootTotal=0;       //单品包装费总价
            let packMealTotal=0        //套餐包装费总价
            if(state.cartData.length>0){
                for(let i=0;i<state.cartData.length;i++){
                    if (state.cartData[i].is_meal == '0') {        //如果非套餐
                        //计算非套餐总价
                        footTotal += parseFloat(state.cartData[i].price) *
                                  parseInt(state.cartData[i].amount);
                         if(state.isPack){                          //如果是自提
                            //计算非套餐包装费
                             packFootTotal += parseFloat(state.cartData[i].pack_
price) * parseInt(state.cartData[i].amount);
                        }
                    }
                    if (state.cartData[i].is_meal == '1') {        //如果是套餐
                        for (let j = 0; j < state.cartData[i].meal_items.length; j++) {
                            //计算套餐总价
                            mealTotal += parseFloat(state.cartData[i].meal_items[j].
price) * parseInt(state.cartData[i].meal_items[j].amount);
                             if(state.isPack){                      //如果是自提
                                //计算套餐包装费
                                 packMealTotal += (parseFloat(state.cartData[i].
meal_items[j].pack_price) * parseInt(state.cartData[i].meal_items[j].amount));
                            }
                        }
                    }
                }
                 packTotal = packFootTotal + packMealTotal;
                 //计算非套餐和套餐包装费总价
                total=footTotal+mealTotal+packTotal;
                //计算非套餐和套餐总价
            }
            if(total>0){
                total=parseFloat((total+packTotal).toFixed(2));    //计算菜品总价
            }
            return total;
        },
        //计算包装费总价
        packTotal(state){
            let packFootTotal=0,packMealTotal=0,packTotal=0;
            if(state.isPack){                                      //如果是自提
                if(state.cartData.length>0){
```

```
                    for(let i=0;i<state.cartData.length;i++){
                        if(state.cartData[i].is_meal=='0'){          //如果非套餐
                            //计算非套餐包装费总价
                            packFootTotal+=parseFloat(state.cartData[i].
                            pack_price)*parseInt(state.cartData[i].amount);
                        }
                        if(state.cartData[i].is_meal=='1'){          //如果是套餐
                            for(let j=0;j<state.cartData[i].meal_items.length;j++){
                                //计算套餐包装费总价
                                packMealTotal+=(parseFloat(state.cartData[i].
                                meal_items[j].pack_price)*parseInt(state.
                                cartData[i].meal_items[j].amount));
                            }
                        }
                    }
                }
            }
            //计算包装费总价
            packTotal=parseFloat((packFootTotal+packMealTotal).toFixed(2));
            return packTotal;
        }
    }
}
```

state内部的cartData属性值为购物车里的菜品；cartCount属性值为购物车里菜品的总数量；isPack属性值为Boolean类型，如果为true表示自提，需要支付包装费，否则表示堂食，不需要支付包装费。mutations内部的ADD_CART_DATA方法可以响应式地增加购物车里的菜品，并且判断是否有相同的菜品，如果相同则增加该菜品的数量，否则新增加菜品。actioins内部的addCart()方法可以将菜品添加到数据库中，添加完成后再调用mutations内部的ADD_CART_DATA方法实现数据的响应。getters内部的total()方法可以计算出菜品总价，packTotal()方法可以计算出自提订餐的包装费总价。

在pages/goods/index.vue文件中显示购物车里菜品的数量和总价，并将菜品添加到Vuex中，新增代码如下：

```
...
<template>
...
<view class="handle-cart">
    <view class="cart-cle">
        <view class="cart-cle-2">
            <view class="cart-icon" id="cart-icon">
            <view class="badge" v-show="cartCount>0?true:false">{{cartCoun
             t}}</view>
```

```
            </view>
        </view>
    </view>
    <view class="total">￥{{total}}</view>
</view>
...
</template>

<script>
    ...
    export default {
        ...
         computed:{
          ...mapState({
            ...
           "cartCount":state=>state.cart.cartCount  //购物车里菜品的数量
          }),
           ...mapGetters({
             "total":"cart/total"                   //购物车里菜品的总价
           }),
          },
         methods:{
        ...mapActions({
            ...
            "addCart":"cart/addCart",               //添加购物车的方法
        }),
        ...
        //添加菜品数量
        incAmount(e,gid,new_meal_items,meal_items){
            //如果会员已登录
            if(this.isLogin){
                //调用static/js/utils/index.js文件中的safeUser()方法,
                //如果token验证成功
                this.$utils.safeUser(this,this.branch_shop_id,this.table_code,()=>{
                    //套餐里菜品明细的gid，用于加入购物车时判断是否有相同的套餐
                    let curNewMealItems=new_meal_items && new_meal_items.
                    length>0?JSON.stringify(new_meal_items):"";
                    for(let i=0;i<this.goods.length;i++){
                        for(let j=0;j<this.goods[i].goods.length;j++){
                            //默认套餐里菜品的gid
                            let newMealItems=this.goods[i].goods[j].new_
                            meal_items.length>0?JSON.stringify(this.
                            goods[i].goods[j].new_meal_items):"";
```

```
                    //如果有相同的菜品，该判断是为了增加this.amount数量
                    if(this.goods[i].goods[j].gid==gid &&
                      newMealItems==curNewMealItems){
                        //增加菜品的数量。这里的数量会传递给菜品展示组件和
                        //菜品详情组件，这样数量就可以同步了
                        this.amount=parseInt(this.goods[i].goods[j].amount)+1;

                        //添加购物车的数据
                        let cartData={
                            "title":this.goods[i].goods[j].title,
                            //菜品名称
                            "branch_shop_id":this.branch_shop_id,
                            //分店的id
                            "table_code":this.table_code,
                            //桌号
                            "gid":gid,
                            //菜品的唯一标识
                            "price":this.goods[i].goods[j].price,
                            //菜品价格
                            "amount":1,
                            //菜品数量默认为1，数量的改变在Vuex中写逻辑
                            "is_meal":this.goods[i].goods[j].is_meal,
                            //是否套餐，1为是，0为否
                                "pack_price":this.goods[i].goods[j].
pack_price?this.goods[i].goods[j].pack_price:0,    //包装费
                            "meal_type":0,
                            //套餐类型，0为普通套餐，1为自定义套餐
                            "meal_items":this.goods[i].goods[j].meal_items,
                            //套餐里菜品的明细
                            "new_meal_items":this.goods[i].goods[j].
                            new_meal_items,
                            //套餐里菜品的gid明细
                            "isAmountInput":false,
                            //是否显示数量input输入框
                            "isAmountInputFocus":false
                            //数量input输入框获取焦点
                        };
                        //添加购物车
                        this.addCart({cartData:cartData});
                        break;
                    }
                }
```

```
                }
            })
        }else{        //如果会员未登录
            //跳转到登录页面
            this.pushPage(`/pages/login/index?branch_shop_id=${this.
                branch_shop_id}&table_code=${this.table_code}`);
        }
      }
    }
  }
</script>
```

加粗代码为新增或修改的代码，incAmount函数()内部实现增加菜品的数量并且可以将菜品添加到购物车里。接下来在套餐页面实现添加购物车功能，在pages/goods/meal.vue文件中新增代码如下：

```
<template>
    <view class="page">
        ...
        <view class="cart-main">
            <view class="total">小计:￥{{total}}</view>
            <view class="line"></view>
            <view class="cart" @click="addCart()">
                <view class="cart-icon"></view>
                <view class="text">加入购物车 ></view>
            </view>
        </view>
        ...
    </view>
</template>

<script>
    import {mapActions,mapState} from 'vuex';
    export default {
       ...
        methods:{
            ...mapActions({
                ...
                "asyncAddCart":"cart/addCart"        //添加购物车
            }),
            ...
            //加入购物车
            addCart(){
```

```
                if(this.isSubmit){
                    this.isSubmit=false;
                    let cartData={
                        "title":this.title,                    //菜品标题
                        "branch_shop_id":this.branch_shop_id, //分店的id
                        "table_code":this.table_code,          //桌号
                        "gid":this.gid,                        //菜品的唯一标识
                        "price":this.total,                    //套餐价格
                        "amount":1,                            //套餐数量
                        "is_meal":1,        //是否套餐，1为是,0为否
                        "meal_type":1,      //套餐类型，0为普通套餐，1为自定义套餐
                        "meal_items":this.mealItems,           //套餐里菜品的明细
                        "new_meal_items":this.simpleMealItems,
                        //套餐里菜品的gid明细
                        "isAmountInput":false,
                        //是否显示数量input输入框
                        "isAmountInputFocus":false
                        //数量input输入框获取焦点
                    };
                    //添加购物车
                    this.asyncAddCart({cartData:cartData,success:()=>{
                            uni.showToast({
                                title: '添加购物车成功！',
                                icon:"success",
                                duration: 2000
                            });
                            setTimeout(()=>{
                                this.isSubmit=true;
                                uni.hideToast();              //隐藏toast
                                uni.navigateBack({            //返回上一页
                                    delta: 1
                                });
                            },2000);
                        }});
                }
            }
        }
    }
</script>
```

以上加粗代码为新增代码，addCart()方法内部的this.asyncAddCart()方法可以将套餐添加到数据库中。注意，只要是从套餐页面加入购物车的菜品，当其cartData对象的is_meal属性的

值为1时，表示固定套餐，当meal_type属性的值为1时，表示自定义套餐。

接下来在菜品展示组件中实现添加购物车功能。在components/goods/index.vue文件中的视图<template>标签内找到以下代码：

```
<view class="inc"></view>
```

替换为

```
<view class="inc" @click.stop="incAmount($event,item2.gid,item2.new_meal_
items,item2.meal_items)"></view>
```

在js代码中添加incAmount()方法，代码如下：

```
methods:{
    //添加菜品数量
    incAmount(e,gid,new_meal_items,meal_items){
        //触发pages/goods/index.vue文件中的incAmount()方法
        this.$emit("incAmount",e,gid,new_meal_items,meal_items)
    },
}
```

incAmount()函数内部使用$emit()方法，可以触发pages/goods/index.vue文件中的incAmount()方法，将菜品添加到购物车中。

11.7.3　开发购物车组件

扫一扫，看视频

在购物车组件中，可以显示加入购物车的菜品、修改菜品数量以及清除购物车里的菜品。首先请求购物车组件相关的服务端接口，在api/cart/index.js文件中新增代码如下：

```
...

//显示购物车
export function getCartData(data){
    return request(config.baseApi+"/v1/user/cart/show","post",data);
}

//更新购物车数量
export function setAmountData(data){
    return request(config.baseApi+"/v1/user/cart/set_amount","post",data);
}

//删除购物车里的单个菜品
export function delData(data){
    return request(config.baseApi+"/v1/user/cart/del","post",data);
```

```
}

//清空购物车里的所有菜品
export function clearCartData(data){
    return request(config.baseApi+"/v1/user/cart/clear","post",data);
}
```

接下来对接到Vuex中，在store/cart/index.js文件中新增代码如下：

```
import Vue from "vue";
import {addCartData,getCartData,setAmountData,delData,clearCartData} from
"../../api/cart";
export default {
    namespaced:true,
    state:{
        cartData:[],                    //购物车里的数据
        ...
    },
    mutations:{
        ...
        //将购物车里的菜品数量同步给菜品展示的数量
        ["SET_GOODS_CART_AMOUNT"](state,payload){
            if(state.cartData.length>0){
                for(let i=0;i<state.cartData.length;i++){
                    for(let j=0;j<payload.goods.length;j++){
                        for(let k=0;k<payload.goods[j].goods.length;k++){
                            let newMealItems=(payload.goods[j].goods[k].
new_meal_items && payload.goods[j].goods[k].new_meal_items.length)>0?JSON.
stringify(payload.goods[j].goods[k].new_meal_items):"";
                            let curNewMealItems=(state.cartData[i].new_meal_
items && state.cartData[i].new_meal_items.length)>0?JSON.stringify(state.
cartData[i].new_meal_items):"";
                            //如果购物车里的菜品和菜品展示相同
                            if(payload.goods[j].goods[k].gid==state.
                              cartData[i].gid && newMealItems==curNewMealItems){
                                //将购物车里菜品的数量赋值给菜品展示
                                payload.goods[j].goods[k].amount=state.cartData[i].amount;
                                break;
                            }
                        }
                    }
                }
            }else{
                //当购物车里没有数据时,将菜品里的数量设置为0
```

```
                    if(payload.goods.length>0){
                        for(let i=0;i<payload.goods.length;i++){
                            for(let j=0;j<payload.goods[i].goods.length;j++){
                                payload.goods[i].goods[j].amount=0;
                            }
                        }
                    }
                }
            }
        },
        //修改购物车里的菜品数量
        ["SET_AMOUNT"](state,payload){
            state.cartCount=0;
            let amount=payload.cartData.amount;
            //如果输入的为非数字，则替换成空
            amount=amount.toString().replace(/[^\d]/g,"");
            //如果输入的数量为空或0，则强行设置为1
            if(!amount || amount=='0'){
                amount=1;
            }
            if(state.cartData.length>0){
                for(let i=0;i<state.cartData.length;i++){
                    let newMealItems=(state.cartData[i]["new_meal_items"]
                    && state.cartData[i]["new_meal_items"].length)>0?JSON.
                    stringify(state.cartData[i]["new_meal_items"]):"";
                    let curNewMealItems=(payload.cartData.new_meal_items
                    && payload.cartData.new_meal_items.length)>0?JSON.
                    stringify(payload.cartData.new_meal_items):"";
                    //如果菜品相同
                    if(state.cartData[i]["gid"]==payload.cartData.gid &&
                    newMealItems==curNewMealItems){
                        //设置菜品数量
                        state.cartData[i].amount=parseInt(amount);
                        //如果套餐里有菜品
                        if((state.cartData[i].meal_items && state.
                        cartData[i].meal_items.length)>0){
                            for(let j=0;j<state.cartData[i].meal_items.length;j++){
                                //设置套餐里菜品的数量
                                state.cartData[i].meal_items[j].amount=state.
                                cartData[i].meal_items[j].dis_amount*parseInt(amount);
                            }
                        }
```

```
                }
                //更新购物车里菜品的总数量
                state.cartCount+=parseInt(state.cartData[i].amount);
            }
        }
    },
    //减少数量和删除菜品
    ["DEL_CART_DATA"](state,payload){
        if(state.cartData.length>0){
            state.cartCount=0;//将购物车里的菜品总数量设置为0
            for(let i=0;i<state.cartData.length;i++){
                let newMealItems=(state.cartData[i]["new_meal_items"]
                && state.cartData[i]["new_meal_items"].length)>0?JSON.
                stringify(state.cartData[i]["new_meal_items"]):"";
                let curNewMealItems=(payload.cartData.new_meal_items
                && payload.cartData.new_meal_items.length)>0?JSON.
                stringify(payload.cartData.new_meal_items):"";
                //如果有相同的菜品
                if(state.cartData[i].gid==payload.cartData.gid &&
                newMealItems==curNewMealItems){
                    //减少菜品数量
                    let cartAmount=parseInt(state.cartData[i].
                    amount)>0?parseInt(state.cartData[i].amount)-1:0;
                    state.cartData[i].amount=cartAmount;

                    //如果数量小于等于0
                    if(cartAmount<=0){
                        state.cartData.splice(i--,1);  //删除该菜品
                    }else{                             //如果数量大于0
                        if(state.cartData[i].meal_items && state.
                        cartData[i].meal_items.length>0){
                            for(let j=0;j<state.cartData[i].meal_
                            items.length;j++){
                                //更改套餐里菜品的数量
                                state.cartData[i].meal_items[j].
                                amount=state.cartData[i].meal_items[j].
                                dis_amount*cartAmount;
                            }
                        }
                    }
                    break;
                }
```

```
                }
                //计算总数量
                for(let i=0;i<state.cartData.length;i++){
                    state.cartCount+=parseInt(state.cartData[i].amount);
                }
            }
        },
        //清除购物车
        ["CLEAR_CART_DATA"](state){
            state.cartData=[];
            state.cartCount=0;
        },
        //设置购物车数据
        ["SET_CART_DATA"](state,payload){
            //将服务端购物车里的菜品同步到state.cartData中
            state.cartData=payload.cartData;
            if(payload.cartData.length>0){
                state.cartCount=0;
                //计算购物车里的菜品总数量
                for(let i=0;i<state.cartData.length;i++){
                    state.cartCount+=parseInt(state.cartData[i].amount);
                }
            }else{
                state.cartCount=0;
            }
        },
        //设置是否自提
        ["SET_PACK"](state,payload){
            state.isPack=payload.isPack;
        },
        //显示数量输入框
        ["SHOW_AMOUNT_INPUT"](state,payload){
            state.cartData[payload.index].isAmountInput=true;
            //显示数量input输入框
            state.cartData[payload.index].isAmountInputFocus=true;
            //数量input输入框获取焦点
                Vue.set(state.cartData,payload.index,state.cartData[payload.index]);
        },
        //隐藏数量输入框
        ["HIDE_AMOUNT_INPUT"](state,payload){
            state.cartData[payload.index].isAmountInput=false;
            //隐藏数量input输入框
```

```
                state.cartData[payload.index].isAmountInputFocus=false;
                //数量input输入框失去焦点
                    Vue.set(state.cartData,payload.index,state.cartData[payload.index]);
            }
        },
        actions:{
            ...
            //减少数量和删除商品
            async delCart(conText,payload){
                //修改数量
                let amount=0;
                let meal_type=payload.meal_type; //1为自定义套餐，0为普通套餐
                if(meal_type!=1){                    //如果是普通套餐
                    amount=payload.cartData.amount;
                }else{        //meal_type=1表示自定义套餐，获取购物车里的数量
                    amount=await getCartData({"uid":conText.rootState.user.
                    uid,"token":conText.rootState.user.token,"platform":conText.
                    rootState.system.platform,branch_shop_id:payload.branch_shop_
                    id,table_code:payload.table_code}).then(res=>{
                        if(res.code==200){
                            for(let i=0;i<res.data.length;i++){
                                let newMealItems=res.data[i].new_meal_items && res.
data[i].new_meal_items.length>0?JSON.stringify(res.data[i].new_meal_items):"";
                                let curMealItems=payload.cartData.new_meal_
                                    items && payload.cartData.new_meal_items.
                                    length>0?JSON.stringify(payload.cartData.
                                    new_meal_items):"";
                                //如果有相同的菜品
                                if(res.data[i].gid==payload.cartData.gid &&
                                  newMealItems==curMealItems){
                                    //减少菜品数量
                                    return res.data[i].amount>0?res.data[i].amount-1:0;
                                }
                            }
                        }
                    })
                }

                let data={
                    "uid":conText.rootState.user.uid,
                    "token":conText.rootState.user.token,
                    "platform":conText.rootState.system.platform,
```

```
                branch_shop_id:payload.branch_shop_id,
                amount:amount,
                meal_items:(payload.cartData.meal_items && payload.cartData.
meal_items.length>0)?JSON.stringify(payload.cartData.meal_items):"",
                new_meal_items:(payload.cartData.new_meal_items && payload.
cartData.new_meal_items.length)>0?JSON.stringify(payload.cartData.
new_meal_items):"",
                gid:payload.cartData.gid
            };
            if(amount>0){               //如果数量大于0，修改数量
                setAmountData(data).then(res=>{
                    if(res.code==200){
                        conText.commit("DEL_CART_DATA",{cartData:payload.cartData});
                    }
                });
            }else{                      //删除购物车里的数据
                delData(data).then(res=>{
                    if(res.code==200){
                        conText.commit("DEL_CART_DATA",{cartData:payload.cartData});
                    }
                })
            }
        },
        //更改购物车数量
        setAmount(conText,payload){
            //客户端购物车的数据
            let data={
                "uid":conText.rootState.user.uid,
                "token":conText.rootState.user.token,
                "platform":conText.rootState.system.platform,
                branch_shop_id:payload.branch_shop_id,
                amount:payload.cartData.amount,
                meal_items:(payload.cartData.meal_items && payload.cartData.
meal_items.length>0)?JSON.stringify(payload.cartData.meal_items):"",
                new_meal_items:(payload.cartData.new_meal_items && payload.
cartData.new_meal_items.length)>0?JSON.stringify(payload.cartData.
new_meal_items):"",
                gid:payload.cartData.gid
            };
            //更新服务端数据库中菜品的数量
            setAmountData(data).then(res=>{
                if(res.code==200){
```

```
                    conText.commit("SET_AMOUNT",{cartData:payload.cartData});
                }
            });
        },
        //清空购物车
        clearCart(conText,payload){
            clearCartData({"uid":conText.rootState.user.uid,"token":conText.
            rootState.user.token,"platform":conText.rootState.system.
            platform,branch_shop_id:payload.branch_shop_id,table_
            code:payload.table_code}).then(res=>{
                if(res.code==200){
                    conText.commit("CLEAR_CART_DATA");
                    if(payload.success){
                        payload.success();
                    }
                }
            });
        },
        //显示购物车数据
        showCartData(conText,payload){
            getCartData({"uid":conText.rootState.user.uid,"token":conText.
            rootState.user.token,"platform":conText.rootState.system.
            platform,...payload}).then(res=>{
                if(res.code==200){              //如果请求成功
                    for(let i=0;i<res.data.length;i++){
                        res.data[i].isAmountInput=false;
                        res.data[i].isAmountInputFocus=false;
                    }
                    conText.commit("SET_CART_DATA",{cartData:res.data});
                    if(payload.success){
                        payload.success();
                    }
                }else{                          //如果没有请求成功
                    conText.commit("SET_CART_DATA",{cartData:[]});
                }
                conText.commit("SET_GOODS_CART_AMOUNT",{goods:conText.rootState.
                    goods.goods});
            })
        }
    },
    ...
}
```

mutations内部的SET_GOODS_CART_AMOUNT方法可以将购物车里的菜品数量同步到菜品展示的数量;SET_AMOUNT方法可以响应式地修改购物车里菜品的数量;DEL_CART_DATA方法可以响应式地减少菜品数量，如果菜品数量小于1，则删除该菜品;CLEAR_CART_DATA方法可以响应式地清除购物车里所有的菜品;SET_CART_DATA方法可以将服务端购物车里的数据同步到state.cartData中，这样就可以在购物车里永久显示添加的菜品;SET_PACK方法可以设置是否自提，如果为true表示自提，需要支付包装费;SHOW_AMOUNT_INPUT方法显示数量输入框;HIDE_AMOUNT_INPUT方法隐藏数量输入框。actions内部的delCart()方法可以对服务端购物车里的菜品数量进行减少和删除;setAmount()方法可以更改服务端购物车里菜品的数量;clearCart()方法可以清空服务端购物车里的菜品;showCartData()方法可以显示购物车里的菜品。

接下来开发购物车组件。在components文件夹中创建cart/index.vue文件，该文件中的代码如下：

```
<template>
    <view>
        <view class="mask" v-show="show" @click="close"></view>
        <view :class="{'cart-main':true, show:show}">
            <view class="nav">
                <view class="nav-title">
                    <view class="icon"></view>
                    <view class="text">已选菜品</view>
                </view>
                <view class="clear-main" @click="clearCart()">
                    <view class="icon"></view>
                    <view class="text">清空</view>
                </view>
            </view>
            <scroll-view scroll-y="true" class="goods-main">
                <view class="scroll-wrap">
                    <view class="goods-list" v-for="(item,index) in cartData":
                     key="index">
                        <view class="title-area">
                            <view class="title">{{item.title}}</view>
                            <view class="price-amount">
                                <view class="price">￥{{item.price}}</view>
                                <view class="handle-amount">
                                    <!--如果数量大于0，显示减号按钮-->
                                                <view class="dec" v-show="item.
amount>0?true:false" @click="decAmount($event,item.gid,item.new_meal_items,item.
meal_items)"></view>
                                    <!--如果数量大于0并且isAmountInput的值为
                                     false，显示文本数量-->
```

```
                                    <view class="text" v-if="!item.isAmountInput
&& item.amount>0" @click="showAmountInput(index)">{{item.amount}}</view>
                                    <!--如果isAmountInput的值为true，显示input
                                    输入框，这里可以修改数量-->
                                    <input v-if="item.isAmountInput"
:focus="item.isAmountInputFocus" :value="item.amount" type="number" @
blur="setAmount($event,item.gid,item.new_meal_items,item.meal_items,index)" />
                                    <!--添加购物车按钮-->
                                    <view class="inc" @click="incAmount
($event,item.gid,item.new_meal_items, item.meal_items)"></view>
                                </view>
                            </view>
                        </view>
                        <!--如果非套餐并且包装费大于0，则显示包装费-->
                        <view class="pack-price" v-if="!item.meal_items &&
item.pack_price>0">包装费:￥{{item.pack_price}}</view>
                        <!--如果是套餐，则显示套餐里菜品的明细-->
                        <view class="goods-desc" v-if="item.meal_items &&
item.meal_items.length>0">
                            <view class="goods-desc-list" v-for="(item2,index2)
in item.meal_items" :key="index2"><view class="goods-title">- {{item2.title}}
x{{item2.amount}}</view><view v-if="item2.pack_price>0">  包 装 费: ￥
{{item2.pack_price}}</view></view>
                        </view>
                    </view>
                </view>
            </scroll-view>
        </view>
    </view>
</template>

<script>
    import {mapMutations} from "vuex";
    export default {
        name: "cart",
        data(){
            return {
                triggered:false
            }
        },
        props:{
            //是否显示购物车组件
            show:{
```

```
            type:Boolean,
            default: false
        },
        //购物车里菜品的数据
        cartData:{
            type:Array,
            default:()=>[]
        },
        //获取桌号和分店的id
        values:{
            type:Object,
            default:()=>{}
        }
    },
    methods:{
        ...mapMutations({
            SHOW_AMOUNT_INPUT:"cart/SHOW_AMOUNT_INPUT",
            //显示数量input输入框
            HIDE_AMOUNT_INPUT:"cart/HIDE_AMOUNT_INPUT"
            //隐藏数量input输入框
        }),
        //隐藏购物车组件
        close(){
            this.$emit("close");
        },
        //增加数量
        incAmount(e,gid,new_meal_items,meal_items){
            //触发pages/goods/index.vue文件中的incAmount()方法
            this.$emit("incAmount",e,gid,new_meal_items,meal_items);
        },
        //减少数量
        decAmount(e,gid,new_meal_items,meal_items){
            //触发pages/goods/index.vue文件中的decAmount()方法
            this.$emit("decAmount",e,gid,new_meal_items,meal_items);
        },
        //更改数量
        setAmount(e,gid,new_meal_items,meal_items,index){
            //触发pages/goods/index.vue文件中的setAmount()方法
            this.$emit("setAmount",e,gid,new_meal_items,meal_items);
            //隐藏数量input输入框
            this.hideAmountInput(index);
        },
        //清除购物车
```

```
            clearCart(){
                if(this.cartData.length>0){
                    uni.showModal({
                        title: '',
                        content: '确认要清空购物车吗?',
                        success: (res)=> {
                            if (res.confirm) {
                                this.$emit("clearCart");
                            } else if (res.cancel) {

                            }
                        }
                    });
                }
            },
            //显示数量输入框
            showAmountInput(index){
                this.SHOW_AMOUNT_INPUT({index});
            },
            //隐藏数量输入框
            hideAmountInput(index){
                this.HIDE_AMOUNT_INPUT({index});
            }
        }
    }
</script>

<style scoped>
    .mask{width:100%;height:100%;position: fixed;z-index:88;bottom:0px;left:
     0px;background-color:rgba(0,0,0,0.6)}
    .cart-main{width:100%;height:994rpx;background-color:#FFFFFF;position:
     absolute;bottom:0px;left:0px;z-index:89;transform: translateY(100%);tra
     nsition:transform 0.3s;}
    .cart-main.nav{width:100%;height:100rpx;border-bottom: 1px solid
     #EFEFEF;display: flex;justify-content: space-between;align-items:
     center;box-sizing: border-box;padding:0px 20rpx;}
    .cart-main.nav-title{width:auto;height:auto;display:flex;}
    .cart-main.nav-title.icon{width:45rpx;height:45rpx;background-image:url("~@/
     static/images/goods/cart.png");background-size:100%;background-repeat: no-
     repeat;background-position: center;margin-right:10rpx;}
    .cart-main.text{font-size:28rpx;color:#333333;}
    .cart-main.clear-main{width:auto;height:auto;display:flex;}
```

```
    .cart-main.clear-main.icon{width:36rpx;height:36rpx;background-
     image:url("~@/static/images/goods/clear.png");background-size:100%;background-
     repeat: no-repeat;background-position: center;margin-right:10rpx;}
    .cart-main.goods-main{width:100%;height:730rpx;overflow:hidden;}
    .cart-main.goods-main.goods-list{width:94%;margin:0 auto;}
    .cart-main.goods-list.title-area{width:100%;height:90rpx;display:flex;
     justify-content: space-between;align-items: center;}
    .cart-main.goods-list.title-area .title{font-size:28rpx;color:#333333;}
    .cart-main.goods-list.goods-desc{width:94%;margin:0 auto;}
    .cart-main.goods-list.goods-desc-list{width:100%;height:auto;font-
     size:24rpx;color:#929292;display: flex;justify-content:
     space-between;margin-bottom:20rpx;}
    .cart-main.goods-list.pack-price{font-size:24rpx;color:#333333;}

    .price-amount{width:auto;height:100%;display:flex;justify-content:
     space-between;align-items: center;}
    .price-amount.points{font-size:28rpx;color:#333333;}
    .price-amount.price{width:auto;height:auto;font-size:28rpx;
     color:#333333;margin-right:20rpx;}
    .price-amount image{width:30rpx;height:30rpx;margin-right:10rpx;}
    .handle-amount{width:auto;height:60rpx;display:flex;align-items:
     center;justify-content:flex-end}
    .handle-amount.dec{width:55rpx;height:55rpx;background-image:url("~@/
     static/images/main/dec.png");background-size:100%;background-position:
     center;background-repeat: no-repeat;}
    .handle-amount.inc{width:55rpx;height:55rpx;background-image:url("~@/
     static/images/main/inc.png");background-size:100%;background-position:
     center;background-repeat: no-repeat;}
    .handle-amount.text{font-size:28rpx;color:#000000;margin-right:
     10rpx;margin-left:10rpx;width:50rpx;text-align:center;}
    .handle-amount.text.disabled{color:#9d9d9d}
    .handle-amountinput{width:50rpx;height:100%;text-align:
     center;margin-right:10rpx;margin-left:10rpx;}
    .cart-main.show{transform: translateY(0%);}
    .goods-main.tip{width:100%;font-size:24rpx;color:#E30019;text-
     align:center;margin-top:30rpx;}
    .goods-main.scroll-wrap{
        width:100%;
        min-height:120%;
    }
</style>
```

incAmount()方法内部使用this.$emit()方法触发pages/goods/index.vue文件中的incAmount方法，可以增加菜品的数量并将菜品添加到购物车；decAmount()方法内部使用this.$emit()方法触发pages/goods/index.vue文件中的decAmount()方法，可以减少菜品的数量和删除菜品；setAmount()方法内部使用this.$emit()方法触发pages/goods/index.vue文件中的setAmount()方法，可以修改菜品的数量；clearCart()方法可以清除购物车里的菜品。为了提高用户体验，在清除前先使用uni.showModal模态框提示用户，当用户点击确认按钮后再使用this.$emit()方法触发pages/goods/index.vue文件中的clearCart()方法，清除购物车里的菜品。

接下来在pages/goods/index.vue文件中使用购物车组件，新增代码如下：

```
<template>
    <view class="page">
        ...
        <!--只有在菜品展示组件显示的情况下，才显示购物车导航-->
        <view class="cart-main" v-show="goodsAction">
            <view class="handle-cart" @click="showCart()">
                <view class="cart-cle">
                    <view class="cart-cle-2">
                        <view class="cart-icon" id="cart-icon">
                            <view class="badge" v-show="cartCount>0?true:false">
                             {{cartCount}}</view>
                        </view>
                    </view>
                </view>
                <view class="total"> ¥{{total}}</view>
            </view>
            <view class="line"></view>
            <view class="submit">
                <template v-if="false">提交订单 &gt;</template>
                <template v-else>未选购商品</template>
            </view>
        </view>
        <cart :show="setIsCart" :cart-data="cartData"  @close="isCart=false"
@incAmount="incAmount"  @decAmount="decAmount"  @setAmount="setAmount"
@clearCart="clearCart()" :values="{branch_shop_id,table_code}"></cart>
    </view>
</template>

<script>
    import {mapState,mapActions,mapMutations,mapGetters} from 'vuex';
    ...
    import Cart from '../../components/cart';
    export default {
```

```
        name: "goods",
        data(){
            return {
                ...
                isCart:false,                           //是否显示购物车组件
            }
        },
        onLoad(opts){
            ...
            //如果有桌号则代表堂食，如果没有桌号则代表自提点餐
            this.SET_PACK({isPack:!this.table_code?true:false});
        },
        onShow(){
            ...
            //显示购物车里的菜品数据
this.showCartData({branch_shop_id:this.branch_shop_id,table_code:this.table_code});
        },
        computed:{
            ...mapState({
                ...
                "cartData":state=>state.cart.cartData        //购物车里的菜品
            }),
            ...
            //是否显示购物车面板
            setIsCart(){
                if(this.cartData.length>0){
                    return this.isCart;
                }else{
                    this.isCart=false;
                    //解决购物车里没有数据自动隐藏以及数据不显示的问题
                    return this.isCart;
                }
            },
        },
        components:{
            ...
            Cart
        },
        ...
        methods:{
            ...mapMutations({
                ...
```

```
        SET_PACK:"cart/SET_PACK"              //设置是否自提
    }),
    ...mapActions({
        ...
        "addCart":"cart/addCart",             //添加购物车的方法
        "delCart":"cart/delCart",             //删除购物车里的菜品
        "showCartData":"cart/showCartData", //显示购物车里的菜品
        "asyncSetAmount":"cart/setAmount",  //修改购物车里菜品的数量
        "asyncClearCart":"cart/clearCart"   //清空购物车
    }),
    ...
    //修改数量
    setAmount(e,gid,new_meal_items,meal_items,index,index2){
        let amount=e.target.value;
        this.amount=amount;
        //设置菜品展示的数量
        this.SET_GOODS_AMOUNT({amount:e.target.value, gid:gid,
        new_meal_items: new_meal_items});
        //更新购物车里菜品的数量
        this.asyncSetAmount({branch_shop_id:this.branch_shop_id,
        cartData: {amount:amount,gid:gid,meal_items:meal_items,new_
        meal_items:new_meal_items}});
        if(index!=undefined && index2!=undefined){
            this.HIDE_AMOUNT_INPUT({index,index2}) //隐藏输入框
        }
    },
    //添加菜品数量
    incAmount(e,gid,new_meal_items,meal_items){
        //如果会员已登录
        if(this.isLogin){
            //调用static/js/utils/index.js文件中的safeUser()方法,
            //如果token验证成功
            this.$utils.safeUser(this,this.branch_shop_id,this.table_code,()=>{
                let isGoodsSame=false;               //是否还有相同的菜品
                //套餐里菜品明细的gid，用于添加购物车时判断是否有相同的套餐
                let curNewMealItems=new_meal_items && new_meal_items.
                    length>0?JSON.stringify(new_meal_items):"";
                for(let i=0;i<this.goods.length;i++){
                    for(let j=0;j<this.goods[i].goods.length;j++){
                        //默认套餐里菜品的gid
                            let newMealItems=this.goods[i].goods[j].new_
meal_items.length>0?JSON.stringify(this.goods[i].goods[j].new_meal_items):"";
```

```
                //如果有相同的菜品，该判断是为了增加this.amount
                //数量以及判断是否为自定义套餐
                if(this.goods[i].goods[j].gid==gid &&
                newMealItems==curNewMealItems){
                  isGoodsSame=true;
                  //增加菜品的数量。这里的数量会传递给菜品展示组件
                  //和菜品详情组件，这样数量可以同步
                  this.amount=parseInt(this.goods[i].
                     goods[j].amount)+1;

                  //添加购物车的数据
                  let cartData={
                    "title":this.goods[i].goods[j].title,
                    //菜品名称
                    "branch_shop_id":this.branch_shop_id,
                    //分店的id
                    "table_code":this.table_code,//桌号
                    "gid":gid,            //菜品的唯一标识
                    "price":this.goods[i].goods[j].price,
                    //菜品价格
                    "amount":1,
                    //菜品数量默认为1，数量的改变在Vuex中写逻辑
                    "image":this.goods[i].goods[j].image,
                    //菜品的图片
                    "is_meal":this.goods[i].goods[j].is_meal,
                    //是否套餐，1为是，0为否
                     "pack_price":this.goods[i].goods[j].
pack_price?this.goods[i].goods[j].pack_price:0,    //包装费
                    "meal_type":0,
                    //套餐类型，0为普通套餐，1为自定义套餐
                    "meal_items":this.goods[i].goods[j].
                    meal_items,
                    //套餐里菜品的明细
                    "new_meal_items":this.goods[i].
goods[j].new_meal_items,                 //套餐里菜品的gid明细
                    "isAmountInput":false,
                    //是否显示数量input输入框
                    "isAmountInputFocus":false
                    //数量input输入框获取焦点
                  };
                  //添加购物车
                  this.addCart({cartData:cartData});
```

```
                        break;
                    }
                }
            }
            //如果没有相同的菜品，表示是自定义套餐
            if(!isGoodsSame){
                //自定义套餐数据
                let cartData={
                    "branch_shop_id":this.branch_shop_id,
                    //分店的id
                    "table_code":this.table_code, //桌号
                    "gid":gid,                    //套餐的唯一标识
                    "amount":1,
                    //菜品数量默认为1，数量的改变在Vuex中写逻辑
                    "is_meal":1,                  //是否套餐，1为是，0为否
                    "meal_type":1,
                    //套餐类型，0为普通套餐，1为自定义套餐
                    "meal_items":meal_items,
                    //套餐里菜品的明细
                    "new_meal_items":new_meal_items,
                    //套餐里菜品的gid明细
                    "isAmountInput":false,
                    //是否显示数量input输入框
                    "isAmountInputFocus":false
                    //数量input输入框获取焦点
                };
                //添加购物车
                this.addCart({cartData:cartData});
            }
        })

    }else{                                        //如果会员未登录
        //跳转到登录页面
        this.pushPage(`/pages/login/index?branch_shop_id=${this.
        branch_shop_id}&table_code=${this.table_code}`);
    }
},
//清空购物车
clearCart(){
    if(this.isLogin){
        this.$utils.safeUser(this,this.branch_shop_id,this.
        table_code,()=>{
```

```
                if(this.goods.length>0){
                    //将菜品展示的数量设置为0
                    for(let i=0;i<this.goods.length;i++){
                        for(let j=0;j<this.goods[i].goods.length;j++){
                            this.goods[i].goods[j].amount=0;
                        }
                    }
                    this.amount=0;
                }
                //清除数据库中的购物车数据
                this.asyncClearCart({branch_shop_id:this.branch_
                    shop_id,table_code:this.table_code});
            });
        }else{
            this.pushPage(`/pages/login/index?branch_shop_id=${this.
                branch_shop_id}&table_code=${this.table_code}`);
        }
    },
    //显示购物车
    showCart(){
        if(this.isLogin){
            this.$utils.safeUser(this,this.branch_shop_id,this.
                table_code,()=>{
                if(this.cartData.length>0){
                    this.isCart=true;          //显示购物车组件
                }
            });
        }else{
            this.pushPage(`/pages/login/index?branch_shop_id=${this.
                branch_shop_id}&table_code=${this.table_code}`);
        }
    },
    //减少购物车里的菜品数量
    decAmount(e,gid,new_meal_items,meal_items){
        if(this.isLogin){
            this.$utils.safeUser(this,this.branch_shop_id,this.
                table_code,()=>{
                let curNewMealItems=new_meal_items && new_meal_items.
                    length>0?JSON.stringify(new_meal_items):"";
                if(this.goods.length>0){
                    let isGoodSame=false;
```

```
                                    for(let i=0;i<this.goods.length;i++){
                                        for(let j=0;j<this.goods[i].goods.
                                            length;j++){
                                                let newMealItems=this.
goods[i].goods[j].new_meal_items.length>0?JSON.stringify(this.
goods[i].goods[j].new_meal_items):"";

                                                if(this.goods[i].goods[j].
gid==gid && newMealItems==curNewMealItems){
                                                    isGoodSame=true;
                                                    //减少菜品展示里的菜品数量
                                                    this.goods[i].goods[j].
amount=parseInt(this.goods[i].goods[j].amount)>0?parseInt(this.goods[i].
goods[j].amount)-1:0;
                                                    //减少菜品的数量
                                                    this.amount = parseInt(this.
                                                    goods[i].goods[j].amount);
                                                        this.$set(this.goods[i].
goods,j,this.goods[i].goods[j]);
                                                    this.delCart({cartData:this.
goods[i].goods[j],branch_shop_id:this.branch_shop_id,table_code:this.
table_code,meal_type:0});
                                                    break;
                                                }
                                            }
                                        }
                                        //如果是自定义套餐
                                        if(!isGoodSame){
                                            this.delCart({cartData:{gid:gid,meal_
items:meal_items,new_meal_items:new_meal_items},meal_type:1,branch_shop_
id:this.branch_shop_id,table_code:this.table_code});
                                        }

                                }
                        });
                    }else{
                        this.pushPage(`/pages/login/index?branch_shop_id=${this.
                        branch_shop_id}&table_code=${this.table_code}`);
                    }
                },
            }
    }
</script>
```

以上加粗代码为新增或修改的代码，到这里对购物车组件的开发基本完成。接下来将菜品展示组件和菜品详情组件与购物车组件关联起来，11.7.2节中在菜品展示组件实现了修改菜品数量和增加菜品数量的功能，现在增加减少菜品数量的功能。在components/goods/index.vue文件中的视图<template>标签内找到以下代码：

```
<view class="dec" v-show="parseInt(item2.amount)>0?true:false"></view>
```

替换为

```
<view class="dec" @click.stop="decAmount($event,item2.gid,item2.new_meal_
items,item2.meal_items)" v-show="parseInt(item2.amount)>0?true:false">
</view>
```

新增click事件，并调用decAmount()方法，在js中实现decAmount()方法，代码如下：

```
methods:{
   //减少菜品数量
   decAmount(e,gid,new_meal_items,meal_items){
       //触发pages/goods/index.vue文件中的decAmount()方法
        this.$emit("decAmount",e,gid,new_meal_items,meal_items)
   },
}
```

以上加粗代码为新增代码。为了提升用户体验，只有在菜品数量大于0时才显示数量。在视图中找到以下代码：

```
<view class="text" v-if="!item2.isAmountInput" @click.stop="showAmountInput(
index,index2)">{{item2.amount}}</view>
```

替换为

```
<view class="text" v-if="!item2.isAmountInput && item2.amount>0" @click.stop
="showAmountInput(index,index2)">{{item2.amount}}</view>
```

上面加粗代码为新增代码。接下来在pages/goods/index.vue文件中找到<goods>组件，添加自定义事件@decAmount，代码如下：

```
<goods :branchShopId="branch_shop_id" :isShow="goodsAction" :gid="gid"
:amount="amount" @setAmount="setAmount" @showAmountInput="showAmountInput"
:classifysData="classifys" :goodsData="goods" @showGoodsDetails="showGoods
Details" @incAmount="incAmount" @decAmount="decAmount"></goods>
```

完善菜品详情组件，在components/goods/details.vue文件中的视图<template>标签内找到以下代码：

```
<view class="handle-amount">
    <view class="dec" v-show="amount>0?true:false"></view>
    <view class="text" v-if="!isAmountInput" @click="showAmountInput">{{amount}}</view>
```

```
    <input v-if="isAmountInput" :focus="isAmountInputFocus" :value="amount"
     type="number" @blur="setAmount" />
    <view class="inc"></view>
</view>
```

替换为

```
<view class="handle-amount">
    <view class="dec" v-show="amount>0?true:false" @click="decAmount"></view>
    <view class="text" v-if="!isAmountInput && amount>0" @click="showAmountI
     nput">{{amount}}</view>
    <input v-if="isAmountInput" :focus="isAmountInputFocus" :value="amount"
     type="number" @blur="setAmount" />
    <view class="inc" @click="incAmount($event,amount)"></view>
</view>
```

以上加粗代码为新增代码。接下来在js中分别实现绑定事件的回调方法decAmount()和incAmount()，代码如下：

```
methods:{
    //增加菜品数量
    incAmount(e){
        //触发components/goods/index.vue文件中的incAmount()方法
        this.$emit("incAmount",e,this.gid,"","")
    },
    //减少菜品数量
    decAmount(e){
        //触发components/goods/index.vue文件中的decAmount()方法
        this.$emit("decAmount",e,this.gid,"","")
    },
}
```

在components/goods/index.vue文件中找到<goods-details>组件，添加自定义事件@incAmount和@decAmount，代码如下：

```
<goods-details :show="isShowGoodsDetails" :branchShopId="branchShopId" :gid="gid"
:amount="amount" @close="isShowGoodsDetails=false" @setAmount="setAmount"
@decAmount="decAmount" @incAmount="incAmount"></goods-details>
```

加粗代码为新增代码，<goods-details>组件的属性show为显示该组件，属性branchShopId为分店的编号，属性gid为菜品的编号，属性amount为菜品的数量，自定义事件close可以隐藏该组件，自定义事件setAmount可以更改菜品的数量，自定义事件decAmount可以减少和删除菜品，自定义事件incAmount可以增加菜品的数量和添加到购物车。至此，购物车功能全部开发完成。

11.8　提交订单

提交订单页面包含购物车里的菜品展示、价格计算、是否打包、在线支付等功能。提交订单页面预览效果如图11-17所示。

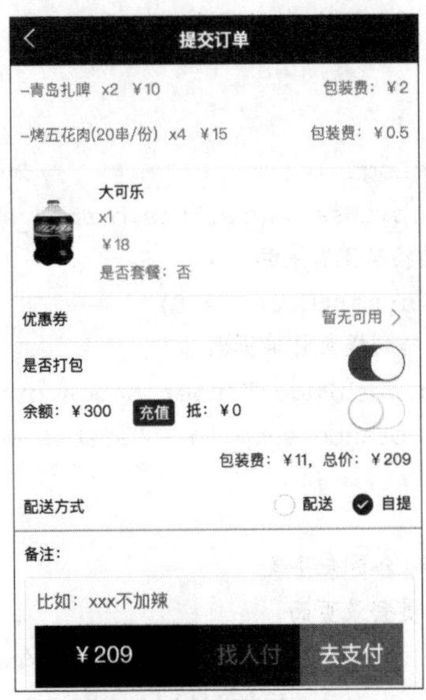

图 11-17　提交订单页面预览效果

图11-17是真实运营的项目界面，其中配送方式、找人付、优惠券、充值等功能是后期根据业务需求扩展开发的，本章的实战项目中并没有这些功能。但是，服务端API接口文档包括这些功能，读者如果想开发，可以根据接口文档自行尝试。

提交订单相关接口在"接口文档"文件夹下的"订单.docx""支付.docx"文件中。

11.8.1　开发提交订单页面

扫一扫，看视频

如果购物车里有菜品，在点餐页面的"未选购商品"区域会显示"提交订单"，点击"提交订单"按钮，可以进入提交订单页面。在pages/goods/index.vue文件中找到以下代码：

```
<view class="submit">
    <template v-if="false">提交订单 &gt;</template>
    <template v-else>未选购商品</template>
</view>
```

替换为

```
<view class="submit" @click="submitOrder()">
    <template v-if="cartCount>0">提交订单 &gt;</template>
```

```
    <template v-else>未选购商品</template>
</view>
```

以上加粗代码为新增代码。接下来在js中的methods对象中添加submitOrder()方法，代码如下：

```
<script>
   methods:{
         //提交订单
          submitOrder(){
             if(this.isLogin) {                //如果会员已登录
           this.$utils.safeUser(this,this.branch_shop_id,this.table_code,()=>{
                //如果购物车里有菜品
                if (this.cartCount > 0) {
                   //跳转到提交订单页面
                   this.pushPage("/user_pages/order/index?branch_shop_id=" +
                   this.branch_shop_id + "&table_code=" + this.table_code);
               }
            });
               }else{//会员未登录
            //跳转到会员登录页面
                                    this.pushPage('/pages/login/index?branch_shop_
                                    id=${this.branch_shop_id}&table_code=${this.
                                    table_code}');
            }
      },
   }
</script>
```

submitOrder()方法可以跳转到提交订单页面。提交订单页面需要会员登录之后才能访问，所以在跳转之前需要使用this.isLogin判断会员是否登录，如果登录则跳转到提交订单页面，否则跳转到会员登录页面。

开发提交订单相关服务端接口。在api文件夹下创建order/index.js文件，该文件中的代码如下：

```
import config from "../../static/js/conf/config";
import {request} from "../../static/js/utils/request";

//提交订单
export function addOrderData(data){
    return request(config.baseApi+"/v1/user/order/add","post",data);
}
```

```
//微信小程序统一下单接口
export function wechatUnifiedorderData(data){
    return request(config.baseApi+"/v1/wxpay/wechat_unifiedorder", "post",data);
}

//提交订单后返回的信息
export function getLastOrderData(data){
    return request(config.baseApi+"/v1/user/order/last","post",data)
}

//直接支付
export function payData(data){
    return request(config.baseApi+"/v1/payorder/notify","post",data)
}
```

addOrderData()函数可以将购物车里的菜品和订单信息添加到数据库中;wechatUnifiedorderData()函数是微信统一下单接口，用来开发微信支付;getLastOrderData()函数可以获取提交订单后的订单编号;payData()函数可以跳过微信支付直接付款成功，一般用于支付金额为0的情况，如使用余额支付或者使用优惠券支付。

接下来对接到Vuex中，在store文件夹下创建order/index.js文件，该文件中的代码如下：

```
import {addOrderData,wechatUnifiedorderData,getLastOrderData} from "../../
api/order";
export default {
    namespaced:true,
    state:{
        lastOrder:{}                    //提交订单后返回的信息
    },
    mutations:{
        //设置提交订单后返回的信息
        ["SET_LAST_ORDER"](state,payload){
            state.lastOrder=payload.lastOrder;
        }
    },
    actions:{
        //提交订单
        addOrder(conText,payload){
            addOrderData({uid:conText.rootState.user.uid,token:conText.rootState.
user.token,platform:conText.rootState.system.platform,...payload}).then(res=>{
                if(payload.success){
                    payload.success(res);
                }
```

```
            })
        },
        //微信小程序统一下单接口
        wechatUnifiedorder(conText,payload){
            return wechatUnifiedorderData({uid:conText.rootState.user.uid,token:conText.
rootState.user.token,platform:conText.rootState.system.platform,...payload}).then(res=>{
                return res;
            })
        },
        //获取提交订单后返回的信息
        getLastOrder(conText,payload){
            getLastOrderData({uid:conText.rootState.user.uid,token:conText.
rootState.user.token,platform:conText.rootState.system.platform}).then(res=>{
                if(res.code==200){
                    conText.commit("SET_LAST_ORDER",{lastOrder:res.data});
                }
            })
        }
    },
    //如果支付金额为0，则直接支付
    payOrder(conText,payload){
        payData({uid:conText.rootState.user.uid,token:conText.rootState.user.
token,platform:conText.rootState.system.platform,...payload}).then(res=>{
            if(res.code==200){
                if(payload.success){
                    payload.success();
                }
            }
        })
    },
}
```

state内部的lastOrder属性值为提交订单后返回的信息，如订单编号。mutations内部的SET_LAST_ORDER方法可以设置提交订单后返回的信息。actions内部的addOrder()方法可以将购物车里的菜品和订单信息提交到数据库；wechatUnifiedorder()方法是微信小程序统一下单接口，开发微信支付时需要使用此接口；getLastOrder()方法可以获取提交订单后返回的信息，如订单编号；payOrder()方法可以跳过微信支付直接付款，如使用优惠券或是余额支付，当支付金额为0时需要使用此方法。

接下来注册到Vuex的模块中，在store/index.js文件中新增代码如下：

```
...
import order from './order';
```

```
Vue.use(Vuex);

const store = new Vuex.Store({
    modules:{
        ...
        order
    }
});

export default store
```

以上加粗代码为新增代码。还需要请求服务端接口获取饭店的区域，在api/business/index.js文件中新增代码如下：

```
...
//获取饭店区域
export function getHotelAreaData(data){
    return request(config.baseApi+"/v1/business/hotel_area","get",data);
}
```

接下来对接到Vuex中，在store/business/index.js文件中新增代码如下：

```
import {getShopData,getShopInfoData,getHotelAreaData} from "../../api/business";
export default {
    namespaced:true,
    state:{
        ...
        hotelArea:""            //饭店区域
    },
    mutations: {
        ...
        //设置饭店区域
        ["SET_HOTEL_AREA"](state,payload){
            state.hotelArea=payload.hotelArea;
        },
    },
    actions:{
        ...
        //获取饭店区域
        getHotelArea(conText,payload){
            getHotelAreaData(payload).then(res=>{
                if(res.code==200){
                    conText.commit("SET_HOTEL_AREA",{hotelArea:res.data.title});
```

```
                }
            })
        },
    }
}
```

以上加粗代码为新增代码。接下来开发提交订单页面，为了提高性能，将对与会员相关的功能页面进行分包处理。在根目录创建user_pages文件夹，在该文件夹下创建order/index.vue文件，并配置路由。在pages.json文件中新增代码如下：

```
{
    ...
    "subPackages": [
        {
            "root": "user_pages",
            "pages": [
                {
                    "path": "order/index",
                    "style": {
                        "navigationBarTitleText": "提交订单",
                        "disableScroll": false
                    }
                }
            ]
        }
    ]
}
```

5.1.5节已经介绍了分包，这里不再赘述。user_pages/order/index.vue文件中的代码如下：

```
<template>
    <view class="page">
        <view class="shop-title">{{shopInfo.branch_shop_name}}</view>
        <view class="order-main">
            <view class="order-list">
                <view class="order-info" v-if="table_code">
                    <view class="area-table">
                        <view class="area">区域:{{hotelArea}}</view>
                        <view class="table-code">桌号:{{table_code}}</view>
                    </view>
                </view>
                <view class="order-desc">
                    <view class="goods-list-main" v-for="(item,index) in
                    cartData" :key="index">
```

```
                <view class="goods-list">
                    <view class="image">
                        <image :src="item.image"></image>
                    </view>
                    <view class="goods-info">
                        <view class="title"><text selectable= "true">
                        {{item.title}}</text></view>
                        <view class="info-text">x{{item.amount}}</view>
                        <view class="info-text">￥{{item.price}}</view>
                        <template v-if="item.place_type=='1'">
                            <view class="info-text" >积分赠送</view>
                            <view class="info-text" >所需积分:{{item.
                            need_points}}</view>
                        </template>
                        <template v-else>
                                <view class="info-text" v-if="item.is_
meal=='0' && item.pack_price>0">包装费:￥{{item.pack_price}}</view>
                            <view class="is-meal">是否套餐:{{item.is_
                            meal=='1'?'是':'否'}}</view>
                        </template>
                    </view>
                </view>
                <view class="meal-items" v-if="item.is_meal=='1'">
                    <view class="item-list" v-for="(item2,index2) in
                    item.meal_items" :key="index2">
                        <view class="item">
                            <view>- <text selectable="true">{{item2.
                            title}}</text></view>
                            <view>x{{item2.amount}}</view>
                            <view>￥{{item2.price}}</view>
                        </view>
                            <view v-if="item2.pack_price>0">包装费:￥
                            {{item2.pack_price}}</view>
                    </view>
                </view>
            </view>
        </view>
        <view class="order-row">
            <view class="order-col-1">
                <view>是否打包</view>
            </view>
            <view class="order-col-2">
```

```
                    <switch :disabled="isDisablePack" :checked="isPack"
                    @change="changePack" color="#E30019" />
                </view>
            </view>
            <view class="price-status">
                <view class="price">包装费:￥{{packTotal}}, 总价:￥{{true_total}}</view>
            </view>
            <view class="order-row" v-if="isDisablePack">
                <view class="order-col-1">
                    <view>配送方式</view>
                </view>
                <view class="order-col-2 active">
                    <label><radio class="radio-style" value="2" color= "#E30019"
                     :checked="true" />自提</label>
                </view>
            </view>
        </view>
        <view class="remarks">
            备注:<br/>
            <view class="content">
                <textarea placeholder="比如:xxx不加辣" v-model="remarks" />
            </view>
        </view>
    </view>
    <view class="pay-wrap">
        <view class="price"> ￥{{true_total}}</view>
        <cover-view class="pay" @click="goPay()">去支付</cover-view>
    </view>
</view>
</template>

<script>
    import {mapState,mapActions,mapGetters,mapMutations} from "vuex";
    export default {
        name: "order-index",
        data(){
            return {
                branch_shop_id:"",     //分店的id
                table_code:"",         //桌号
                remarks:"",            //备注
                isDisablePack:false,  //是否打包
            }
        },
```

```
onLoad(opts){
    this.branch_shop_id=opts.branch_shop_id?opts.branch_shop_id:"";
    //分店的id
    this.table_code=opts.table_code?opts.table_code:'';  //桌号
    this.$utils.safeUser(this,this.branch_shop_id,this.table_code);
    //验证会员登录token是否合法
    this.isPay=true;                      //防止出现网速慢时多次重复提交的问题
    if(!this.table_code){                 //如果没有桌号，则代表自提点餐
        this.isDisablePack=true;          //如果是自提点餐，则需要支付包装费
    }
    //获取饭店的区域
    this.getHotelArea({branch_shop_id:this.branch_shop_id,table_
        code:this.table_code});
    //如果Vuex中的购物车数据为空
    if(!this.cartData || this.cartData.length<=0){
        //重新获取购物车里的数据
        this.showCartData({branch_shop_id:this.branch_shop_
            id,table_code:this.table_code});
    }
    //设置是否打包
    this.SET_PACK({isPack:!this.table_code?true:false});
    //获取商铺信息
    this.showShopInfo({branch_shop_id:this.branch_shop_id});
},
methods:{
    ...mapActions({
        showShopInfo:"business/showShopInfo",       //获取商铺信息
        getHotelArea:"business/getHotelArea",       //获取饭店的区域
        showCartData:"cart/showCartData",           //获取购物车里的菜品
        addOrder:"order/addOrder",                  //提交订单
        payOrder:"order/payOrder",              //如果支付金额为0，则直接支付
        wechatUnifiedorder:"order/wechatUnifiedorder"  //微信支付统一下单接口
    }),
    ...mapMutations({
        SET_PACK:"cart/SET_PACK"                    //设置是否打包
    }),
    //跳转页面
    pushPage(url){
        uni.navigateTo({
            url:url
        })
    },
    //是否自提
```

```
        changePack(e){
            this.SET_PACK({isPack:e.detail.value});
        },
        //去支付
        goPay(){
            if(this.isPay) {
                this.isPay = false;
                //如果为堂食
                if (!this.isDisablePack) {
                    this.distribution_type = "0";
                } else {    //如果不是堂食，则设置为自提
                    this.distribution_type = "2";
                }
                //提交订单
                this.addOrder({
                    branch_shop_id: this.branch_shop_id,
                    table_code: this.table_code,
                    remarks: this.remarks,
                    is_pack: this.isPack ? '1' : '0',//是否打包，1为是，0为否
                    distribution_type: this.distribution_type,
                    //配送方式，0为堂食,2为自提
                    success: async (res) => {
                        let true_total = 0;                 //实际支付金额
                        let ordernum = "";                  //订单编号
                        if (res.code == 200) {
                            true_total = res.data.true_total;
                            ordernum = res.data.ordernum;

                        } else {
                            uni.showToast({
                                title: '' + res.data + '',
                                icon: "none",
                                duration: 2000
                            })
                        }
                        this.isPay = true;
                    }
                });
            }
        }
    },
    computed:{
        ...mapState({
```

```
            shopInfo:state=>state.business.shopInfo,    //商铺信息
            hotelArea:state=>state.business.hotelArea, //饭店的区域
            cartData:state=>state.cart.cartData,        //购物车数据
            openId:state=>state.user.openId,            //会员的openId
            isPack:state=>state.cart.isPack             //是否打包
        }),
        ...mapGetters({
            total:"cart/total",                     //购物车里菜品的总价
            packTotal:"cart/packTotal"              //购物车里菜品的包装费总价
        }),
        //实际付款金额
        true_total(){
            return parseFloat(this.total.toFixed(2));
        }
    }
  }
</script>

<style scoped>
    .page{width:100%;min-height:100vh;background-color:#FFFFFF;overflow:hidden;
    margin-bottom:160rpx;}
    .shop-title{width:100%;height:80rpx;background-color:#FFFFFF;border-
    bottom: 1px solid #EFEFEF;font-size: 28rpx; text-align: center;
    line-height:80rpx;}
    .order-main{width:100%;}
    .order-main.order-list{width:100%;border-bottom:1px solid #f17f1f;}
    .order-main.order-list.order-info{width:100%;border-bottom: 1px solid
    #EFEFEF;box-sizing: border-box;padding:20rpx;}
    .order-main.order-list.area-table{width:100%;display:flex;font-
    size:28rpx;margin-top:20rpx;}
    .order-main.order-list.area-table .area{margin-right:40rpx;}

    .order-main.order-desc{width:100%;}
    .order-main.goods-list-main{width:100%;box-sizing: border-
    box;padding:20rpx;border-bottom: 1px solid #EFEFEF;}
    .order-main.order-desc.goods-list{width:100%;display:flex;justify-
    content: space-between;align-items: center;}
    .order-main.order-desc.goods-list.image{width:120rpx;height:120rpx;}
    .order-main.order-desc.goods-list.image image{width:100%;height:100%;
    border-radius: 5px;}
    .order-main.order-desc.goods-list.goods-info{width: 80%; height:
    auto;font-size:28rpx;overflow:hidden;}
    .order-main.order-desc.goods-list.goods-info .title{width:100%;height:
    45rpx;overflow:hidden;text-overflow: ellipsis;white-space: nowrap;}
```

```
.order-main.order-desc.goods-list.goods-info .info-text{width:auto;height:
auto;color:#909090;margin-bottom:10rpx;}
.order-main.order-desc.goods-list.goods-info.is-meal {width:auto;height:
auto;color:#909090;}
.order-main.order-desc.meal-items{width:100%;margin-top:20rpx;}
.order-main.order-desc.meal-items.item-list{width:100%;height:80rpx;
display:flex;justify-content:space-between;align-items:
center;font-size:28rpx;color:#909090}
.order-main.order-desc.meal-items.item-list.item{display:flex;}
.order-main.order-desc.meal-items.item-list.item view{margin-right:20rpx;}
.order-main.order-desc.meal-items.item-list.refund-btn{width:auto;
height:auto;padding:8rpx 15rpx;color:#FFFFFF;font-size:28rpx;background-
color:#007aff;border-radius: 4px;}
.order-main.order-list.price-status{width:100%;height:auto;box-sizing:
border-box;padding:20rpx;display:flex;justify-content: flex-end;align-
items: center;}
.order-main.order-list.price-status.price{font-size: 28rpx;
color:#333333}

.order-main.order-row{width:100%;height:auto;border-bottom:
1px solid #EFEFEF;justify-content: space-between;align-items:
center;display:flex;box-sizing: border-box;padding:20rpx 20rpx;}
.order-main.order-row.order-col-1{font-size:28rpx;display:flex;}
.order-main.order-row.order-col-2{display:flex;font-size:28rpx;color:#909
090;height:40rpx;align-items: center;}
.order-main.order-row.order-col-2.active{color:#000000}
.order-main.order-row.radio-style{transform:scale(.8)}
.order-main.order-row.order-col-2 label{margin-left:30rpx;}

.order-main.remarks{font-size:28rpx;width:100%;margin-top:20rpx;box-
sizing: border-box;padding:0px 20rpx;}
.order-main.remarks.content{width:100%;height:300rpx;overflow:hidden;margin-
top:20rpx;border-radius: 5px;border:1px solid #EFEFEF;}
.order-main.remarks.content textarea{width:100%;height:100%;padding:20rpx;}
.order-main.remarks.address{width:100%;height:70rpx;overflow:hidden;
margin-top:20rpx;border-radius: 5px;border:1px solid #EFEFEF;box-
sizing: border-box;color:#717171;padding:13rpx;}

.pay-wrap{width:90%;height:110rpx;background-color: #000000; position: fixed;
bottom:10rpx;left:50%;transform: translateX(-50%);display:flex;z-index:10;}
```

```
    .pay-wrap.price{font-size:40rpx;color:#FFFFFF;width:auto;box-sizing:
    border-box;background-color:#000000;position: absolute;z-index:10;
    left:10%;top:25rpx;}
    .pay-wrap.pay{width:190rpx;height:100%;background-color:#f17f1f;text-
    align:center;line-height:110rpx;font-size:40rpx;color:#FFFFFF;position:
    fixed;z-index:10;right:0%;bottom:0rpx;}
</style>
```

该页面的核心就是调用Vuex中的方法获取商铺信息和购物车里的菜品。点击“去支付”按钮，调用goPay()方法，在该方法内部使用this.addOrder()方法，将订单信息提交到服务端数据库，提交完成后，服务端返回实付金额和订单编号，获取金额和订单编号，接下来即可开发微信支付。

注意：

提交订单时无须将金额传递给服务端接口，服务端会根据购物车里的菜品计算金额。由于客户端的金额不安全，可以随意改动，因此需要在服务端计算支付金额，提交订单后最终将金额和订单编号返回给客户端。

11.8.2 实现微信支付

扫一扫，看视频

10.18节已经介绍了微信支付及其支付流程，本小节在this.addOrder()的success回调函数中实现微信支付，代码如下：

```
//提交订单
this.addOrder({
    branch_shop_id: this.branch_shop_id,
    table_code: this.table_code,
    remarks: this.remarks,
    is_pack: this.isPack ? '1' : '0',                //是否打包，1为是，0为否
    distribution_type: this.distribution_type,       //配送方式，0为堂食,2为自提
    success: async (res) => {
        let true_total = 0;                          //实际支付金额
        let ordernum = "";                           //订单编号
        if (res.code == 200) {
            true_total = res.data.true_total;
            ordernum = res.data.ordernum;
            if (true_total <= 0) {
            //如果实际付款金额小于等于0，则直接调用付款成功接口
                this.payOrder({
                    ordernum: ordernum, pay_type: 1, success: () => {
                        this.pushPage('/user_pages/order/pay_success?branch_
shop_id=' + this.branch_shop_id + '&table_code=' + this.table_code)
```

```
                    }
                });
            } else {

                //#ifdef MP-WEIXIN
                //微信支付成功的回调地址
                let notifyUrl = this.$config.domain + "/api/home/wxpay/wechat_notify";
                //微信支付统一下单接口
                let unifOrder = await this.wechatUnifiedorder({
                    open_id: this.openId,           //用户登录的openId
                    notify_url: notifyUrl,          //支付成功的回调地址
                    ordernum: ordernum,             //订单编号
                    price: true_total,              //支付金额
                    body: "点餐"                    //商品描述
                });
                let timeStamp = unifOrder.data['timeStamp'];        //时间戳
                let nonceStr = unifOrder.data['nonceStr'];          //随机字符串
                let packages = unifOrder.data['package'];
                //统一下单接口返回的prepay_id参数值
                let sign = unifOrder.data['paySign'];               //签名
                uni.requestPayment({
                    provider: 'wxpay',
                    timeStamp: timeStamp.toString(),
                    nonceStr: nonceStr,
                    package: packages,
                    signType: 'MD5',
                    paySign: sign,
                    success: (res) => {
//跳转到支付成功页面
this.pushPage('/user_pages/order/pay_success?branch_shop_id=' + this.
branch_shop_id + '&table_code=' + this.table_code)
                    },
                    fail: (err) => {
                    }
                });
                //#endif
            }
        } else {
            uni.showToast({
                title: '' + res.data + '',
```

```
                icon: "none",
                duration: 2000
            })
        }
        this.isPay = true;
    }
});
```

以上加粗代码为新增代码。注意，在进行微信支付之前需要判断支付金额是否大于0，如果大于0再调用微信支付接口，否则直接调用this.payOrder方法支付即可，因为传递给微信支付接口的金额必须大于0。

11.8.3　开发支付成功页面

扫一扫，看视频

微信支付完成后跳转到支付成功页面，显示用户支付状态，在此页面可以跳转到订单管理页面查看订单明细。支付成功页面预览效果如图11-18所示。

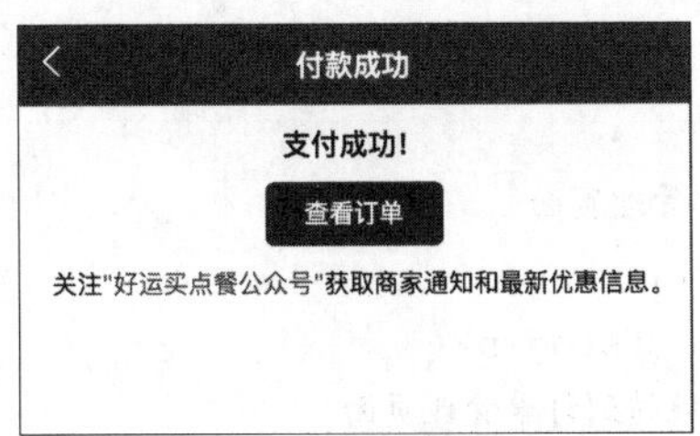

图11-18　支付成功页面预览效果

首先在user_pages/order文件夹下创建pay_success.vue文件，该文件中的代码如下：

```
<template>
    <view class="page">
        <view class="order-main">
            <view class="pay-success">支付成功! </view>
            <button class="view-order" @click="goMyOrder()">查看订单</button>
            <view class="msg">关注<text decode="true" class="official">&quot;好
             运买点餐公众号&quot;</text>获取商家通知和最新优惠信息。</view>
        </view>
    </view>
</template>

<script>
    import {mapState} from 'vuex';
    export default {
        name: "pay-success",
        data(){
```

```
            return {
                branch_shop_id:"",              //分店的id
                table_code:""                   //桌号
            }
        },
        onLoad(opts){
            this.branch_shop_id=opts.branch_shop_id?opts.branch_shop_id:"";
            //分店的id
            this.table_code=opts.table_code?opts.table_code:"";  //桌号
            //验证会员登录的token是否合法
            this.$utils.safeUser(this,this.branch_shop_id,this.table_code);
        },
        computed:{
            ...mapState({
                isLogin:state=>state.user.isLogin  //是否登录会员
            })
        },
        methods:{
            //跳转到订单管理页面
            goMyOrder(){
                if(this.isLogin){                          //如果已登录
                    //跳转到订单管理页面
                    this.pushPage('/pages/myorder/index?branch_shop_id='+this.
                        branch_shop_id+'&table_code='+this.table_code+'');
                }else{
                    //如果未登录，则跳转到会员登录页面
                    this.pushPage('/pages/login/index?branch_shop_id=${this.
                        branch_shop_id}&table_code=${this.table_code}');
                }
            },
            //跳转页面
            pushPage(url){
                uni.navigateTo({
                    url:url
                })
            },
        }
    }
</script>
```

```
<style scoped>
    .page{width:100%;min-height:100vh;background-color:#FFFFFF;
    overflow:hidden;}
    .order-main{width:100%;}
    .order-main.pay-success{width:auto;font-size:32rpx;color:#000000;text-
    align: center;margin-top:20rpx;font-weight: bold;}
    .order-main.view-order{width:200rpx;height:auto;font-
    size:28rpx;color:#FFFFFF;border-radius: 5px; background-color: #007aff;
    margin-top:20rpx;}
    .order-main.msg{font-size:28rpx;margin-top:20rpx;margin-left:20rpx;text-
    align: center;}
    .order-main.msg.official{color:#E30019;}
</style>
```

该页面非常简单，只需提示用户支付成功即可，并且用户可以从该页面跳转到订单管理页面。注意，不要忘记配置路由，在pages.json文件中新增代码如下：

```
{
    ...
    "subPackages": [
        {
            "root": "user_pages",
            "pages": [
                ...
                {
                    "path": "order/pay_success",
                    "style": {
                        "navigationBarTitleText": "付款成功",
                        "disableScroll": false
                    }
                }
            ]
        }
    ]
}
```

支付成功页面后期不需要放到tabBar中，该页面属于会员登录后才能访问的页面。为了提高性能的可优化性和可维护性，将该页面放到user_pages分包中。

11.9　个人中心

扫一扫，看视频

个人中心可以显示会员的头像、积分、余额等，可以进入我的订单、个人资料、绑定手机等页面。个人中心页面预览效果如图11-19所示。

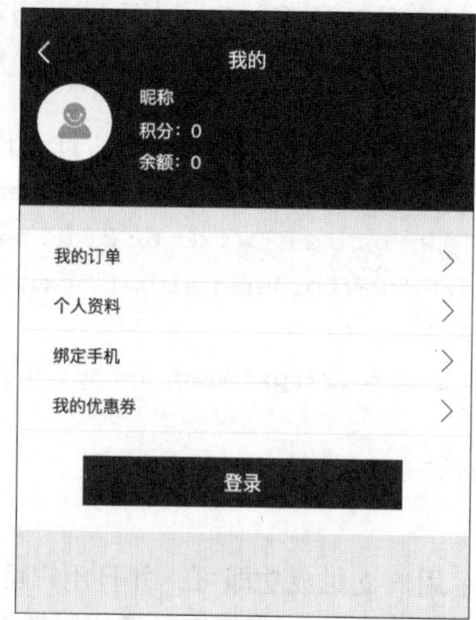

图 11-19　个人中心页面预览效果

个人中心页面的接口在“接口文档”文件夹下的“会员登录、注册、信息.docx”文件中。下面讲解如何开发个人中心的页面。

在点餐页面点击图11-20所示的按钮进入个人中心页面。

图 11-20　点击按钮

进入个人中心页面后，首先请求服务端接口，在api/user/index.js文件中新增代码如下：

```
...
//获取用户信息
export function getUserInfoData(data){
    return request(config.baseApi+"/v1/user/userinfo","post",data);
}

//安全退出
export function outLoginData(data){
    return request(config.baseApi+"/v1/outlogin","post",data);
}
```

getUserInfoData()函数可以从服务端接口获取用户信息，outLoginData()函数可以重新设置服务端数据库中的token值，实现安全退出。

接下来对接到Vuex中，在store/user/index.js文件中新增代码如下：

```
import {getWeChatOpenIdData,setWeChatUserData,getDewxbizdataData,bindWechatLoginBind
PhoneNumberData,safeUserData,getUserInfoData,outLoginData} from '../../api/user';
export default {
    namespaced:true,
    state:{
        ...
        //会员信息
        userInfo:{}
    },
    mutations: {
        ...
        //设置会员信息
        ["SET_USERINFO"](state,payload){
            state.userInfo=payload.userInfo;
        },
        //退出会员
        ["OUT_LOGIN"](state,payload){
            state.uid="";
            state.token="";
            state.openid="";
            state.isLogin=false;
            uni.removeStorageSync('uid');
            uni.removeStorageSync('token');
            uni.removeStorageSync('isLogin');
            uni.removeStorageSync('openid');
        }
    },
    actions:{
        ...
        //获取会员信息
        getUserInfo(conText,payload){
                getUserInfoData({uid:conText.rootState.user.uid,token:conText.
rootState.user.token,platform:conText.rootState.system.platform,...payload}).
then(res=>{
                if(res.code==200){
                    conText.commit("SET_USERINFO",{userInfo:res.data});
                    if(payload.success){
                        payload.success(res.data)
                    }
```

```
                }else{
                    conText.commit("SET_USERINFO",{userInfo:{}});
                }
            })
        },
        //安全退出
        outLogin(conText,payload){
            outLoginData({uid:conText.rootState.user.uid,platform:conText.
            rootState.system.platform}).then(res=>{
                if(res.code==200){
                    conText.commit("OUT_LOGIN");
                }
            })
        }
    }
}
```

以上加粗代码为新增代码，state内部的userInfo属性用来存储用户信息。mutations内部的SET_USERINFO方法可以设置用户信息，并将用户信息赋值给state内部的userInfo属性；OUT_LOGIN方法可以实现当退出会员时将会员信息清空的功能。actions内部的getUserInfo()方法可以从服务端接口获取用户信息并提交给mutations内部的SET_USERINFO方法；outLogin()方法内部调用outLoginData()方法，可以将服务端数据库中的token值重置，实现安全退出。

接下来在pages文件夹下创建my/index.vue文件，并配置路由。在pages.json文件中新增代码如下：

```
{
    "pages": [
        ...
        {
            "path": "pages/my/index",
            "style": {
                "navigationBarTitleText": "我的",
                "disableScroll": true,
                "navigationStyle": "custom"
            }
        }
    ],
    ...
}
```

个人中心页面后期有可能被放到tabBar中，所以不能放到分包中。配置完路由后，开发个人中心页面，在pages/my/index.vue文件中的代码如下：

```
<template>
    <view class="page">
        <view class="status_bar bg-color"></view>
        <view :class="{header:true,ipx:isIpx}">
            <view class="back" @click="goBack()">
                <view class="back-icon"></view>
            </view>
            <view class="title">我的</view>
            <view :class="{'user-info':true,ipx:isIpx}">
                <view class="head"><image :src="userInfo.head?userInfo.
                head:'../../static/images/user/head.png'"></image></view>
                <view class="nickname">{{userInfo.nickname?userInfo.
                nickname:"昵称"}}</view>
                <view class="show-total">
                    <view class="text">积  分:{{userInfo.points?userInfo.
                    points:"0"}}</view>
                </view>
                <view class="show-balance">
                    <view class="text">余  额:{{userInfo.balance?userInfo.
                    balance:"0"}}</view>
                </view>
            </view>
        </view>
        <view class="pannel-main">
            <view class="list" @click="userPushPage('/pages/myorder/
            index?branch_shop_id='+branch_shop_id+'&table_code='+table_code+'')">
                <view class="text">我的订单</view>
                <view class="arrow"></view>
            </view>
            <view class="list" @click="userPushPage('/user_pages/profile/
            index?branch_shop_id='+branch_shop_id+'&table_code='+table_
            code)">
                <view class="text">个人资料</view>
                <view class="arrow"></view>
            </view>
            <view class="list" @click="userPushPage('/user_pages/bind_
            cellphone/index?branch_shop_id='+branch_shop_id+'&table_
            code='+table_code)">
                <view class="text">绑定手机</view>
                <view class="arrow"></view>
            </view>
```

```
            <button type="button" class="out-btn" @click="safePage">
             {{isLogin?' 安全退出':'登录'}}</button>
        </view>
    </view>
</template>

<script>
    import {mapState,mapActions} from 'vuex';
    export default {
        name: "my",
        data(){
            return {
                branch_shop_id:"",
                table_code:""
            }
        },
        onLoad(opts){
            this.branch_shop_id=opts.branch_shop_id;
            this.table_code=opts.table_code;
        },
        onShow(){
            this.getUserInfo({branch_shop_id:this.branch_shop_id});
        },
        computed:{
            ...mapState({
                isIpx:state=>state.system.isIpx,
                isLogin:state=>state.user.isLogin,
                "userInfo":state=>state.user.userInfo        //会员信息
            })
        },
        methods:{
            ...mapActions({
                "asyncOutLogin":"user/outLogin",              //Vuex中的安全退出
                getUserInfo:"user/getUserInfo"                //获取会员信息
            }),
            //安全退出，在safePage()方法里调用
            outLogin(){
                uni.showModal({
                    title: '',
                    content: '确认要退出吗？ ',
                    success: (res)=> {
                        if (res.confirm) {
```

```
                    this.asyncOutLogin();
                    this.userPushPage("/pages/login/index?branch_
                        shop_id=" + this.branch_shop_id + "&table_
                        code=" + this.table_code);
                } else if (res.cancel) {

                }
            }
        });
    },
    //安全退出与登录
    safePage(){
        if(this.isLogin){     //如果会员是已登录状态
            this.outLogin()   //安全退出
        }else {               //如果未登录
            //跳转到登录页面
            this.userPushPage("/pages/login/index?branch_shop_id=" +
this.branch_shop_id + "&table_code=" + this.table_code);
        }
    },
    //判断是否登录跳转页面
    userPushPage(url) {
        if(this.isLogin){      //如果已登录
            uni.navigateTo({
                url:url
            })
        }else{                 //如果未登录，则跳转到登录页面
            uni.navigateTo({
                url:"/pages/login/index?branch_shop_id="+this.
                    branch_shop_id+"&table_code="+this.table_code
            })
        }
    },
    //返回上一页
    goBack(){
        uni.navigateBack({
            delta: 1
        });
    },
    //跳转页面
    pushPage(url){
        uni.navigateTo({
```

```
                url:url
            })
        }
    }
}
</script>

<style scoped>
    .page{width:100%;height:100vh;overflow:hidden;background-color:#EFEFEF;}
    .status_bar.bg-color{background-color:#E30019;}
    .header{width:100%;height:320rpx;background-color:#E30019;position:
     relative;z-index:1;overflow:hidden;}
    .header.back{width:80rpx;height:80rpx;position: absolute; z-index:1;
     left:20rpx;top:55rpx;}
    .header.back.back-icon{width:40rpx;height:40rpx;background-image:
     url("~@/static/images/common/back.png");background-size:100%;background-
     repeat:no-repeat;background-position: center;}
    .header.ipx.back{top:100rpx;}
    .header.ipx{height:425rpx;}
    .header.title{width:100%;height:auto;text-align: center; font-size:
     32rpx;color:#FFFFFF;margin-top:60rpx;}
    .header.ipx.title{margin-top:105rpx;}
    .header.user-info{width:100%;height:180rpx;position: relative; z-
     index:1; margin-top:20rpx;}
    .header.ipx.user-info{margin-top:40rpx;}
    .header.head{width:125rpx;height:125rpx;position:absolute;z-index: 1;
     left:30rpx;top:0rpx;}
    .header.head image{width:100%;height:100%;border-radius: 100%;}

    .header.nickname{width:auto;height:auto;position: absolute;z-index:1;left:
     200rpx;top:0rpx;font-size:28rpx;color:#FFFFFF}
    .header.show-total{width:auto;height:auto;position: absolute;z-index:1;
     left:200rpx;top:55rpx;display:flex;}
    .header.show-total.text{font-size:28rpx;color:#FFFFFF;margin-right:20rpx;}

    .header.show-balance{width:auto;height:auto;position: absolute;z-index:1
     ;left:200rpx;top:105rpx;display:flex;}
    .header.show-balance.text{font-size:28rpx;color:#FFFFFF;margin-right:20rpx;}

    .pannel-main{width:100%;height:auto;background-color:#FFFFFF;margin-
     top:40rpx;background-color:#FFFFFF;padding-bottom:40rpx;}
```

```
    .pannel-main.list{width:100%;height:80rpx;border-bottom: 1px solid
     #EFEFEF;display:flex;justify-content: space-between;align-items:
     center;box-sizing: border-box;padding-left:60rpx;padding-right:20rpx;}
    .pannel-main.list.text{font-size:28rpx;color:#333333}
    .pannel-main.list.arrow{width:40rpx;height:40rpx;background-image:
     url("~@/static/images/user/right_arrow.png");background-size:100%;
     background-repeat: no-repeat;background-position: center;}

    .out-btn{width:526rpx;height:80rpx;margin:0 auto;background-
     color:#E30019;color:#FFFFFF;font-size:32rpx;margin-top:40rpx;}
</style>
```

此页面的核心功能是将用户的昵称、头像、积分和余额显示在视图中。点击“安全退出”按钮，实现退出会员功能。使用userPushPage方法()可以跳转到我的订单、个人资料、绑定手机页面。由于这些页面必须登录会员后才能访问，因此需要在跳转前使用this.isLogin判断会员是否登录，如果未登录，则跳转到会员登录页面。其具体实现逻辑请读者阅读代码仔细体会，代码注释很清晰，不再赘述。

11.10　订单管理

订单管理页面包括查看已付款的订单、订单详情、退款等功能。订单管理页面的接口在“接口文档”文件夹下的“订单.docx”文件中。

11.10.1　开发已付款订单页面

扫一扫，看视频

已付款订单页面预览效果如图 11-21 所示。

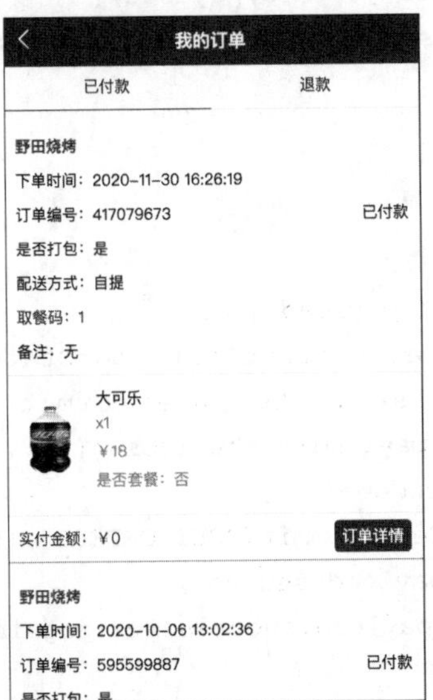

图 11-21　已付款订单页面预览效果

首先请求已付款订单相关的服务端接口，在api/order/index.js文件中新增代码如下：

```
...
//查看已付款订单
export function getOrderData(data){
    return request(config.baseApi+"/v1/user/order/show","post",data);
}
```

getOrderData()函数可以从服务端接口获取已付款的订单列表。接下来对接到Vuex中，在store/order/index.js文件中新增代码如下：

```
import {addOrderData,wechatUnifiedorderData,getLastOrderData,payData,getOrderD
ata} from "../../api/order";
export default {
    namespaced:true,
    state:{
        ...
            orders:[],          //已付款订单数据
    },
    mutations:{
        ...
        //设置已付款订单数据
        ["SET_ORDERS"](state,payload){
            state.orders=payload.orders;
        },
        //设置已付款订单分页数据
        ["SET_ORDERS_PAGE"](state,payload){
            state.orders.push(...payload.orders);
        },
    },
    actions:{
        ...
        //查看已付款订单数据
        getOrder(conText,payload){
            getOrderData({uid:conText.rootState.user.uid,token:conText.
            rootState.user.token,platform:conText.rootState.system.
            platform,...payload}).then(res=>{
                if(res.code==200){
                    conText.commit("SET_ORDERS",{orders:res.data});
                    if(payload.success){
                        payload.success(res.pageinfo.pagenum);
                    }
                }else{
```

```
                    conText.commit("SET_ORDERS",{orders:[]});
                }
                if(payload.complete){
                    payload.complete()
                }
            })
        },
        //查看已付款订单分页数据
        getOrderPage(conText,payload){
            getOrderData({uid:conText.rootState.user.uid,token:conText.rootState.
            user.token,platform:conText.rootState.system.platform,...payload}).
            then(res=>{
                if(res.code==200){
                    conText.commit("SET_ORDERS_PAGE",{orders:res.data});
                }
            })
        },
    }
}
```

以上加粗代码为新增代码，state内部的orders属性为已付款订单的数据列表。mutations内部的SET_ORDERS方法可以设置state内部的orders的值;SET_ORDERS_PAGE方法内部使用push()方法，可以对state.orders数组添加新的值，用于上拉加载数据。actions内部的getOrder()方法可以从服务端接口获取已付款订单数据;getOrderPage()方法可以从服务端接口获取已付款订单的数据，使用commit提交到SET_ORDERS_PAGE，为state.orders数组添加新的值。

接下来创建已付款订单文件。在pages文件夹下创建myorder/index.vue文件，创建完成后配置路由。在pages.json文件中新增代码如下：

```
{
    "pages": [
        ...
        {
            "path": "pages/myorder/index",
            "style": {
                "navigationBarTitleText": "我的订单",
                "disableScroll": false,
                "enablePullDownRefresh": true
            }
        },
    ],
    ...
}
```

继续开发已付款订单页面，在pages/myorder/index.vue文件中的代码如下：

```
<template>
    <view class="page">
        <!--
        Tags:订单状态页面切换组件
        status:订单状态
        branch_shop_id:分店的id
        table_code:桌号
        -->
        <Tags :status="status" :branch_shop_id="branch_shop_id" :
         table_code="table_code"></Tags>
        <view class="order-main" v-if="orders.length>0">
            <view class="order-list" @click="pushPage('/pages/myorder/
             details?branch_shop_id='+branch_shop_id+'&table_code='+table_
             code+'&ordernum='+item.ordernum+'')" v-for="(item,index) in
             orders" :key="index">
                <view class="order-info">
                    <view class="shop-name">{{item.branch_shop_name}}</view>
                    <view class="order-time">下单时间:{{item.order_time}}</view>
                    <view class="ordernum-status">
                        <view class="ordernum">订单编号:{{item.ordernum}}</view>
                        <view class="status">{{item.status=='0'?'待 付 款
                        ':item.status=='1'?'已付款':""}}</view>
                    </view>
                    <view class="area-table" v-if="item.table_code && item.
                     is_pack=='0'">
                        <view class="area">区域:{{item.hotel_area_title}}</view>
                        <view class="table-code">桌号:{{item.table_code}}</view>
                    </view>
                    <view class="remarks">是否打包:{{item.is_pack=='1'?'是':'
                     否'}}</view>
                    <view class="remarks" v-if="item.is_pack=='1'">配送方式:
                     {{item.distribution_type=='1'?'配送':'自提'}}</view>
                    <view class="pick-code">取餐码:{{item.pick_code}}</view>
                    <view class="remarks">备注:{{item.remarks?item.remarks:"
                     无"}}</view>
                </view>
                <view class="order-desc">
                    <view class="goods-list-main" v-for="(item2,index2) in
                     item.order_desc" :key="index2">
                        <view class="goods-list">
                            <view class="image">
                                <image :src="item2.image"></image>
```

```
                    </view>
                    <view class="goods-info">
                        <view class="title">{{item2.title}}</view>
                        <view class="info-text">x{{item2.amount}}</view>
                        <view class="info-text">¥{{item2.price}}</view>
                        <view class="info-text" v-if="item2.pack_
                         price>0">包装费:¥{{item2.pack_price}}</view>
                        <view class="is-meal">是 否 套 餐:{{item2.is_
                         meal=='1'?'是':'否'}}</view>
                    </view>
                </view>
                <view class="refund-content" v-if="item2.refund_state=='-2'">
                    失败原因:{{item2.refund_failed}}
                </view>
                <view class="meal-items" v-if="item2.is_meal=='1'">
                      <block v-for="(item3,index3) in item2.omi_data"
                        :key="index3">
                        <view class="item-list">
                            <view class="item">
                                <view>- {{item3.title}}</view>
                                <view>x{{item3.amount}}</view>
                                <view> ¥{{item3.price}}</view>
                            </view>
                        </view>
                        <view class="pack-price" v-if="item3.pack_
                         price>0">包装费:¥{{item3.pack_price}}</view>
                        <view class="refund-content" v-if="item3.
                         refund_state=='-2'">
                            失败原因:{{item3.refund_failed}}
                        </view>
                    </block>
                </view>
            </view>
        </view>
        <view class="price-status">
            <view class="price">实付金额:¥{{item.true_total}}</view>
            <view class="status">
                <view class="status-btn">订单详情</view>
            </view>
        </view>
    </view>
</view>
<view class="no-data" v-if="orders.length<=0">没有订单! </view>
```

```
        </view>
    </template>

    <script>
        import {mapActions,mapState} from "vuex";
        //切换订单状态页面切换组件
        import Tags from "../../components/myorder/tags";
        export default {
            name: "order",
            data(){
                return {
                    branch_shop_id:0,        //分店的id
                     status:0,               //订单状态，1为已付款，-2为已退款
                    table_code:"",           //桌号
                }
            },
            onLoad(opts){
                this.branch_shop_id=opts.branch_shop_id;
                this.table_code=opts.table_code;
                this.status=opts.status?opts.status:"1";
                this.$utils.safeUser(this,this.branch_shop_id,this.table_code);
                this.curPage=1;          //当前页码
                this.maxPage=0;          //总页码
                this.getOrder({page:1,status:this.status,success:(pageNum)=>{
                        this.maxPage=parseInt(pageNum);
                    }});
            },
            //下拉刷新
            onPullDownRefresh(){
                //获取已付款订单数据
                this.getOrder({page:1,status:this.status,success:(pageNum)=>{
                        this.maxPage=parseInt(pageNum);
                    },
                    complete:()=>{
                        uni.stopPullDownRefresh();
                    }
                });
            },
            //上拉加载数据
            onReachBottom(){
                //如果当前页码小于总页码数
                if(this.curPage<this.maxPage){
                  this.curPage++;
```

```
          //执行获取分页数据
          this.getOrderPage({page:this.curPage,status:this.status});
        }
      },
      components:{
        Tags                                     //订单状态页面切换组件
      },
      computed:{
        ...mapState({
          orders:state=>state.order.orders,  //已付款订单数据
        })
      },
      methods:{
        ...mapActions({
          getOrder:"order/getOrder",             //获取已付款订单数据
          getOrderPage:"order/getOrderPage"      //获取已付款订单分页数据
        }),
        //跳转页面
        pushPage(url){
          uni.navigateTo({
            url:url
          })
        }
      }
    }
</script>

<style scoped>
    .page{width:100%;min-height:100vh;background-color:#FFFFFF;margin-
     top:80rpx;overflow:hidden;}
    .order-main{width:100%;}
    .order-main.order-list{width:100%;border-bottom:1px solid #f17f1f;}
    .order-main.order-list.order-info{width:100%;border-bottom: 1px solid
     #EFEFEF;box-sizing: border-box;padding:20rpx;}
    .order-main.order-list.ordernum-status{width:100%;display:flex;font-
     size:28rpx;justify-content: space-between;margin-top:20rpx;}
    .order-main.order-list.area-table{width:100%;display:flex;font-
     size:28rpx;margin-top:20rpx;}
    .order-main.order-list.area-table .area{margin-right:40rpx;}
    .order-main.order-list.order-time{font-size:28rpx;}
    .order-main.order-list.pick-code{font-size:28rpx; color:#E30019;
     margin-top:20rpx;}
    .order-main.order-list.remarks{font-size:28rpx;margin-top:20rpx;}
```

```
    .order-main.order-list.shop-name{width:100%;font-size: 28rpx; margin-top:20rpx;margin-bottom:20rpx;font-weight: bold;color:#E30019}
    .order-main.order-desc{width:100%;}
    .order-main.goods-list-main{width:100%;box-sizing: border-box;padding: 20rpx;border-bottom: 1px solid #EFEFEF;}
    .order-main.order-desc.goods-list{width:100%;display:flex;justify-content: space-between;align-items: center;margin-bottom:20rpx;}
    .order-main.order-desc.goods-list.image{width:120rpx;height:120rpx;}
    .order-main.order-desc.goods-list.image image{width:100%;height:100%;border-radius: 5px;}
    .order-main.order-desc.goods-list.goods-info{width:80%;height:auto;font-size: 28rpx;overflow:hidden;position: relative;}
    .order-main.order-desc.goods-list.goods-info.title{width:100%;height:45rpx;overflow:hidden;text-overflow: ellipsis;white-space: nowrap;}
    .order-main.order-desc.goods-list.goods-info.info-text{width:auto;height:auto;color:#909090;margin-bottom:10rpx;}
    .order-main.order-desc.goods-list.goods-info.is-meal {width:auto;height:auto;color:#909090;}
    .order-main.order-desc.goods-list.goods-info.handle-status{}
    .order-main.order-desc.goods-list.goods-info.refund-btn{width:auto;height:auto;position: absolute;right:0rpx;bottom:0rpx;padding:8rpx 15rpx;color:#FFFFFF;font-size:28rpx;background-color:#f17f1f;border-radius: 4px;}
    .order-main.order-desc.meal-items{width:100%;margin-top:20rpx;}
    .order-main.order-desc.meal-items.item-list{width:100%;height:80rpx;display:flex;justify-content: space-between;align-items:center;font-size:28rpx;color:#909090}
    .order-main.order-desc.meal-items.item-list.item{display:flex;}
    .order-main.order-desc.meal-items.item-list.item view{margin-right:20rpx;}
    .order-main.order-desc.goods-list-main.refund-content{width:100%;font-size:28rpx;color:#E30019;}
    .order-main.order-list.price-status{width:100%;height:80rpx;box-sizing: border-box;padding:20rpx;display:flex;justify-content: space-between;align-items: center;}
    .order-main.order-list.price-status.price{font-size:28rpx;color:#333333}
    .order-main.order-list.price-status.status{width:auto;height:auto;display:flex;justify-content: flex-end;}
    .order-main.order-list.price-status.status.status-btn{width:auto;height:auto;padding:8rpx 15rpx;border-radius: 4px;background-color:#E30019;font-size:28rpx;color:#FFFFFF;margin-left:20rpx;}
    .order-main.pack-price{font-size:28rpx;color:#909090;margin-left:24rpx;}
</style>
```

上面代码中，Tags为已付款和已退款页面的切换。由于微信小程序没有主路由和子路由之分,因此可以用组件代替以实现类似的效果。在components文件夹下创建myorder/tags.vue文件，该文件中的代码如下：

```
<template>
    <view class="tags">
        <view :class="{item:true, active:status==1?true:false}"
         @click="pushPage('/pages/myorder/index?branch_shop_id='+branch_
         shop_id+'&table_code='+table_code+'&status=1')" >已付款</view>
        <view :class="{item:true, active:status==-2?true:false}"
         @click="pushPage('/pages/myorder/refund?branch_shop_id='+branch_
         shop_id+'&table_code='+table_code+'&status=-2')">退款</view>
    </view>
</template>

<script>
    export default {
        name: "tags",
        props:{
            //订单状态
            status:{
                type:String,
                default:"0"
            },
            //分店的id
            branch_shop_id:{
                type:String,
                default: ""
            },
            //桌号
            table_code:{
                type:String,
                default:""
            }
        },
        methods:{
            pushPage(url){
                uni.redirectTo({
                    url:url
                })
            }
        }
    }
</script>
```

```
<style scoped>
    /* #ifdef H5 */
    .tags{width:100%;height:80rpx;background-color:#FFFFFF;border-
     bottom:1px solid #EFEFEF;display:flex;position: fixed;z-index:10;
     left:0px;top:85rpx;}
    /* #endif */
    /* #ifndef H5 */
    .tags{width:100%;height:80rpx;background-color:#FFFFFF;border-
     bottom:1px solid #EFEFEF;display:flex;position: fixed; z-index:10;
     left:0px;top:0px;}
    /* #endif */
    .tags.item{flex:1;height:100%;font-size:28rpx;color:#333333;text-
     align:center;line-height:80rpx;}
    .tags.item.active{border-bottom:1px solid #E30019;}
</style>
```

props内部的status属性为订单状态，1表示已付款，-2表示退款；branch_shop_id属性为分店的编号；table_code属性为桌号。这些属性将作为URL参数传递给pushPage()方法，pushPage()方法内部使用uni.redirectTo实现已付款订单页面与已退款订单页面之间的跳转。

扫一扫，看视频

11.10.2 开发订单详情页面

订单详情页面预览效果如图11-22所示。

图 11-22 订单详情页面预览效果

首先请求订单详情相关服务端接口，在api/order/index.js文件中新增代码如下：

```
...
//订单详情
export function getOrderDetailsData(data){
    return request(config.baseApi+"/v1/user/order/details","post",data);
}
```

getOrderDetailsData()函数可以从服务端接口获取订单详情数据。接下来对接到Vuex中，在store/order/index.js文件中新增代码如下：

```
import {addOrderData,wechatUnifiedorderData,getLastOrderData,payData,
getOrderData,getOrderDetailsData} from "../../api/order";
export default {
    namespaced:true,
    state:{
        ...
        orderDetails:{},                //订单详情数据
    },
    mutations:{
        ...
        //设置订单详情数据
        ["SET_ORDER_DETAILS"](state,payload){
            state.orderDetails=payload.orderDetails;
        },
    },
    actions:{
        ...
        //查看订单详情
        getOrderDetails(conText,payload){
            getOrderDetailsData({uid:conText.rootState.user.uid,token:conText.
            rootState.user.token,platform:conText.rootState.system.
            platform,...payload}).then(res=>{
                if(res.code==200){
                    conText.commit("SET_ORDER_DETAILS",{orderDetails:res.data});
                }
                if(payload.complete){
                    payload.complete();
                }
            })
        },
    }
}
```

以上加粗代码为新增代码，state内部的orderDetails属性可以存储订单详情数据；mutations内部的SET_ORDER_DETAILS方法可以设置state内部的orderDetails属性值；actions内部的getOrderDetails()方法可以从服务端接口获取订单详情数据，并使用commit提交SET_ORDER_DETAILS，将订单详情数据赋值给state.orderDetails。

在pages/myorder文件夹下创建details.vue文件，该文件中的代码如下：

```
<template>
    <view class="page">
        <view class="shop-title">{{orderDetails.branch_shop_name}}</view>
        <view class="order-main">
            <view class="order-list">
                <view class="order-info">
                    <view class="order-time">下单时间:{{orderDetails.order_time}}</view>
                    <view class="ordernum-status">
                        <view class="ordernum">订单编号:{{ordernum}}</view>
                        <view class="status">{{orderDetails.status=='0'?'待付
                        款':orderDetails.status=='1'?'已付款':""}}</view>
                    </view>
                    <!--如果是堂食订单-->
                    <view class="area-table" v-if="orderDetails.table_code
                    && orderDetails.is_pack=='0'">
                        <view class="area">区 域:{{orderDetails.
                        hotel_area_title}}</view>
                        <view class="table-code">桌号:{{orderDetails.table_code}}</view>
                    </view>
                    <view class="pick-code">取餐码:{{orderDetails.pick_code}}</view>
                    <view class="remarks">是 否 打 包:{{orderDetails.is_
                    pack=='1'?'是':'否'}}</view>
                    <view class="remarks" v-if="orderDetails.is_pack=='1'">
                    配送方式：自提</view>
                    <view class="remarks">备 注:{{orderDetails.remarks?
                    orderDetails.remarks:"无"}}</view>
                </view>
                <view class="order-desc">
                    <!--订单明细-->
                    <view class="goods-list-main" v-for="(item2,index2) in
                    orderDetails.order_desc" :key="index2">
                        <view class="goods-list">
                            <view class="image">
                                <image :src="item2.image"></image>
                            </view>
                            <view class="goods-info">
```

```
                    <view class="title">{{item2.title}}</view>
                    <view class="info-text">x{{item2.amount}}</view>
                    <view class="info-text">¥{{item2.price}}</view>
                    <view class="info-text" v-if="item2.pack_
                    price>0">包装费:¥{{item2.pack_price}}</view>
                    <view class="is-meal">是 否 套 餐:{{item2.is_
                    meal=='1'?'是':'否'}}</view>
                    <view :class="{'refund-btn':true,success:item2.
refund_state=='1',fail:item2.refund_state=='-2',handle:item2.refund_state=='-1'}
"v-if="item2.is_meal=='0' && orderDetails.status=='1' && (orderDetails['true_
total']>0 || orderDetails['balance_total']>0)" @click="applyRefund(item2.order_
item_id,item2.refund_state,item2.is_meal,item2.title,item2.amount)">{{item2.
refund_state=='1'?'退 款 成 功':item2.refund_state=='-2'?'退 款 失 败':item2.refund_
state=='-1'?'处理中':'申请退款'}}</view>
                </view>
            </view>
            <!--退款失败原因-->
            <view class="refund-content" v-if="item2.refund_state=='-2'">
                失败原因:{{item2.refund_failed}}
            </view>
            <!--套餐里菜品的明细-->
            <view class="meal-items" v-if="item2.is_meal=='1'">
                <block v-for="(item3,index3) in item2.omi_data" :key="index3">
                    <view class="item-list">
                        <view class="item">
                            <view>- {{item3.title}}</view>
                            <view>x{{item3.amount}}</view>
                            <view>¥{{item3.price}}</view>
                        </view>
                            <view v-if="orderDetails.status=='1' &&
(orderDetails['true_total']>0)" :class="{'refund-btn':true,success:item3.refund_
state=='1',fail:item3.refund_state=='-2',handle:item3.refund_state=='-1'}"
@click="applyRefund(item3.order_item_id,item3.refund_state,item2.is_meal,item3.
title,item3.amount)">{{item3.refund_state=='1'?'退 款 成 功':item3.refund_
state=='-2'?'退款失败':item3.refund_state=='-1'?'处理中':'申请退款'}}</view>
                    </view>
                        <view class="pack-price" v-if="item3.pack_
price>0">包装费:¥{{item3.pack_price}}</view>
                    <view class="refund-content" v-if="item3.
                    refund_state=='-2'">
                        失败原因:{{item3.refund_failed}}
                    </view>
```

```
                        </block>
                    </view>
                </view>
            </view>
            <view class="price-status">
                <view class="price">菜品金额:￥{{orderDetails['total']}}</view>
                  <view class="price active">包装费:￥{{orderDetails['pack_
                  total']}} </view>
                <view class="price active">实付金额:￥{{orderDetails['true_
                total']}} </view>
            </view>
        </view>
    </view>
    <view class="again-order" @click="replacePage('/pages/goods/index?branch_
     shop_id='+branch_shop_id+'&table_code='+table_code)">再来一单</view>
    <!--申请退款组件-->
    <refund-order :show="showRefund" @cancel="cancelRefund()"
     :orderItemId="orderItemId" :isMeal="isMeal" :title="title"
     :amount="amount"></refund-order>
  </view>
</template>

<script>
    import RefundOrder from "../../components/refund_order";
    import {mapActions,mapState} from "vuex";
    export default {
        name: "order-details",
        data(){
            return {
                branch_shop_id:"",     //分店的id
                ordernum:"",           //订单编号
                table_code:"",         //桌号
                showRefund:false,      //是否显示申请退款组件
                orderItemId:"",        //菜品订单明细的id
                isMeal:"",             //是否套餐
                title:"",              //菜品名称
                amount:1               //菜品数量
            }
        },
        onLoad(opts){
            this.branch_shop_id=opts.branch_shop_id;
            this.table_code=opts.table_code;
```

```
        this.ordernum=opts.ordernum;
        this.$utils.safeUser(this,this.branch_shop_id,this.table_code);
        //订单详情
        this.getOrderDetails({ordernum:this.ordernum});
    },
    //下拉刷新
    onPullDownRefresh(){
        this.getOrderDetails({ordernum:this.ordernum,complete:()=>{
            uni.stopPullDownRefresh();
        }});
    },
    components:{
        RefundOrder         //申请退款组件
    },
    computed:{
        ...mapState({
            orderDetails:state=>state.order.orderDetails,//订单详情数据
        })
    },
    methods:{
        ...mapActions({
            getOrderDetails:"order/getOrderDetails"    //获取订单详情
        }),
        replacePage(url){
            uni.redirectTo({
                url:url
            })
        },
        //申请退款
        applyRefund(orderItemId,refund_state,isMeal,title,amount){
            //如果退款的订单状态不是申请状态，也不是申请失败状态，也不是退款成功状态
            //refund_state的值:-1表示申请状态，-2表示退款失败，1表示退款成功
            if(refund_state!="-1" && refund_state!='-2' && refund_state!='1'){
                this.showRefund=true;              //显示退款组件
                this.orderItemId=orderItemId;   //菜品订单明细的id
                this.isMeal=isMeal;                //是否套餐
                this.title=title;                  //菜品名称
                this.amount=parseInt(amount);   //退菜数量
            }
        },
        //取消申请退款
        cancelRefund(){
```

```
                this.showRefund=false;          //隐藏申请退款组件
                this.amount=1;                  //数量设置为1
            }
        }
    }
</script>

<style scoped>
    .page{width:100%;min-height:100vh;background-color:#FFFFFF;overflow:hidden;}
    .shop-title{width:100%;height:80rpx;background-color:#FFFFFF;border-bottom:
     1px solid #EFEFEF;font-size:28rpx;text-align:center;line-height:80rpx;}
    .order-main{width:100%;}
    .order-main.order-list{width:100%;border-bottom:1px solid #f17f1f;}
    .order-main.order-list.order-info{width:100%;border-bottom: 1px solid
     #EFEFEF;box-sizing: border-box;padding:20rpx;}
    .order-main.order-list.ordernum-status{width:100%;display:flex;font-
     size:28rpx;justify-content: space-between;margin-top:20rpx;}
    .order-main.order-list.area-table{width:100%;display:flex;font-
     size:28rpx;margin-top:20rpx;}
    .order-main.order-list.area-table .area{margin-right:40rpx;}
    .order-main.order-list.order-time{font-size:28rpx;}
    .order-main.order-list.pick-code{font-size:28rpx;color:#E30019;margin-top:20rpx;}
    .order-main.order-list.remarks{font-size:28rpx;margin-top:20rpx;}

    .order-main.order-desc{width:100%;}
    .order-main.goods-list-main{width:100%;box-sizing: border-
     box;padding:20rpx;border-bottom: 1px solid #EFEFEF;}
    .order-main.order-desc.goods-list{width:100%;display:flex;justify-
     content: space-between;align-items: center;margin-bottom:20rpx;}
    .order-main.order-desc.goods-list.image{width:120rpx;height:120rpx;}
    .order-main.order-desc.goods-list.image image{width:100%;height:100%;
     order-radius: 5px;}
    .order-main.order-desc.goods-list.goods-info{width:80%;height:auto;font-size:
     28rpx;overflow:hidden;position: relative;}
    .order-main.order-desc.goods-list.goods-info.title{width:100%;height:45r
     px;overflow:hidden;text-overflow: ellipsis;white-space: nowrap;}
    .order-main.order-desc.goods-list.goods-info.info-text{width:auto;height:
     auto;color:#909090;margin-bottom:10rpx;}
    .order-main.order-desc.goods-list.goods-info.is-meal{width:auto;
     height:auto;color:#909090;}
    .order-main.order-desc.goods-list.goods-info.refund-btn{width:auto;height:auto;
     position: absolute;right:0rpx;bottom:0rpx;padding:8rpx 15rpx;color:#FFFFFF;font-
     size:28rpx;background-color:#f17f1f;border-radius: 4px;}
```

```
    .order-main.order-desc.goods-list.goods-info.refund-btn.success
    {background-color:#007aff;}
    .order-main.order-desc.goods-list.goods-info.refund-btn.fail
    {background-color:#E30019;}
    .order-main.order-desc.goods-list.goods-info.refund-btn.handle
    {background-color:#CCCCCC;}
    .order-main.order-desc.meal-items{width:100%;margin-top:20rpx;}
    .order-main.order-desc.meal-items.item-list{width:100%;height:
    80rpx;display:flex;justify-content: space-between;align-items:
    center;font-size:28rpx;color:#909090}
    .order-main.order-desc.meal-items.item-list.item{display:flex;}
    .order-main.order-desc.meal-items.item-list.item view{margin-right:20rpx;}
    .order-main.order-desc.meal-items.item-list.refund-btn{width:auto;
    height:auto;padding:8rpx 15rpx;color:#FFFFFF;font-size:28rpx;background-
    color:#f17f1f;border-radius: 4px;}
    .order-main.order-desc.meal-items.item-list.refund-btn.success
    {background-color:#007aff;}
    .order-main.order-desc.meal-items.item-list.refund-btn.fail
    {background-color:#E30019;}
    .order-main.order-desc.meal-items.item-list.refund-btn.handle
    {background-color:#CCCCCC}
    .order-main.order-desc.goods-list-main.refund-content {width:100%;
    font-size:28rpx;color:#E30019;}
    .order-main.order-list.price-status{width:100%;height:auto;box-sizing:
    border-box;padding:20rpx;}
    .order-main.order-list.price-status.price{font-size:28rpx; color:
    #333333;margin-bottom:10rpx;}
    .order-main.order-list.price-status.price.active{color:#E30019;}

    .again-order{width:200rpx;height:60rpx;font-size:28rpx;color:#FFFF
    FF;background-color:#E30019;margin:0 auto;text-align:center;line-
    height:60rpx;border-radius: 4px;margin-top:30rpx;margin-bottom:30rpx;}

    .order-main.pack-price{font-size:28rpx;color:#909090;margin-left:24rpx;}
</style>
```

在onLoad()钩子函数内调用Vuex中的getOrderDetails()方法传入订单编号，从服务端接口获取订单详情数据，保存到orderDetails中，并渲染到视图上。使用onPullDownRefresh()钩子函数可以实现下拉刷新，在订单详情页面可以申请退款，视图中的<refund-order>组件为申请退款组件。接下来开发申请退款组件。

11.10.3 开发申请退款组件

申请退款组件预览效果如图11-23所示。

扫一扫，看视频

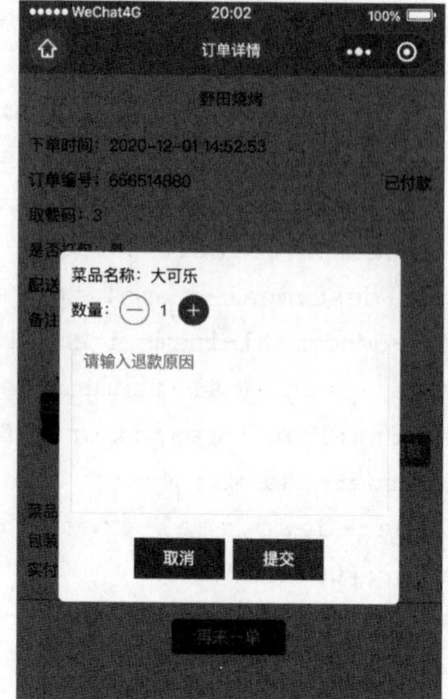

图11-23 申请退款组件预览效果

首先请求申请退款相关的服务端接口，在api/order/index.js文件中新增代码如下：

```
...
//申请退款
export function applyRefundOrderData(data){
    return request(config.baseApi+"/v1/user/order/refund","post",data);
}
```

applyRefundOrderData()函数可以调用服务端接口申请退款。接下来对接到Vuex中，在store/order/index.js文件中新增代码如下：

```
import Vue from "vue";
import {addOrderData,wechatUnifiedorderData,getLastOrderData,payData,
getOrderData,getRefundOrderData,getOrderDetailsData,applyRefundOrderData}
from "../../api/order";
export default {
    ...
    mutations:{
        ...
        //改变订单详情退款状态
        ["SET_ORDER_DETAILS_REFUND_STATE"](state,payload){
            if(state.orderDetails.order_desc){
```

```
                if(payload.is_meal=='1'){//套餐
                    level1:
                        for(let i=0;i<state.orderDetails.order_desc.length;i++){
                            level2:
                                for(let j=0;j<state.orderDetails.order_
                                desc[i].omi_data.length;j++){
                                    if(state.orderDetails.order_desc[i].omi_
                                    data[j].order_item_id==payload.order_
                                    item_id) {
                                        state.orderDetails.order_desc[i].omi_
                                        data[j].refund_state='-1';
                                        Vue.set(state.orderDetails.order_
                                        desc[i].omi_data,j,state.orderDetails.
                                        order_desc[i].omi_data[j]);
                                        break level1;

                                    }
                                }
                        }
                }else{//非套餐
                    for(let i=0;i<state.orderDetails.order_desc.length;i++){
                        if(state.orderDetails.order_desc[i].order_item_
                        id==payload.order_item_id){
                            state.orderDetails.order_desc[i].refund_state='-1';
                            Vue.set(state.orderDetails.order_desc,i,state.
                            orderDetails.order_desc[i]);
                             break;
                        }
                    }
                }
            }
        },
    },
    actions:{
        ...
        //申请退款
        applyRefundOrder(conText,payload){
            applyRefundOrderData({uid:conText.rootState.user.uid,token:conText.
rootState.user.token,platform:conText.rootState.system.platform,...payload}).
then(res=>{
                if(payload.success){
                    payload.success(res);
```

```
                    }
                    if(res.code==200){
                        //改变订单状态
                        conText.commit("SET_ORDER_DETAILS_REFUND_STATE",
                        {order_item_id: payload.order_item_id,is_meal:payload.
                        is_meal});
                    }
                })
            },
        }
    }
```

以上加粗代码为新增代码。mutations内部的SET_ORDER_DETAILS_REFUND_STATE方法可以响应式地改变订单的状态，actions内部的applyRefundOrder()方法可以向服务端申请退款并改变订单的状态。

接下来开发申请退款组件。在components文件夹下创建refund_order/index.vue文件，该文件中的代码如下：

```
<template>
    <view class="mask" v-show="show">
        <view class="refund-win">
            <view class="title">菜品名称:{{title}}</view>
            <view class="amount-wrap">
                <view class="text">数量:</view>
                <view class="handle-amount">
                    <view class="dec" @click="refundAmount>1?--refundAmount:1"></view>
                    <text>{{refundAmount}}</text>
                    <view class="inc" @click="incAmount()"></view>
                </view>
            </view>
            <view class="refund-content"><textarea fixed="true" placeholder="
            请输入退款原因" v-model="refundContent"></textarea></view>
            <view class="handle-btn">
                <button class="cancel" @click="cancel()">取消</button>
                <button class="submit" @click="submitData()">提交</button>
            </view>
        </view>
    </view>
</template>

<script>
    import {mapActions} from "vuex";
```

```
export default {
    name: "refund-order",
    data(){
        return {
            refundAmount:1,        //退菜数量
            refundContent:""       //退款原因
        }
    },
    props:{
        //是否显示组件
        show:{
            type:Boolean,
            default:false
        },
        //是否套餐
        isMeal:{
            type:String,
            default:""
        },
        //订单明细的id
        orderItemId:{
            type:String,
            default:""
        },
        //菜品名称
        title:{
            type:String,
            default:""
        },
        //菜品数量
        amount:{
            type:Number,
            default:0
        }
    },
    mounted(){
        this.isSubmit=true;
    },
    methods:{
        ...mapActions({
            applyRefundOrder:"order/applyRefundOrder"  //申请退款
        }),
```

```
//提交申请退款
submitData(){
    if(this.isSubmit){
        this.isSubmit=false;
        this.applyRefundOrder({refund_amount:this.refundAmount,order_
            item_id:this.orderItemId,refund_content:this.
            refundContent,is_meal:this.isMeal,success:(res)=>{
               if(res.code==200){
                   uni.showToast({
                       title:"申请退款成功，我们会尽快处理！ ",
                       icon:"none",
                       duration:2000
                   });
                   setTimeout(()=>{
                       this.isSubmit=true;
                       this.cancel();
                   },2000);
               }else{
                   this.isSubmit=true;
                   uni.showToast({
                       title:""+res.data+"",
                       icon:"none",
                       duration:2000
                   });
               }
            }});
    }

},
//隐藏组件
cancel(){
    this.refundAmount=1;
    this.refundContent="";
    this.$emit("cancel");
},
//增加菜品退款数量
incAmount(){
    if(this.refundAmount<this.amount){
        this.refundAmount+=1;
    }else{
        uni.showToast({
            title:"退款数量不能大于购买数量",
            icon:"none",
            duration:2000
```

```
                    });
                }
            }
        },
        watch:{
            //监听父组件菜品数量的变化
            amount(val){
                this.refundAmount=val;
            }
        }
    }
</script>

<style scoped>
    .mask{width:100%;height:100%;position: fixed;left:0px; top:0px;
    z-index:90;background-color:rgba(0,0,0,0.6)}
    .refund-win{width:80%;height:600rpx;background-color:#FFFFFF;position:
    absolute;left:0px;top:0px;bottom:0px;right:0px;margin:auto;z-index:1;border-
    radius: 5px;box-sizing: border-box;padding:20rpx;font-size:28rpx;}

    .amount-wrap{width:100%;height:80rpx;display:flex;align-items: center;}
    .handle-amount{width:auto;height:60rpx;display:flex;align-items:
    center;justify-content:flex-end}
    .handle-amount.dec{width:55rpx;height:55rpx;background-image:url("~@/
    static/images/main/dec.png");background-size:100%;background-position:
    center;background-repeat: no-repeat;margin-right:20rpx}
    .handle-amount.inc{width:55rpx;height:55rpx;background-image:url("~@/
    static/images/main/inc.png");background-size:100%;background-position:
    center;background-repeat: no-repeat;}
    .handle-amount text{font-size:28rpx;color:#333333;margin-right:20rpx;}

    .refund-content{width:100%;height:300rpx;border:1px solid
    #EFEFEF;border-radius: 4px;margin-top:10rpx;padding:20rpx;box-sizing:
    border-box;}
    .refund-content textarea{width:100%;height:100%;font-size:28rpx;}
    .handle-btn{width:350rpx;height:auto;margin:0 auto;margin-top:
    40rpx;display:flex;}
    .handle-btn.submit{width:160rpx;height:70rpx;background-color:
    #E30019;font-size:28rpx;color:#FFFFFF;}
    .handle-btn.cancel{width:160rpx;height:70rpx;background-color:
    #007aff;font-size:28rpx;color:#FFFFFF;}
</style>
```

申请退款组件可以更改退菜的数量并填写退款原因，最终提交给服务端接口完成退款申请。该组件需要从父组件接收几个必要的属性，props内部的show属性可以显示该组件；isMeal属性表示是否为套餐，值为1表示是套餐，0表示非套餐；orderItemId属性为订单明细的编号；title属性为菜品的名称；amount属性为退菜的数量。submitData()方法可以提交退款申请，如果退款申请成功，则调用cancel()方法隐藏该组件。

11.10.4 查看已退款的订单

扫一扫，看视频

用户申请退款完成后，在商家端单击“同意”按钮，则可将款项原路退回用户的账户中。由于本实战项目没有开发商家端，因此可以进入平台后台管理系统进行操作与测试，后台管理地址和账号在11.1节获取。进入后台，单击订单管理按钮，进入订单管理页面，单击左侧“退款”按钮，找到申请退款的菜品，单击“确认现金退款”按钮，即可实现退款，退款成功后即可在客户端显示。客户端退款页面预览效果如图11-24所示。

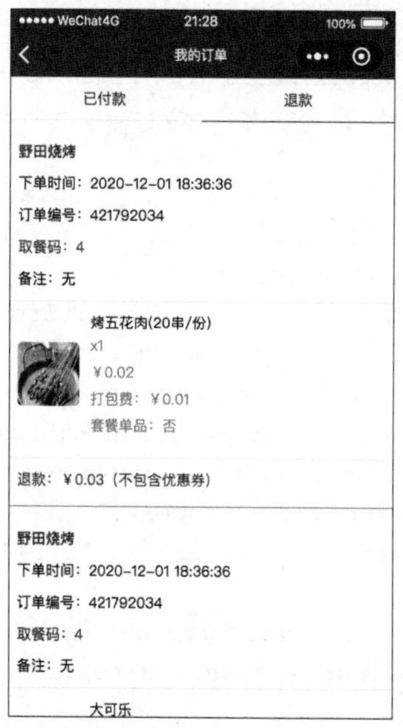

图 11-24 客户端退款页面预览效果

首先请求退款相关的服务端接口，在api/order/index.js文件中新增代码如下：

```
...
//查看已退款订单
export function getRefundOrderData(data){
    return request(config.baseApi+"/v1/user/order/show_refund","post",data);
}
```

getRefundOrderData()函数可以从服务端接口获取已退款订单列表数据。接下来对接到

Vuex中，在store/order/index.js文件中新增代码如下：

```
import Vue from "vue";
import {addOrderData,wechatUnifiedorderData,getLastOrderData,payData,
getOrderDetailsData,applyRefundOrderData,getOrderData,getRefundOrderData}
from "../../api/order";
export default {
    namespaced:true,
    state:{
        ...
        refundOrders:[],                //已退款订单数据
    },
    mutations:{
        ...
        //设置已退款订单数据
        ["SET_REFUND_ORDERS"](state,payload){
            state.refundOrders=payload.refundOrders;
        },
        //设置已退款订单分页数据
        ["SET_REFUND_ORDERS_PAGE"](state,payload){
            state.refundOrders.push(...payload.refundOrders);
        }
    },
    actions:{
        ...
        //查看已退款订单数据
        getRefundOrder(conText,payload){
            getRefundOrderData({uid:conText.rootState.user.uid,token:conText.
rootState.user.token,platform:conText.rootState.system.platform,...payload}).
then(res=>{
                if(res.code==200){
                    conText.commit("SET_REFUND_ORDERS",{refundOrders:res.data});
                    if(payload.success){
                        payload.success(res.pageinfo.pagenum);
                    }
                }else{
                    conText.commit("SET_REFUND_ORDERS",{refundOrders:[]});
                }
                if(payload.complete){
                    payload.complete();
                }
            })
        },
```

```
        //查看退款订单分页数据
        getRefundOrderPage(conText,payload){
            getRefundOrderData({uid:conText.rootState.user.uid,token:conText.
rootState.user.token,platform:conText.rootState.system.platform,...payload}).
then(res=>{
                if(res.code==200){
                    conText.commit("SET_REFUND_ORDERS_PAGE",{refundOrders:res.data});
                }
            })
        }
    }
}
```

以上加粗代码为新增代码，state内部的refundOrders属性可以存储已退款订单数据。mutations内部的SET_REFUND_ORDERS方法可以设置state内部的refundOrders属性值；SET_REFUND_ORDERS_PAGE方法内部使用push()方法，可以为state.refundOrders数组添加新的值，用于上拉加载数据。actions内部的getRefundOrder()方法可以从服务端接口获取已退款订单数据，并使用commit提交SET_REFUND_ORDERS，将获取的数据存储到state.refundOrders中；getRefundOrderPage()方法可以从服务端接口获取已退款订单分页的数据，并使用commit提交SET_REFUND_ORDERS_PAGE，将数据添加到state.refundOrders中。

接下来开发已退款订单页面。在pages/myorder文件夹下创建refund.vue文件，该文件中的代码如下：

```
<template>
    <view class="page">
        <Tags :status="status" :branch_shop_id="branch_shop_id"
        :table_code="table_code"></Tags>
        <view class="order-main" v-if="refundOrders.length>0">
            <view class="order-list" v-for="(item,index) in refundOrders" :key="index">
                <view class="order-info">
                    <view class="shop-name">{{item.branch_shop_name}}</view>
                    <view class="order-time">下单时间:{{item.order_time}}</view>
                    <view class="ordernum-status">
                        <view class="ordernum">订单编号:{{item.ordernum}}</view>
                    </view>
                    <view class="area-table" v-if="item.is_pack=='0'">
                        <view class="area">区域:{{item.hotel_area_title}}</view>
                        <view class="table-code">桌号:{{item.table_code}}</view>
                    </view>
                    <view class="pick-code">取餐码:{{item.pick_code}}</view>
                    <view class="remarks">备 注:{{item.remarks?item.remarks:"
                    无"}}</view>
                </view>
```

```
            <view class="order-desc">
                <view class="goods-list-main">
                    <view class="goods-list">
                        <view class="image">
                            <image :src="item.image"></image>
                        </view>
                        <view class="goods-info">
                            <view class="title">{{item.title}}</view>
                            <view class="info-text">x{{item.refund_amount}}</
                             view>
                            <view class="info-text"> ¥{{item.price}}</view>
                            <view class="info-text" v-if="item.pack_
                             price>0">打包费:¥{{item.pack_price}}</view>
                            <view class="is-meal">套餐单品:{{item.is_meal=='1'
                             ?'是':'否'}}</view>
                        </view>
                    </view>
                </view>
            </view>
            <view class="price-status">
                <view class="price">退款:¥{{item.refund_total}}</view>
            </view>
        </view>
    </view>
    <view class="no-data" v-if="refundOrders.length<=0">没有退款订单! </view>
  </view>
</template>

<script>
    import {mapActions,mapState} from "vuex";
    import Tags from "../../components/myorder/tags";
    export default {
        name: "order",
        data(){
            return {
                branch_shop_id:0,       //分店的id
                status:0,               //订单状态，1为已付款，-2为已退款
                table_code:"",          //桌号
            }
        },
        onLoad(opts){
            this.branch_shop_id=opts.branch_shop_id;
            this.table_code=opts.table_code;
```

```
            this.status=opts.status?opts.status:"-2";
            this.$utils.safeUser(this,this.branch_shop_id,this.table_code);
            this.curPage=1;              //当前页码
            this.maxPage=0;              //总页码
            this.getRefundOrder({page:1,success:(pageNum)=>{
                    this.maxPage=parseInt(pageNum);
                }})
        },
        //下拉刷新
        onPullDownRefresh(){
            this.getRefundOrder({page:1,success:(pageNum)=>{
                    this.maxPage=parseInt(pageNum);
                },
                complete:()=>{
                    uni.stopPullDownRefresh();
                }
            })
        },
        //上拉加载数据
        onReachBottom(){
            if(this.curPage<=this.maxPage){
                this.curPage++;
                //执行获取分页数据
                this.getRefundOrderPage({page:this.curPage});
            }
        },
        components:{
            Tags
        },
        computed:{
            ...mapState({
                refundOrders:state=>state.order.refundOrders  //退款数据
            })
        },
        methods:{
            ...mapActions({
                getRefundOrder:"order/getRefundOrder",       //获取退款数据
                getRefundOrderPage:"order/getRefundOrderPage"
                //获取退款分页数据
            })
        }
    }
</script>
```

```
<style scoped>
    .page{width:100%;min-height:100vh;background-color:#FFFFFF;margin-
     top:80rpx;overflow:hidden;}
    .order-main{width:100%;}
    .order-main.order-list{width:100%;border-bottom:1px solid #f17f1f;}
    .order-main.order-list.order-info{width:100%;border-bottom: 1px solid
     #EFEFEF;box-sizing: border-box;padding:20rpx;}
    .order-main.order-list.ordernum-status{width:100%;display:flex;font-
     size:28rpx;justify-content: space-between;margin-top:20rpx;}
    .order-main.order-list.area-table{width:100%;display:flex;font-size:
     28rpx;margin-top:20rpx;}
    .order-main.order-list.area-table .area{margin-right:40rpx;}
    .order-main.order-list.order-time{font-size:28rpx;}
    .order-main.order-list.pick-code{font-size: 28rpx;color:#E30019;
     margin-top:20rpx;}
    .order-main.order-list.remarks{font-size:28rpx;margin-top:20rpx;}
    .order-main.order-list.shop-name{width:100%;font-size:28rpx;margin-
     top:20rpx;margin-bottom:20rpx;font-weight: bold;color:#E30019}

    .order-main.order-desc{width:100%;}
    .order-main.goods-list-main{width:100%;box-sizing: border-
     box;padding:20rpx;border-bottom: 1px solid #EFEFEF;}
    .order-main.order-desc.goods-list{width:100%;display:flex;justify-
     content: space-between;align-items: center;margin-bottom:20rpx;}
    .order-main.order-desc.goods-list.image{width:120rpx;height:120rpx;}
    .order-main.order-desc.goods-list.image image{width:100%;height:100%;
     border-radius: 5px;}
    .order-main.order-desc.goods-list.goods-info{width:80%;height:auto;font-size:
     28rpx;overflow:hidden;position: relative;}
    .order-main.order-desc.goods-list.goods-info.title{width:100%;height:
     45rpx;overflow:hidden;text-overflow: ellipsis;white-space: nowrap;}
    .order-main.order-desc.goods-list.goods-info.info-text{width:auto;height:
     auto;color:#909090;margin-bottom:10rpx;}
    .order-main.order-desc.goods-list.goods-info.is-meal{width:auto; height:
     auto;color:#909090;}

    .order-main.order-list.price-status{width:100%;height:auto;box-sizing:
     border-box;padding:20rpx;}
    .order-main.order-list.price-status.price{font-size:28rpx;color:
     #333333;margin-bottom:10rpx;}
    .order-main.order-list.price-status.price.active{color:#E30019;}
</style>
```

在onLoad()钩子函数内调用Vuex中的getRefundOrder()方法，可以从服务端接口获取已退款订单列表数据并存储到refundOrders中，使用v-for将数据渲染到视图中；onPullDownRefresh()钩子函数内部的代码可以实现下拉刷新；onReachBottom()钩子函数内部的代码可以实现上拉加载数据。请读者仔细阅读代码，注释很清晰，不再详细讲解。

11.11 修改个人资料

扫一扫，看视频

修改个人资料包括修改头像、昵称、姓名、性别等信息，相关接口在“接口文档”文件夹下的“会员登录、注册、信息.docx”文件中。

个人资料页面预览效果如图 11-25 所示。

图 11-25 个人资料页面预览效果

首先请求个人资料相关的服务端接口，在api/user/index.js文件中新增代码如下：

```
...

//修改用户信息
export function updateUserData(data){
    return request(config.baseApi+"/v1/user/update","post",data);
}
```

updateUserData()函数可以调用服务端接口修改用户信息。接下来对接Vuex，在store/user/index.js文件中新增代码如下：

```
import {getWeChatOpenIdData,setWeChatUserData,getDewxbizdataData,
bindWechatLoginBindPhoneNumberData,safeUserData,getUserInfoData,updateUserDat
a} from '../../api/user';
export default {
    ...
    actions:{
        ...
        //修改用户信息
```

```
        updateUser(conText,payload){
            updateUserData({uid:conText.rootState.user.uid,token:conText.
            rootState.user.token,platform:conText.rootState.system.
            platform,...payload}).then(res=>{
                if(res.code==200){
                    if(payload.success){
                        payload.success(res);
                    }
                }
            })
        },
    }
}
```

以上加粗代码为新增代码，actions内部的updateUser()方法内部调用updateUserData()函数，可以调用服务端接口修改用户信息。

接着在user_pages文件夹下创建profile/index.vue文件，创建完成后配置路由。在pages.json文件中新增代码如下：

```
{
    ...
    "subPackages": [
        {
            "root": "user_pages",
            "pages": [
                ...
                {
                    "path": "profile/index",
                    "style": {
                        "navigationBarTitleText": "个人资料",
                        "disableScroll": true
                    }
                }
            ]
        }
    ]
}
```

以上加粗代码为新增代码，修改个人资料页面时，必须登录会员后才能访问与会员相关的页面，所以将其放在user_pages分包下。user_pages/profile/index.vue文件中的代码如下：

```
<template>
    <view class="page">
        <view class="head-wrap">
```

```
            <view class="text">头像</view>
            <view class="head" @click="handleHead"><image :src="showHead">
             </image></view>
        </view>
        <view class="list">
            <view class="text">昵称</view>
            <view class="input-wrap">
                <input type="text" class="input" v-model="nickname"
                 placeholder="请输入昵称" />
            </view>
        </view>
        <view class="list">
            <view class="text">姓名</view>
            <view class="input-wrap">
                <input type="text" class="input" v-model="name"
                 placeholder="请输入姓名" />
            </view>
        </view>
        <view class="list" @click="selectGender">
            <view class="text">性别</view>
            <view class="input-wrap">
                <view class="text-tip">{{gender=="1"?'男':gender=='2'?'女':
                 '请选择'}}</view>
                <view class="arrow"></view>
            </view>
        </view>
        <button type="button" @click="submit()" class="submit-btn">保存</button>
    </view>
</template>

<script>
    import {mapActions,mapState} from "vuex";
    export default {
        name: "profile",
        data(){
            return {
                nickname:"",    //昵称
                gender:"",      //性别，1为男，2为女
                name:"",        //姓名
                showHead:"../../static/images/user/head.png",
                //在视图中显示的头像图片的地址
                head:""         //保存到数据库的头像图片名称
```

```
        }
    },
    onLoad(opts){
        this.branch_shop_id=opts.branch_shop_id;        //分店的id
        this.table_code=opts.table_code;                //桌号
        this.$utils.safeUser(this,this.branch_shop_id,this.table_code);
        this.isSubmit=true;
        //获取用户信息
        this.getUserInfo({branch_shop_id:this.branch_shop_id,success:(data)=>{
            this.nickname=data.nickname;                //昵称
            this.gender=data.gender;                    //性别
            this.name=data.name;                        //姓名
            this.showHead=data.head;                    //头像
        }});
    },
    computed:{
        ...mapState(
            {
                uid:state=>state.user.uid,
                token:state=>state.user.token,
                platform:state=>state.system.platform
            }
        )

    },
    methods:{
        ...mapActions({
            getUserInfo:"user/getUserInfo",             //获取用户信息
            updateUser:"user/updateUser"                //修改用户信息
        }),
        //选择性别
        selectGender(){
            uni.showActionSheet({
                itemList: ['男', '女'],
                success:  (res)=> {
                    this.gender=res.tapIndex+1; //性别，1为男，2为女
                },
                fail: (res)=> {
                    // console.log(res.errMsg);
                }
            });
        },
```

```
//头像操作
handleHead(){
    uni.showActionSheet({
        itemList: ['拍照','从相册选择' ],
        success:  (res)=> {
            let sourceType=[];
            let index=res.tapIndex;
            if(index==0){                    //拍照
                sourceType=["camera"];
            }else if(index==1){              //从相册选择
                sourceType=["album"];
            }
            //选择图片
            uni.chooseImage({
                count: 1,                    //默认为9
                sizeType: ['compressed'],
                //可以指定是原图还是压缩图，默认两者都有
                sourceType: sourceType, //从相册选择
                success: (res)=> {
                    let tmpFilePath=res.tempFilePaths[0];
                    this.showHead=tmpFilePath;
                    //上传图片
                    uni.uploadFile({
                        url: this.$config.baseApi+'/v1/user/uploadhead',
                        //上传图片接口地址
                        filePath: tmpFilePath,          //文件
                        name: 'head',   //服务端接口字段名称
                        formData:{       //数据参数
                            uid:this.uid,
                            token:this.token,
                            platform:this.platform
                        },
                        //上传成功
                        success: (uploadFileRes) => {
                            let res=JSON.parse(uploadFileRes.data);
                            //文件名称
                            this.head=res.data.msbox;
                        },
                        fail:(uploadFileRes)=>{
                          // console.log(uploadFileRes);
                        }
                    });
```

```
                }
            });
        },
        fail: (res)=> {
            // console.log(res.errMsg);
        }
    });
},
//提交保存
submit(){
    if(this.isSubmit){
        if(this.nickname.match(/^\s*$/)){
            uni.showToast({
                title: '请输入昵称',
                icon:"none",
                duration: 2000
            });
            return;
        }
        if(this.name.match(/^\s*$/)){
            uni.showToast({
                title: '请输入姓名',
                icon:"none",
                duration: 2000
            });
            return;
        }
        this.isSubmit=false;
        //修改用户信息并保存到数据库
        this.updateUser({branch_shop_id:this.branch_shop_id,name:
this.name,nickname:this.nickname,gender:this.gender,head:this.head,success:()=>{
                uni.showToast({
                    title: '修改成功',
                    icon:"none",
                    duration: 2000
                });
                setTimeout(()=>{
                    uni.navigateBack({
                        delta: 1
                    });
                },2000)
            }});
```

```
                }

            }
        }
    }
</script>

<style scoped>
    .page{width:100%;height:100vh;overflow:hidden;background-color:#FFFFFF;}
    .head-wrap{width:100%;height:120rpx;border-bottom: 1px solid #EFEFEF;
     display:flex;justify-content: space-between;align-items: center;box-
     sizing: border-box;padding:0px 30rpx;}
    .head-wrap.text,.list.text{font-size:28rpx;color:#333333;}
    .head-wrap.head{width:100rpx;height:100rpx;}
    .head-wrap.head image{width:100%;height:100%;border-radius: 100%;}
    .list{width:100%;height:80rpx;border-bottom: 1px solid EFEFEF;display:flex;
     justify-content: space-between;align-items: center;box-sizing: border-box;
     padding:0px 30rpx;}
    .list.input-wrap{width:auto;height:100%;display:flex;align-items: center;}
    .list.input-wrap.input{width:380rpx;height:66rpx;font-size:28rpx;text-align:right;}
    .list.arrow{width:40rpx;height:40rpx;background-image: url("~@/static/
     images/user/right_arrow.png");background-size:100%;background-repeat:
     no-repeat;background-position: center;margin-left:20rpx;}
    .list.text-tip{font-size:28rpx;color:#333333;}
    .submit-btn{width:200rpx;height:60rpx;margin:0 auto;background-
     color:#E30019;color:#FFFFFF;font-size:28rpx;margin-top:40rpx;line-
     height:60rpx;border-radius: 4px;}
</style>
```

在onLoad()钩子函数内调用Vuex内部的getUserInfo()方法，可以从服务端接口获取会员的昵称、性别、姓名和头像，分别赋值给data()方法返回的属性nickname、gender、name和showHead，并将数据渲染到视图中。selectGender()方法内部使用uni.showActionSheet API（见10.6.4节）实现选择性别；handleHead()方法（见10.2节）可以实现头像上传；submit()方法内部的this.updateUser()方法可以将输入的用户信息保存到服务端数据库中。

11.12 绑定手机号

扫一扫，看视频

在绑定手机号页面中可以修改手机号，相关接口在“接口文档”文件夹下的“会员登录、注册、信息.docx”和“验证码.docx”文件中。

绑定手机号页面预览效果如图11-26所示。

图 11-26 绑定手机号页面预览效果

在正式运行的项目中点击“获取验证码”按钮，可以接收到短信通知获取验证码。要实现发送短信的功能，必须有一个短信运营商，如阿里云短信服务商，使用服务端语言（如PHP、Node、Java等），利用短信服务端提供的SDK或API接口发送短信到用户的手机上。因为本章的项目是测试版本，没有对接短信服务商，所以不会收到短信通知，但是可以从服务端接口获取数字验证码来模拟短信验证码。

首先请求绑定手机号的相关服务端接口，在api/user/index.js文件中新增代码如下：

```
...

//绑定手机号
export function bindCellphoneData(data){
    return request(config.baseApi+"/v1/user/wechat_bind_mobile","post",data);
}

//是否绑定过手机号
export function existCellphoneData(data){
    return request(config.baseApi+"/v1/exist_cellphone","post",data);
}
```

bindCellphoneData()函数调用服务端接口，可以将用户的手机号保存到数据库中；existCellphoneData()函数可以查询服务端数据库是否绑定过手机号。接下来在api文件夹下创建vcode/index.js文件，该文件中的代码如下：

```
import config from "../../static/js/conf/config";
import {request} from "../../static/js/utils/request";

//发送短信验证码
export function sendCodeData(data){
    return request(config.baseApi+"/v1/vcode/send_code","post",data);
}
```

sendCodeData()函数调用服务端接口，可以实现发送短信验证码功能。接下来对接到Vuex

中，在store/user/index.js文件中新增代码如下：

```
import {getWeChatOpenIdData,setWeChatUserData,getDewxbizdataData,
bindWechatLoginBindPhoneNumberData,safeUserData,getUserInfoData,updateUserData,
bindCellphoneData,existCellphoneData} from '../../api/user';
export default {
    ...
    actions:{
        ...
        //绑定手机号
        bindCellphone(conText,payload){
            return bindCellphoneData({uid:conText.rootState.user.uid,token:conText.
rootState.user.token,platform:conText.rootState.system.platform,...payload}).
then(res=>{
                return res;
            })
        },
        //是否存在手机号
        existCellphone(conText,payload){
            return existCellphoneData(payload).then(res=>{
                return res;
            })
        },
    }
}
```

以上加粗代码为新增代码，actions内部的bindCellphone()方法内部调用bindCellphoneData()函数，可以将手机号保存到服务端的数据库中；existCellphone()方法内部调用existCellphoneData()函数，可以查询服务端数据库是否绑定过手机号。

接着开发短信验证码相关的Vuex。在store文件夹下创建vcode/index.js文件，该文件中的代码如下：

```
import {sendCodeData} from "../../api/vcode";
export default {
    namespaced:true,
    actions:{
        //发送短信验证码
        sendCode(conText,payload){
            return sendCodeData(payload).then(res=>{
                return res;
            })
        }
    }
}
```

actions内部的sendCode()方法调用sendCodeData()函数，可以实现发送短信验证码功能。接下来对接到Vuex的模块中，在store/index.js文件中新增代码如下：

```
...
import vcode from '../vcode';

Vue.use(Vuex);

const store = new Vuex.Store({
    modules:{
        vcode
    }
});

export default store
```

以上加粗代码为新增代码。接着开发绑定手机号页面，在user_pages文件夹下创建bind_cellphone/index.vue文件，创建完成后配置路由。在pages.json文件中新增代码如下：

```
{
    ...
    "subPackages": [
        {
            "root": "user_pages",
            "pages": [
                ...
                {
                    "path": "bind_cellphone/index",
                    "style": {
                        "navigationBarTitleText": "绑定手机号",
                        "disableScroll": true
                    }
                }
            ]
        }
    ]
}
```

以上加粗代码为新增代码。在user_pages/bind_cellphone/index.vue文件中的代码如下：

```
<template>
    <view class="page">
        <view class="tip-wrap">
```

```
            <view class="tip-icon"></view>
            <view class="text">新手机号验证后，即可绑定成功！ </view>
        </view>
        <view class="input-wrap">
            <input type="text" placeholder="手机号" class="cellphone" v-model="cellphone" />
        </view>
        <view class="input-wrap">
            <view class="msg-code">
                <input type="text" placeholder="请输入短信验证码" v-model="msgCode"  />
            </view>
            <view :class="{'msg-code-btn':true, active:isSendCode}" @click="
             getMsgCode()">{{msgText}}</view>
        </view>
        <button type="button" class="submit-btn" @click="submit()">确认绑定</button>
    </view>
</template>

<script>
    import {mapActions} from "vuex";
    export default {
        name: "bind-cellphone",
        data(){
            return {
                cellphone:"",           //手机号
                msgCode:"",             //短信验证码
                msgText:"获取验证码",
                isSendCode:true         //是否可以发送短信验证码
            }
        },
        onLoad(opts){
            this.timer=null;
            this.isSubmit=true;
            this.branch_shop_id=opts.branch_shop_id;
            this.table_code=opts.table_code;
            this.$utils.safeUser(this,this.branch_shop_id,this.table_code);
        },
        onUnload(){
            //页面离开时清除定时器
            clearInterval(this.timer);
        },
```

```
methods: {
    ...mapActions({
        existCellphone:"user/existCellphone",       //手机号是否绑定过
        bindCellphone:"user/bindCellphone",         //绑定手机号
        sendCode:"vcode/sendCode"                   //发送短信验证码
    }),
    //确认绑定
    async submit() {
        if(this.isSubmit){
            if (this.cellphone.match(/^\s*$/)) {
                uni.showToast({
                    title: '请输入手机号',
                    icon: "none",
                    duration: 2000
                });
                return;
            }
            if (!this.cellphone.match(/^1[0-9][0-9]\d{8}$/)) {
                uni.showToast({
                    title: '您输入的手机号格式不正确',
                    icon: "none",
                    duration: 2000
                });
                return;
            }
            //校验手机号是否绑定过
            let existCellphone=await this.existCellphone({cellphone:this.
            cellphone});
            if(existCellphone.data.exist=='1'){
                uni.showToast({
                    title: '手机号已存在，请更换',
                    icon:"none",
                    duration: 2000
                });
                return;
            }
            if (this.msgCode.match(/^\s*$/)) {
                uni.showToast({
                    title: '请输入短信验证码',
                    icon: "none",
```

```
                duration: 2000
            });
            return;
        }
        this.isSubmit=false;
        //绑定手机号
        let bcData=await  this.bindCellphone({cellphone:this.
            cellphone,sms_code:this.msgCode});
        if(bcData.code==200){
            uni.showToast({
                title: "绑定成功! ",
                icon: "none",
                duration: 2000
            });
            setTimeout(()=>{
                uni.navigateBack({
                    delta: 1
                });
            },2000);
        }else{
            this.isSubmit=true;
            uni.showToast({
                title: "" + bcData.data + "",
                icon: "none",
                duration: 2000
            });
            return;
        }
    }


},
//发送短信验证码
async getMsgCode() {
    if (this.isSendCode) {
        if (this.cellphone.match(/^\s*$/)) {
            uni.showToast({
                title: '请输入手机号',
                icon: "none",
                duration: 2000
            });
```

```
        this.isCodeSend = true;
        return;
    }
    if (!this.cellphone.match(/^1[0-9][0-9]\d{8}$/)) {
        uni.showToast({
            title: '您输入的手机号格式不正确',
            icon: "none",
            duration: 2000
        });
        this.isSendCode = true;
        return;
    }
    //校验手机号是否绑定过
    let existCellphone=await this.existCellphone
        ({cellphone:this.cellphone});
    if(existCellphone.data.exist=='1'){
        uni.showToast({
            title: '手机号已存在，请更换',
            icon:"none",
            duration: 2000
        });
        this.isCodeSend = true;
        return;
    }
    //发送短信验证码
    let smsCode = await this.sendCode({cellphone: this.cellphone});
    if (smsCode.code == 200) {
        //smsCode.data.sms_code获取短信验证码
        uni.showToast({
            title: "" + smsCode.data.sms_code + "",
            icon: "none",
            duration: 2000
        });
        this.isSendCode = false;
        let time = 60;
        this.msgText = "重新获取(" + time + "s)";
        this.timer = setInterval(() => {
            if (time <= 0) {
                clearInterval(this.timer);
                this.isSendCode = true;
                this.msgText = "获取验证码";
```

```
                                    } else {
                                        time--;
                                        this.msgText = "重新获取(" + time + "s)";
                                    }
                                    }, 1000);
                            }
                        }
                    }
                }
            }
</script>

<style scoped>
    .page{width:100%;min-height:100vh;overflow:hidden;background-
     color:#FFFFFF;}
    .tip-wrap{width:100%;height:100rpx;background-color:#f3f5c4;display:flex;box-
     sizing: border-box;align-items: center;padding-left:40rpx;margin-bottom:40rpx;}
    .tip-wrap.tip-icon{width:40rpx;height:40rpx;background-image:url("~@/
     static/images/user/tip.png");background-size:100%;background-position:
     center;background-repeat: no-repeat;margin-right:20rpx;}
    .tip-wrap.text{font-size:28rpx;color:#ac7700;}

    .input-wrap{width:90%;height:80rpx;background-color:#FFFFFF;border:1px
     solid #EFEFEF;margin:0 auto;display:flex;align-items: center;box-sizing:
     border-box;padding-left:20rpx;border-radius: 5px;margin-bottom:20rpx;}
    .input-wrap.cellphone{width:90%;height:80%;font-size:28rpx;}

    .input-wrap.msg-code{width:70%;height:100%;border-right: 1px solid #EFEFEF}
    .input-wrap.msg-code input{width:100%;height:100%;font-size:28rpx;}

    .input-wrap.msg-code-btn{width:29%;height:100%;text-align:center;line-
     height:80rpx;font-size:28rpx;color:#717376;}
    .input-wrap.msg-code-btn.active{color:#E30019}

    .input-wrap.msg-code-btn image{width:80%;height:80%;margin-top:9rpx;}

    .submit-btn{width:200rpx;height:60rpx;margin:0 auto;background-
     color:#E30019;color:#FFFFFF;font-size:28rpx;margin-top:40rpx;line-
     height:60rpx;border-radius: 4px;}
</style>
```

以上代码逻辑不是很复杂，注释很清晰，请读者仔细阅读。这里简单讲解发送短信验证码的逻辑。点击“获取验证码”按钮，触发getMsgCode()方法，可以获取短信验证码。在该方法内部，this.isSendCode变量的作用是判断是否可以发送短信验证码，值为true表示可以发送短信验证码，否则不能发送短信验证码。如果该变量的值为true，则先校验手机号是否为空，再校验手机号格式是否正确。如果校验通过，调用existCellphone则方法判断手机号是否绑定过。如果未绑定，则可以获取短信验证码。

接下来调用sendCode()方法，从服务端接口获取短信验证码。由于没有对接短信服务商接口，因此无法发送到手机上，可以使用uni.showToast将短信验证码显示出来进行测试，再将this.isSendCode变量设置为false，这样再点击“获取验证码”按钮时，就无法再次获取短信验证码，便可以防止60s内重复获取短信验证码的问题。创建变量time，赋值为60，将变量msgText的值设置为“重新获取(" + time + "s)”，再将“获取验证码”按钮的值设置为msgText，这样就可以动态地改变“获取验证码”按钮的文字。创建定时器setInterval()函数并赋值给变量this.timer，间隔时间设置为1000ms，即1s。在setInterval()函数内部进行判断，如果time小于等于0，表示60s倒计时执行完毕，使用clearInterval(this.timer)清除定时器，this.isSendCode的值设置为true，这样点击“获取验证码”按钮就可以再次获取短信验证码，将msgText变量的值设置为“获取验证码”；如果time不小于等于0，则将time的值递减，将msgText的值设置为“重新获取(" + time + "s)”。

至此，获取短信验证码的功能开发完毕。注意，最后要在onUnload()钩子函数内清除定时器，因为uni-app和Vue一样是单页面应用，在页面离开时需要手动清除定时器。

11.13 最终发布上线

扫一扫，看视频

项目开发完成后，使用微信开发者工具上传代码。可以使用体验版测试发行版本的程序，测试完成后提交审核，审核成功后发布即可。

微信小程序发布流程如图11-27所示。

图 11-27 微信小程序发布流程

首先使用HBuilder X发行小程序，接下来在微信开发者工具中单击“上传”按钮，如图11-28所示。

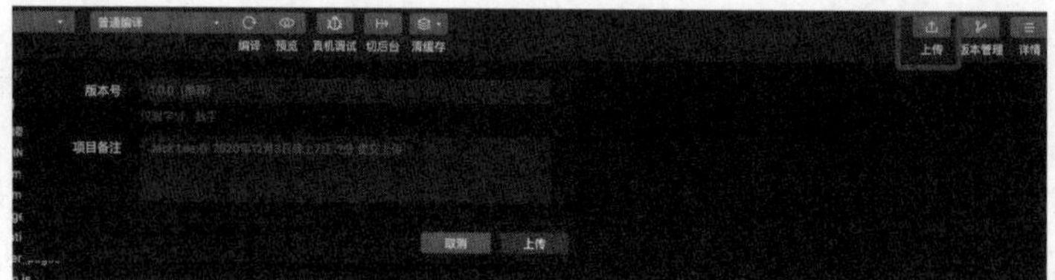

图 11-28 上传小程序

填写版本号，单击“上传”按钮进行上传。上传成功后，进入微信小程序管理后台（https://mp.weixin.qq.com），单击左侧导航的“版本管理”按钮，进入版本管理页面，可以看到开发版本信息。单击“提交审核”下拉按钮并选择“选为体验版本”，这样便可以在真机下测试发行版本的程序，如图 11-29 所示。

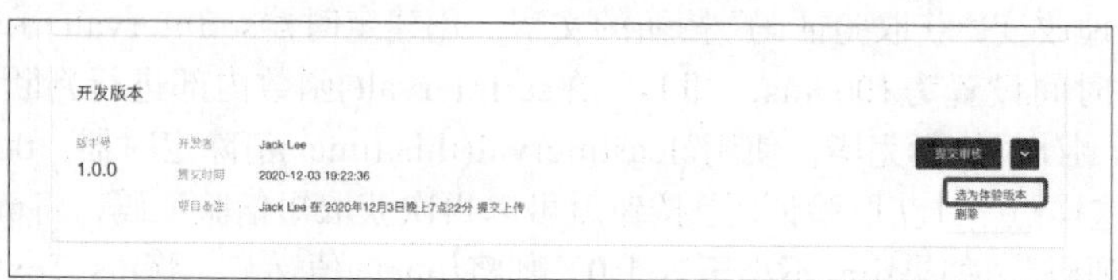

图 11-29 选择“选为体验版本”

选择体验版本后，会设置扫码后访问小程序的默认路径，如图 11-30 所示。

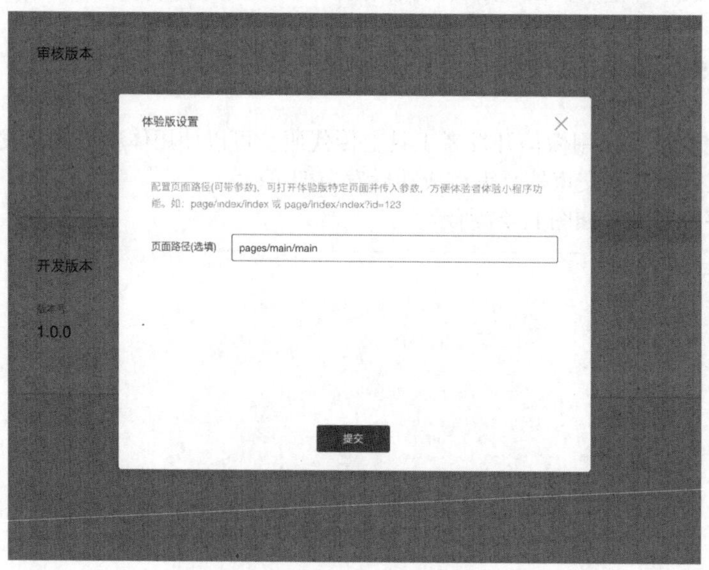

图 11-30 设置默认路径

设置完成后，单击“提交”按钮，出现如图 11-31 所示的界面。

图 11-31　体验版二维码

扫描二维码即可进行测试。体验版只能供项目成员和体验成员访问，单击左侧导航栏的“成员管理”按钮，进入成员管理页面添加即可，按照提示操作即可，比较简单不做详解。测试完成后，单击图11-29中的“提交审核”按钮，按照提示填写审核信息后，进行审核，如图11-32所示。

图 11-32　审核版本

如果是不加急审核，预计审核时间为1~7天，通常1天之内就能审核完成。审核完成后会出现“发布”按钮，单击“发布”按钮后有两个选项：全量发布和灰度发布(分阶段发布)，其中全量发布是指所有用户访问程序时都会使用当前最新的发布版本，灰度发布是指分不同时间段控制部分用户使用最新发布版本。如果在使用高峰期时发布，则推荐选择灰度发布。通常，在点餐小程序发布成功5分钟后就能在微信客户端里面搜索到。

11.14　小结

本章是uni-app的实战部分，也是本书的重中之重，实现了仿美团微信点餐小程序的核心功能，包括点餐、菜品搜索、会员登录、购物车、提交订单、个人中心、订单管理、修改个人资定手机号、发布上线等，同时提供了真实的企业级完整项目接口文档，从真机测试到最发布上线，和在企业中实际开发无异。本章的实战项目同时复习了组件的使用、分包的应用、微信授权登录、微信支付、如何看懂接口文档与服务端通信等综合知识。通过练习并完成该实战项目，读者既能巩固前面学到的理论知识点，同时在做项目的过程中也可以提高自己的实战能力、逻辑思维能力和解决问题能力，这样在工作中才能游刃有余。